D0164289

AN INTRODUCTION TO
MOLECULAR ORBITALS

AN INTRODUCTION TO MOLECULAR ORBITALS

by
YVES JEAN and FRANÇOIS VOLATRON

translated and edited by
Jeremy Burdett

New York   Oxford
OXFORD UNIVERSITY PRESS
1993

Oxford University Press

Oxford   New York   Toronto
Delhi   Bombay   Calcutta   Madras   Karachi
Kuala Lumpur   Singapore   Hong Kong   Tokyo
Nairobi   Dar es Salaam   Cape Town
Melbourne   Auckland
and associated companies in
Berlin   Ibadan

Library of Congress Cataloging-in-Publication Data
Jean, Yves, 1948–
[Orbitales moleculaires. English]
An introduction to molecular orbitals / by Yves Jean and François Volatron ;
translated and edited by Jeremy Burdett.
p. cm.   Includes bibliographical references and index.
ISBN-13 978 0-19-506918-1
ISBN 0-19-506918-8
1. Molecular orbitals.   I. Volatron, François.
II. Burdett, Jeremy K., 1947–   III. Title.
QD461.J4313   1993
514.2'24—dc20   92-45676

9 8 7 6 5
Printed in the United States of America
on acid-free paper

Translator's and Editor's Note

Some years ago during a summer at the Laboratoire de Chimie Théorique at The Université de Paris-Sud, Yves Jean told me about a plan that he and François Volatron had for writing a book on Molecular Orbital Theory to be called 'Les Orbitales Moléculaires en Chimie: Introduction et Applications'. This was to be developed from their highly successful lecture course for students at the university and, unlike its predecessors would focus on the nuts and bolts of drawing molecular orbital diagrams. Importantly it should be aimed at beginners. At that early stage, and knowing the didactic skill of the authors, I volunteered to translate and edit an English language version. Since then, as the book took shape, we have had many discussions during my subsequent visits to Paris. The preparation of the English language edition took place alongside the final version of the French edition, and some of the editorial changes have made their way back into the original. Much of the work was done during the tenure of a John Simon Guggenheim Memorial Fellowship. The manuscript was effortlessly typed by Nita Yack, to whom I owe a thousand thanks.

JEREMY BURDETT
Chicago, Illinois
March 1992

Contents

II BUILDING UP MOLECULAR ORBITALS AND ELECTRONIC STRUCTURE

III INTRODUCTION TO THE STUDY OF THE GEOMETRY AND REACTIVITY OF MOLECULES

Introduction

Molecular Orbital Theory is one of the methods available to us for the study of the electronic structure of molecules and molecular properties, such as geometry and reactivity which it determines. When teaching molecular orbital theory, one can stress the numerical precision of the model or one can try to generate a qualitative understanding of the results. It goes without saying that these two approaches can (and indeed should) coexist, since they are effectively two sides of the same coin. In this book we have opted for a qualitative, non-mathematical, treatment with which to understand some of the results generated by numerical models. Its aim is to give the student a theoretical method of analysis which relies on some simple ideas, applicable to chemical problems of varying complexity. The reason for our choice is two-fold. First of all there is no book of this available at a sufficiently introductory level for the student who has had little exposure to theoretical chemical methods. The second is a pedagogical one. The mathematical apparatus for the rigorous treatment of the electronic structure of molecules is important, but its applications most often stop at very small systems, invariably containing two electrons at most. It is then quite attractive for the student to discover and use another approach, which while not leading to exact numerical values of observables, nevertheless provides a way to interpret experimental results using a language understandable by all chemists. This qualitative orbital description of electronic structure is currently routinely applied to the study of organic, inorganic, organometallic and solid-state systems. It is not without merit that this is a theoretical approach which can be usefully applied to a very broad range of chemical problems. It provides a sense of unity for the whole area of chemistry and encourages chemists to think beyond the traditional divisions of the field. During the past twenty-five years many people have contributed to the development and applications of molecular orbital theory which have been used in this book. We shan't even start to compile a list but special homage must be paid to Roald Hoffmann and Kenichi Fukui, Nobel laureates in chemistry for 1981, for the decisive role which they played in the development of an interpretive theory of the electronic structure of molecules, their structures and reactivity.

The goal of this book is to present an introduction to molecular orbital theory which will illustrate the major areas of its application, namely the electronic structure, geometry and reactivity of molecules. It is divided into four parts; (I) Introduction to atomic and molecular structure, (II) Building up molecular orbitals and electronic structure, (III) Introduction to the study of the geometry and reactivity of molecules and (IV) Problems. Each chapter is followed by a set of problems with answers.

We would like to thank J. K. Burdett, professor of chemistry at The University of Chicago, who has actively supported our project to write a book on molecular orbital theory and its applications to problems of structure and reactivity at an elementary level. His encouragement and advice have been very useful during our efforts.

A number of our colleagues have taken the responsibility of reading a part or all of the manuscript, and their comments have been important in shaping the scientific content. We thank R. Abouaf (Paris XI), B. Bigot (ENS Lyon), D. Cauchy (Amiens), C. Iung (Paris XI), Y. Justum (Paris XI), J.-M. Leclerq (Paris VI), J.-M. Lefour (Ecole Polytechnique), A. Lledos (Barcelona), Nguyen Trong Anh (Ecole Polytechnique), P. Perrot (Villeneuve d'Asq), M. Picard (Paris VI), J.-C. Rayez (Bordeaux), J.-Y. Saillard (Rennes I), L. Salem (Paris XI) and A. Stritch (Strasbourg I). We also thank the members of the Theoretical Chemistry Laboratory at Orsay where this book was written, for their help in many different ways. We thank in particular, O. Eisenstein the director of the laboratory and P. Hiberty who carried out the calculations presented in the appendix to Chapter 7. B. Housse and I. El Rachidi worked on the preparation of the manuscript and M. and P. Jean helped with the writing style.

Finally we dedicate this book to Françoise, Marie, Jeanne, Frédéric and Denis for their unfailing moral support.

Orsay, France
1991

Part I
Introduction to atomic and molecular structure

1 From the periodic table to molecules

A knowledge of the electronic structure of molecules is essential for understanding their geometry, their physico-chemical characteristics and their reactivity. The challenge which then faces chemists is one of providing an approach to describe how molecules hold together in a way which allows access to such properties. At the end of the nineteenth century it became clear that classical models of atoms, and therefore of molecules too, were not able to account for several important properties, such as the existence of specific absorption and emission spectra. Very little insight was available into the structure and properties of molecules. This led in 1900 to the postulate of a theory in which energy was quantized. In 1912 such an approach was used to describe the structure and spectra of atoms. This 'old' quantum theory was superseded around 1925, by a new theory, that of quantum mechanics, which we recognize today as being the only way at present to satisfactorily describe systems made up of light particles such as electrons. Quantum mechanics therefore forms the basis for the development of accurate theories of the electronic structure of atoms and molecules. Among these are the ideas of molecular orbital theory which form the basis of this book.

There exist, however, simpler models which are often used to describe bonding in molecules. Prominent among them are the theory of Lewis, with which we are able to rapidly draw out a bonding scheme for a molecule, and that developed by Gillespie and others which allows prediction of its geometry. These two models, which take up most of the first chapter, often give important information about the electronic structure of a molecule and its reactivity and as a result, are used by a wide spectrum of chemists. In spite of their limitations they are an indispensable precondition to deeper studies using models which at the same time are more precise yet more complex. To be able to make use of them necessitates a general knowledge of the electronic properties of atoms which are briefly reviewed at the beginning of this chapter.

1.1. The periodic classification

More than a century ago Mendeleev proposed a classification of the chemical elements (63 were known at this time) in a tabular form which collected in each column elements having similar chemical properties, a chemical family. Remember that at the time there was no information available concerning the structure of the

Table 1.1. The first three rows of the periodic table. In each case the elemental symbol and its atomic number are given.

	1							2
1st period	H							He
	3	4	5	6	7	8	9	10
2nd period	Li	Be	B	C	N	O	F	Ne
	11	12	13	14	15	16	17	18
3rd period	Na	Mg	Al	Si	P	S	Cl	Ar

atom, (nucleus and electrons) neither of course a theoretical model capable of describing their properties. It is then remarkable to note that the table proposed by Mendeleev is, with a few exceptions the one which is used today, and that the spaces left free at the time have been filled in little by little by the elements which have since been discovered.

1.1.1. Short description of the first three periods

Today we are on firmer ground when it comes to describing the electronic structure of atoms. Recall briefly that the atom consists of two parts. (i) a nucleus containing protons and neutrons. The protons, of which there are Z, the atomic number, carry a positive charge ($+e = 1.6 \times 10^{-19}$ coulomb), but the neutrons are electrically neutral. The total nuclear charge is therefore $+Ze$. (ii) Z electrons which move around the nucleus each carrying a negative charge of $-e$. Overall the atom is neutral.

Coming back to the periodic classification, the first three periods, or rows, are given above in Table 1.1. The table is read line by line and from left to right. (H, He; Li, . . . , Ne; Na, . . . , Ar). The place of an element is directly determined by Z, the number of protons or electrons, which increases by one each time we move from one element to the next along the table. There is therefore one electron for hydrogen (H), two for helium (He), three for lithium (Li), ten for neon (Ne), fourteen for silicon (Si), etc. If we remember that the elements of the same family are found in the same column then we can see that elemental properties vary in a periodic way with Z. Accordingly the properties of fluorine (with nine electrons) are extremely different from those of neon which has an extra electron, but those of neon and argon are very similar even though they differ by eight electrons.

1.1.2. Core and valence electrons

This periodicity in chemical properties controlled by Z or the number of electrons in the neutral atom leads to a shell structure for the electrons. Each time we move from one row to the next we move to a different shell of electrons. On moving from the first to the second row the first shell becomes full. It can only hold two electrons. Similarly, moving from the second to the third row leaves the second shell full of electrons. It may accommodate eight electrons since there are eight elements in the second row. We may proceed in the same general way down the table but in Chapter 2 will see how things get a little more complicated with the filling associated with the fourth row of the table. The electrons which occupy the last shell of an atom are called the *valence electrons* and those in earlier shells *core electrons*. Notice that elements from the same column (and therefore belonging to the same family) have the same number of valence electrons. So carbon (C) and silicon (Si) both have four valence electrons but carbon has two core electrons (the first shell is full) whereas silicon has ten (the first two shells are full). We note too that the column on the far right of the table contains elements with both two (He) and eight (Ne, Ar) valence electrons. A common feature of these elements is that they have a completely filled valence shell. The similarity in the properties of the elements from the same column leads us to think that the electrons responsible for determining chemical properties are those which are located in the outermost shell, namely the valence electrons.

This organization by family is well used by chemists who refer to certain of them by name. The elements of the first family (the left-hand column of the periodic table) possess a single electron in their valence shell. These are the alkali metals. (Hydrogen, although it too only has one electron is usually considered to lie on its own. It is the only notable exception to this idea of chemical families within the periodic classification.) The second column with two valence electrons are the alkaline earths. The penultimate family with seven valence electrons are the halogens and finally the last family with a filled valence shell are called by one of three names, either the rare, inert, or noble gases.

1.1.3. Systems with eight valence electrons

Among the different families of the periodic classification, that of the inert gases stands out because all the elements in this group have very little chemical reactivity. They have a very limited tendency to form strong chemical bonds with other atoms. The origin of this chemical inertness is the saturation of the valence shell by two electrons in the case of He or, more generally by eight electrons (Ne, Ar). Such chemical inertness is found too in the monatomic ions which also have a saturated valence shell. These ions may be obtained by adding one electron to a halogen atom which already has seven valence electrons (for example $Cl + e^- \rightarrow Cl^-$) or removal of the only valence electron from an alkali metal atom (for example $Na \rightarrow Na^+ + e^-$). These ions have a considerably reduced chemical reactivity when compared to the atoms from which they are derived.

1.1.4. The electronegativity concept

As we have described, the halogens can attain a position of chemical stability by the addition of a single electron to their valence shell. This property, seen for the isolated atom, also manifests itself in molecules, so a halogen atom will always have a tendency to attract electrons from a neighboring atom in order to complete its valence shell. Contrarily an alkali metal will tend to give up an electron. There is therefore in a molecule a reorganization of the electrons between those atoms with an affinity for electrons and those for which the opposite is true. All of the elements may be, by and

Table 1.2. The electronegativities (Pauling's scale) of the elements of the first three rows of the periodic table.

H 2.2						
Li 1.0	Be 1.6	B 2.0	C 2.6	N 3.0	O 3.4	F 4.0
Na 0.9	Mg 1.3	Al 1.6	Si 1.9	P 2.2	S 2.6	Cl 3.2

large, put into one of the two categories. The concept of electronegativity is used to describe this. *Electronegativity is the tendency of an atom in a molecule to attract electrons to itself.*

Electronegativity measures a 'tendency' and is not defined in an unequivocal way, but several definitions have been proposed and are described in the following chapter. The variation in electronegativity across the periodic table is, however, always the same, independent of definition. It increases from bottom to top and from left to right across the periodic table (Table 1.2). This last trend is easy to explain. The elements to the far left of the table with a small number of electrons will have a tendency to give them up to reach a closed shell configuration (elements with low electronegativities) while those situated at the right-hand side (with the exception of the inert gases which already have a closed shell) will have a tendency to accept electrons (elements with high electronegativities). For the set of elements given in Table 1.2 the most electronegative is therefore fluorine, and the least electronegative is sodium. We will return later to the quantitative scale (due to Pauling) used here.

1.2. Lewis' theory and Lewis structures

1.2.1. Bond pairs and lone pairs

The simplest description of a chemical bond, that of a pair of electrons shared between two atoms was proposed by Lewis around 1915. Each bonded atom can either provide one electron, or both electrons can come from the same atom. Consider, for example the dihydrogen molecule. Here each atom contains one proton and one electron and the chemical bond is formed by the sharing of the two (in this case the only two) electrons of the system. The bond is commonly represented as in **1-1** either as a pair of dots or more frequently as a line between the two atoms. The two electrons forming the bond are referred to as a *bond pair* or *bonding pair* of electrons.

1-1   H• + •H ⟶ H ⦂ H   or   H—H

If the atom possesses more than one valence electron it can form several bonds with different atoms. A very **simple** example is that of the BeH_2 molecule. The beryllium atom possesses two valence electrons (labeled with a cross in **1-2**) which can each form a bond with one atom of hydrogen. Similarly carbon, which has four valence electrons, is able to form four bonds with four hydrogen atoms in this way, leading (**1-3**) to the methane molecule (CH_4). Note that this scheme only describes

1-2   H• + ×Be× + •H ⟶ H ⚬× Be ×⚬ H   or   H—Be—H

1-3

$$\begin{array}{ccc} & H & & H \\ & \times\bullet & & | \\ H\; \substack{\bullet\\\times}\; C\; \substack{\times\\\bullet}\; H & & \text{or} & H—C—H \\ & \times\bullet & & | \\ & H & & H \end{array}$$

the arrangement of the bonds in the molecule, but does not give any indication of its spatial geometry. It is important not to confuse this picture of the molecule with any three-dimensional description. We will look at the geometry problem later (Section 1.4).

Sometimes an atom forms a smaller number of bonds than expected by considering the number of valence electrons. There then remain on the atom, electrons which do not participate in chemical bonding, and for this reason are called *nonbonding electrons*. These electrons, arranged in pairs, form one or more *lone pairs* (to distinguish them from bond pairs) and remain localized on the atom. For example in the ammonia molecule, NH_3, nitrogen has five valence electrons (×) and is linked to three hydrogen atoms. Three of the five electrons are used to form three N—H bonds, and the remaining two which are not involved in bonding form a lone pair. This is conventionally written as a bar or pair of crosses at the side of the nitrogen atom (**1-4**).

$$\begin{array}{ccc} & H & & H \\ & \times\bullet & & | \\ H\; \substack{\bullet\\\times}\; N\; \substack{\times\\} & & \text{or} & H—N| \\ & \times\bullet & & | \\ & H & & H \end{array}$$

1-4

An analogous bonding scheme can be constructed for the water molecule, H_2O. Oxygen, with its six valence electrons (×) can share two with the two hydrogen atoms, leaving four nonbonding electrons which form two lone pairs (**1-5**).

1-5 $$H\; \substack{\bullet\\\times}\; \overset{\times\times}{\underset{\times\times}{O}}\; \substack{\times\\\bullet}\; H \quad \text{or} \quad H—\underline{\overline{O}}—H$$

In the same way in hydrogen fluoride (HF), there is a bond between fluorine and hydrogen and therefore three lone pairs (**1-6**). This scheme which shows the way

1-6 $$H\; \substack{\bullet\\\times}\; \overset{\times\times}{\underset{\times\times}{F}}\; \substack{\times\\} \quad \text{or} \quad H—\overline{\underline{F}}|$$

electrons are distributed between bond pairs and lone pairs is called the *Lewis structure*.

Up until now we have only considered the case where an atom with more than one valence electron forms bonds to hydrogen. The bonding scheme between two atoms in general is often just as easy to treat. In the ethane molecule, C_2H_6, each carbon uses three electrons to form three C—H bonds. There then remain two electrons (one on each carbon) which come together and form a C—C bond (**1-7**).

1-7

In other molecules, more than one pair of electrons may be shared between two atoms. Depending upon whether there are two or three electron pairs to be shared, the two atoms are said to be doubly or triply bonded respectively. Let us take as an example the ethylene molecule, C_2H_4. Two electrons per carbon are used in the formation of C—H bonds. The remaining four electrons (two on each carbon) are shared to form two bonds. A double line between these atoms symbolizes the existence of the two bonds (**1-8**). According to the same principle the two carbon atoms are triply bonded in the acetylene molecule, C_2H_2 (**1-9**).

1-8

1-9

The number of bonds which are formed according to these ideas directly determines the attractive force between the two atoms; the more bonds the stronger the atoms are bonded to each other. From a geometrical point of view this translates as a shorter distance between the two atoms. So, for example, the C≡C triple bond in acetylene (120 pm) is shorter than the double bond in ethylene (134 pm) which is in turn shorter than the single bond in ethane (154 pm). Analogously the energy needed to break apart the two carbon atoms (the bond energy) increases with the number of bonds. It is about 351 kJ mol^{-1} for a single bond (C—C), 623 kJ mol^{-1} for a double bond (C=C) and finally 834 kJ mole^{-1} for a triple bond (C≡C).

1.2.2. The octet rule

Even for very simple molecules the construction of Lewis structures runs into a number of difficulties. For example in the nitrogen molecule, N_2, one can envisage three different bonding schemes which differ in the way the ten electrons (five pairs) are distributed between bond pairs and lone pairs. The two nitrogen atoms could be singly bonded with two lone pairs, each triply bonded with one lone pair each, or even quintuply bonded (i.e., five bonds) with no lone pairs at all. The octet rule determines the choice of the structure which most faithfully describes the molecule. It may be stated as follows: *The maximal stability of a molecule is found when each atom of the second or third row is surrounded by four electron pairs (bonded or nonbonded). The atoms of the first row (hydrogen and helium) only need a single pair of electrons.*

The justification for the rule comes from the observation of the chemical stability of the inert gases which possess eight valence electrons (two for helium). Returning to the nitrogen molecule the quintuply bonded structure is not a suitable one since each atom is associated with five electron pairs (ten electrons). In the singly bonded structure where each nitrogen possesses two lone pairs, there are only a total of three electron pairs (six electrons) around each atom. It is therefore the triply bonded structure where each atom has one lone pair and a total of four electron pairs (eight electrons) which best represents the structure of N_2 (**1-10**).

1-10 $|N{\equiv}N|$

The octet rule is practically a universal one for the atoms of the second and third rows, but it does, however, have some exceptions. There are in fact perfectly stable compounds where there are a smaller number of electrons, and compounds where there are a larger number.

(a) Compounds deficient in electrons

It is sometimes impossible to surround each atom with four pairs of electrons. The molecule trimethylborane, $B(CH_3)_3$, is an illustration of this. Boron has three valence electrons which allows formation of a bond with each of the carbon atoms. It is then surrounded by three pairs of electrons, one short of the number required for the octet rule. This situation is sometimes represented with an empty box at the side of the electron-deficient atom (**1-11**). Some molecules though are perfectly stable even

$$\begin{array}{c} CH_3 \\ | \\ H_3C-B-CH_3 \\ \square \end{array}$$

1-11

though they are electron deficient in this sense. They include many compounds based on clusters of atoms and several simple molecules. An example of the latter is the structure of the boron halides BX_3 (X = halogen). Here, boron with its three valence electrons forms three bonds, one to each halogen. Whereas the halogen atoms do satisfy the eight-electron rule only six electrons are present at the boron atom.

(b) Hypervalent compounds

The elements of the third row are able to form compounds where there are more than four pairs of electrons around the central atom. This is essentially limited to the elements silicon to chlorine in the third row, but is found too for the heavier elements of these families. So, for example in PCl_5 and SF_6, there are respectively five and six pairs of electrons around the central phosphorus or sulfur atom (**1-12**). This behavior is sometimes called *expanding the octet*.

1-12

1.2.3. Formal charges

Bond formation in molecules can arise in two ways. Either each atom contributes one electron or both electrons come from the same atom. Although each atom contributes one electron to the bond in **1-13a**, neither partner formally gains or loses an electron. Contrawise the two electrons in **1-13b** are contributed by the same

1-13a   A• + •B   ⟶   A—B

1-13b   A + ⦂B   ⟶   A—B

atom so that on forming the bond, one atom loses an electron and the other gains an electron. In order to trace this type of electron transfer we attribute to each atom in the molecule a number of electrons calculated in the following way. A bond pair is distributed equally between the two bonded atoms (one electron per atom). The two electrons of a lone pair are assigned to the atom on which they reside. Such an electronic assignment is a formal one since it apportions the electrons in a chemical bond equally and ignores the relative electronegativities of the atoms. Once this assignment has been done however, one can compare the number of electrons attributed to an atom (n_e) with the number of valence electrons (n_v). The difference is the *formal charge*, positive if there are fewer electrons associated with the atom in the molecule compared to the isolated atom, and negative if there are more. According to this scheme the molecule HCN is represented by a Lewis structure in which each atom is neutral, but the Lewis structure for the molecule CNH shows a charge

		Bond pairs	Lone pairs	n_e	n_v	Formal charge
H—C≡N⦂	H	1	0	1	1	0
	C	4	0	4	4	0
	N	3	1	5	5	0
⦂C≡N—H	H	1	0	1	1	0
	C	3	1	5	4	−1
	N	4	0	4	5	+1

separation. The molecule is neutral overall since the formal charges sum to zero. This is a general result; the sum of the formal charges is equal to the charge carried by the species.

1.2.4. Classification of reactants

The determination of the Lewis structure allows in general a simple description of the electronic structure for a molecule which may be used to understand aspects of its chemical reactivity. Three broad categories of reactants may be simply described starting off from these structures.

(a) Radicals

It is not always possible for all molecules to arrange the electrons in pairs (bonded or non-bonded) as we have described up until now. This is most obvious for molecules with an odd number of electrons. For example, in the CH_3 species which has seven valence electrons, carbon uses three of its four electrons in bonds to the hydrogen atoms. The fourth electron, unable to find a partner to make a lone pair or another bond pair, remains on the carbon atom (**1-14**). Molecules which contain such unpaired electrons are called *radicals*.

$$
\begin{array}{c}
H \\
| \\
H-C\cdot \\
| \\
H
\end{array}
$$

1-14

The radical center is generally surrounded by seven valence electrons and, following the octet rule, needs to complete its valence shell by the acquisition of one more electron, for chemical stability. The simplest route is that of *dimerization* of the radical so that the unpaired electrons on each center form a new bond pair viz:

$$H_3C\cdot + \cdot CH_3 \rightarrow H_3C-CH_3$$

This reaction though is not very common. Since there is usually a low concentration of radicals in a reaction medium the chances of them meeting to react in this fashion is low. In general a radical will react with another molecule where all the electrons are paired, to produce a new radical which then reacts in the same way with yet another stable species. In this way we get a *chain reaction* where the number of radicals present is conserved. The chlorination of methane, CH_4, using molecular chlorine, Cl_2, proceeds by this type of mechanism, initiated by a chlorine radical.

$$Cl\cdot + CH_4 \rightarrow \cdot CH_3 + HCl$$
$$\cdot CH_3 + Cl_2 \rightarrow CH_3Cl + \cdot Cl$$

In the same way the polymerization of ethylene $H_2C=CH_2$ can occur via a radical chain reaction where a radical $R \cdot$ initiates the reaction

$$R \cdot + H_2C{=}CH_2 \rightarrow R{-}CH_2{-}CH_2 \cdot$$

$$R{-}CH_2{-}CH_2 \cdot + H_2C{=}CH_2 \rightarrow R{-}CH_2{-}CH_2{-}CH_2{-}CH_2 \cdot$$

etc.

The chain reaction keeps going until all of the reactive species are consumed. It stops when reactions between two radicals (not very frequent) have consumed all the radicals present. In general terms such chain reactions consist of the three steps we have described; *initiation* (by a radical species), *propagation* (where the chain gets longer) and *termination* (where the reaction stops because the radicals are consumed).

(b) Lewis acids and bases

Lewis acids are species which contain an atom which is short of a pair of electrons to make an octet (or doublet for the first-row case). Like radicals these Lewis acids react in a way so as to complete ther electron shell. The ideal partner is a molecule with a lone pair which it can use to make a new bond. Molecules with such lone pairs are called *Lewis bases*. We can write an acid–base reaction in the Lewis sense as in **1-15**.

1-15 $A\square + |B \longrightarrow \overset{\ominus}{A}{-}\overset{\oplus}{B}$

The H^+ ion is a special example of a Lewis acid, since it has no electrons at all in its valence shell. Since it can be saturated by a single pair the proton reacts with a number of Lewis bases.

$$H^+ + NH_3 \rightarrow NH_4{}^+$$

$$H^+ + H_2O \rightarrow H_3O^+$$

In the special case of H^+ there is a nice overlap between the Lewis and Brønsted notions of acids and bases. However the Lewis ideas are more general than the Brønsted one since this is strictly limited to reactions of H^+. For example BF_3 (containing a vacant site) reacts with NH_3 (containing a nitrogen-located lone pair) to form a stable adduct via an acid–base reaction in the Lewis (but not Brønsted) sense

$$BF_3 + NH_3 \rightarrow F_3B^-{-}NH_3{}^+$$

Such reactions are used in organic chemistry in the synthesis of unstable intermediates such as the cations Cl^+ and $CH_3{}^+$ which are not easily obtained by other routes.

$$AlCl_3 + Cl{-}Cl \rightarrow AlCl_4{}^- + Cl^+$$

$$AlCl_3 + CH_3{-}Cl \rightarrow AlCl_4{}^- + CH_3{}^+$$

A particularly interesting example of the Lewis acid/base classification is the family of carbenes CX_2. These species are at the same time Lewis bases, since they have a lone pair on carbon, and Lewis acids since they only have six valence electrons. This electronic structure confers a special reactivity. For example dichlorocarbene can dimerize via a double acid/base reaction to give tetrachloroethylene (1-16). Carbenes and their silicon and germanium homologues form a class of reactive molecules which currently are often used in organic synthesis.

1-16
$$Cl_2C \quad + \quad CCl_2 \longrightarrow Cl_2C{=}CCl_2$$

1.2.5. Dipole moments of diatomic molecules

The *dipole moment* (μ) of a system which overall is electrically neutral stems from the existence of charges, $+q$ and $-q$ localized at two points separated in space. It is a vector quantity whose magnitude is equal to the product of the charge q and the distance, d, between these two points ($|\mu| = qd$). Conventionally one represents the dipole moment by an arrow which points from the positive end to the negative end. In general the unit used to measure dipole moments is the Debye (D) although this is not an SI unit. (1 D $= 3.34 \times 10^{-30}$ coulomb meters.)

An isolated atom does not possess a permanent dipole moment since the barycenter ('mass center') of the negative charges is located at the nucleus, where the positive charge is also centered. However in many molecules the barycenters of the positive and negative charges do not coincide. The result is the creation of a dipole moment. This quantity is extremely important in chemistry, being both readily determined experimentally and giving information about the electronic and geometrical structure of polyatomic molecules (Section 1.4.4).

In a symmetric diatomic molecule, such as H_2 or Cl_2, the two bonding electrons are equally distributed between the two nuclei since the atoms have the same electronegativity. These molecules do not have a dipole moment therefore and we say that their bonding is purely covalent. This is not the case if the two bonded atoms are different. The more electronegative atom preferentially attracts the electrons in the bond and leads to an asymmetry in the charge distribution. If we consider the electrons in the bond moving around both atoms then this asymmetry may be understood in terms of an electron pair which spends more time around one atom than the other. Overall there is an excess of electrons ($-\delta$) on the more electronegative atom and a deficiency of electrons ($+\delta$) on the less electronegative atom. We say such a bond is polar. For example, the larger electronegativity of hydrogen compared to lithium (Table 1.2) leads to the existence of a dipole moment in the Li—H molecule (1-17) directed from Li towards H. When the electronegativity

1-17
$$\overset{+\delta}{Li} - \overset{-\delta}{H}$$
$$\underset{\mu}{\longrightarrow}$$

difference between the atoms is very large the bond pair is monopolized by just one atom, and the molecule is described by the interaction between two ions, one positive and the other negative. We say such a bond is purely ionic. Such a situation probably does not exist in practice but ions held together largely by Coulombic forces are found in the gas phase, as solvated ion-pairs in solution and in what are called 'ionic' solids. A classical example of the latter is solid sodium chloride in which we can envisage the presence of Na^+ and Cl^- ions which, via their mutual coulombic attraction, hold the solid together.

Purely ionic bonds, where the two electrons of the bonding pair are found on a single atom, and purely covalent bonds, where the two electrons are equally shared between the two atoms, are the two extremes of the chemical bond (1-18). More

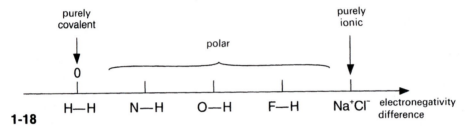

1-18

often, the two atoms do not have the same electronegativity, but the polarization of the bond only leads to a partial electron transfer from one atom to the other. The partial ionic character of a polar bond is simply determined as the ratio of the experimental dipole movement (μ_{exp}) to that which would be expected if the charge transfer were complete (μ_{ion}). (To compute the latter we need the experimental bond length.) This ratio varies from 0 (a purely covalent bond with $\mu_{exp} = 0$) to 1 (a purely ionic bond with $\mu_{exp} = \mu_{ion}$). As an example consider the HF molecule. The experimentally determined dipole movement of this species is $\mu_{exp} = 1.82$ D and the distance between the two atoms is 92 pm. If the bond were purely ionic (H^+F^-) the dipole moment would be

$$\mu_{ion} = q \times d = 1.6 \times 10^{-19} \text{ (coulombs)} \times 0.92 \times 10^{-10} \text{ (meters)}$$

$$= 1.472 \times 10^{-29} \text{ C m}$$

$$= 4.41 \text{ D}$$

The ionic character of the H—F bond is then equal to $\mu_{exp}/\mu_{ion} = 1.82/4.41 = 0.41$ or 41%. We could represent the bond as $H^{+\delta}F^{-\delta}$ with $\delta = 0.41$.

1.3. Resonance or mesomerism

In all of the cases we have studied up until now, a single Lewis structure gives a satisfactory description of the molecule. The bonding scheme in which the electrons are distributed between lone pairs and bond pairs was in accord with experimental observables such as the correlation of bond lengths with bond multiplicity, acid/base properties and radical chemistry. There are however, many molecules for which a single Lewis structure does not give a correct qualitative description and which leads to erroneous predictions.

1.3.1. Examples of mesomeric structures: carbonate ion ($CO_3{}^{2-}$) and benzene (C_6H_6)

In the carbonate ion, $CO_3{}^{2-}$, the carbon atom can form four bonds with three atoms of oxygen (**1-19**) which lead to two single bonds (C—O_1 and C—O_2) and one double bond (C=O_3). The two extra electrons contributed by the two negative charges are used to complete the octet (i.e., fill the valence shell) of the two oxygen atoms O_1 and O_2. The octet rule is satisfied for all four atoms and the description is thus apparently a satisfactory one. However in this Lewis structure the oxygen atoms are not all equivalent. Two (O_1 and O_2) are singly bonded and one (O_3) is doubly bonded to the central carbon atom, a description which runs counter to the experimental observation that all three C—O distances are of the same length (129 pm). *Thus the representation of the carbonate ion by a single Lewis structure is not correct.*

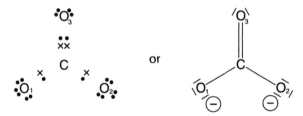

1-19

A way around this problem is the following. We may construct three Lewis structures which differ only in the location of the C=O double bond (**1-20**), the oxygen lone pairs and the formal charges on the oxygen atoms. The carbonate ion is then described as a superposition, or a mixture of the three forms. We say that there is *resonance* between three *mesomeric* forms, represented by a double-headed arrow. It is important not to confuse this notation with that of a chemical equilibrium (⇌). In the latter case the species connected by the arrows are real chemical entities which can in many cases be observed or isolated. Here though the arrow indicates the inability to represent the structure of a single species by just one Lewis structure. So the experimental structure in which all the CO distances are equal is represented by a mixture of these three structures, each contributing equally to the overall structure. If the weights of the three mesomeric forms were not the same then the three bonds would not have equal lengths.

1-20

In the construction of these mesomeric forms the 'skeleton' of the molecule, namely the collection of nuclei and the single bonds remains invariant. The conversion of one mesomeric form to another is usually represented by curved arrows to describe the movement of the electron pairs from oxygen to CO bond and vice-versa (1-21).

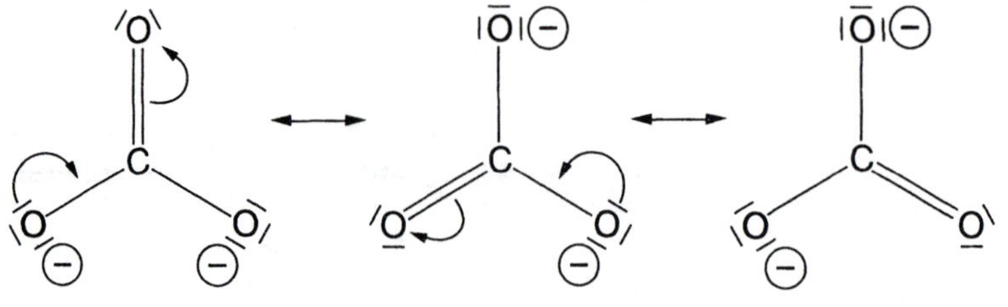

1-21

A particularly important application of resonance is in the description of the benzene molecule, C_6H_6. Experimentally it has been shown that the molecule is perfectly symmetrical and in particular that all of the CC bonds are equal. Just as in the previous example writing a single Lewis structure (1-22) does not lead to such a feature. This single structure implies an alternation between long and short CC distances around the ring appropriate for single and double bonds respectively. In order to reproduce the observed structure with equal CC distances another meso-meric form, related to the first, needs to be introduced. These two mesomeric forms (1-23) are known as the *Kékulé structures* of benzene, and need to be included with equal weight. One result of such a description is that the bonds between the carbon atoms lie somewhere between single (C—C) and double (C=C) bonds. This is confirmed by the experimental value of the C—C distance in benzene. At 140 pm it is effectively intermediate between that of a single bond (154 pm) and that of a double bond (134 pm). Besides the geometrical consequences of the resonance we have just described, there is another interesting result. This is that the three double bonds are not exactly localized between three pairs of carbon atoms but are delocalized

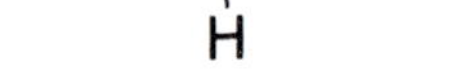

1-22

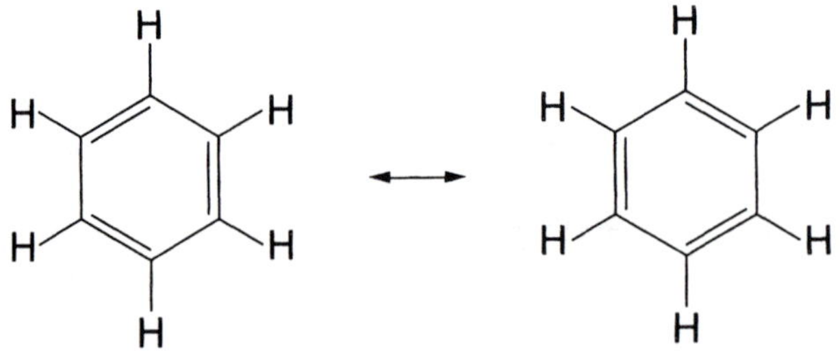

1-23

over the whole ring. One way of describing this particular electronic state of affairs is to draw a circle within the hexagonal benzene skeleton (**1-24**) to represent these six electrons delocalized over six C—C linkages. This delocalization has a number of consequences, notably that the chemical reactivity of benzene is very different from that of compounds such as ethylene where the double bond is localized between a pair of carbon atoms.

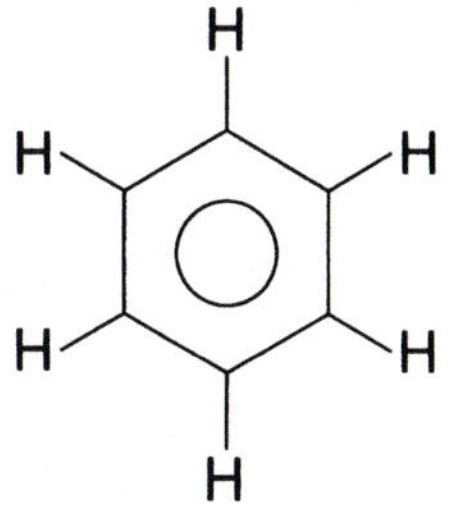

1-24

1.3.2. Selection of mesomeric or resonance structures

For the average molecule, without the symmetry constraints of the carbonate ion or benzene, it is often possible to write several mesomeric forms. There often exists a single structure which is reasonably satisfactory, but the addition of other less important structures (which accordingly receive a smaller weight in the overall structure) can often improve the electronic description of the molecule. Since it is usually possible to write a large number of resonance structures, it is important to eliminate those whose contribution is negligible. The following three rules are good ones to follow:

(i) As far as possible respect the octet rule.
(ii) Maximize the number of bonds.
(iii) Avoid too many atoms which have formal charges. Especially avoid the localization of two charges on the same atom.

1.3.3. Application to the structure of aniline

Consider the aniline molecule $C_6H_5NH_2$. Like benzene this molecule has to be represented by the two Kékulé structures shown in **1-25**. In addition to these we can write the three other mesomeric forms of **1-26** by using the nitrogen lone pair to

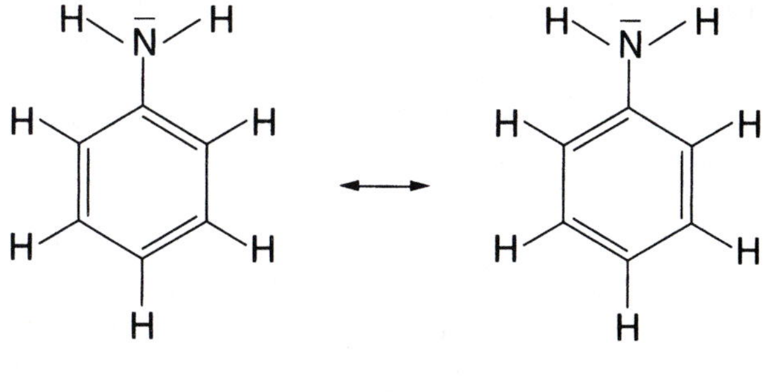

1-25

1-26a   **1-26b**   **1-26c**

form a double $N=C_1$ bond. (In **1-26** we have not shown the hydrogen atoms attached to the ring carbon atoms. This is a frequently used shorthand for cyclic molecules such as these.) At the same time we have to get rid of the double bond $C_1=C_2$ and put a lone pair on C_2 to satisfy the octet rule (**1-26a**). Only two double bonds remain in the benzene ring. It is easy to see that nitrogen carries a formal charge of $+1$ (four electrons in place of five) and that the carbon atom carrying the lone pair has a formal charge of -1 (five electrons in place of four). The other two mesomeric forms (**1-26b, c**) are simply obtained by distributing the two double bonds of the ring in a different way. The negative charge is then found on C_4 (**1-26b**) or on C_6 (**1-26c**). These three structures are less important than those of **1-25** since they come about as a result of charge separation. However they contribute in a non-negligible way to the description of the molecule. This may be seen by consideration of the length of the C—N bond found experimentally for this molecule. At 140 pm it is somewhat shorter than a typical C—N single bond (147 pm) but significantly longer than a typical C=N double bond (132 pm). This implies admixture of the resonance forms **1-26** with their formal C=N double bonds into the actual electronic description of the molecule. Another consequence of the involvement of these structures will be examined in Section 1.4.3b.

1.4. Molecular geometry

1.4.1. Spatial representation of molecules: Cram's model

The representation on a piece of paper of a three-dimensional object is facilitated by the use of some drawing conventions. Several, including those of Newman and Fischer are used in specific areas of chemistry but Cram's is of general utility and is described here.

(i) The bond between two atoms lying in the plane of the page is represented by a straight line (**1-27**).

1-27   A——B

(ii) The bond between an atom A lying in the plane of the page and an atom B

1-28   B ➤ A

lying in front of it is represented by a solid wedge (**1-28**). To encourage a
perspective view the thick part of the wedge is next to the atom which lies
closer to the viewer (i.e., B).

(iii) The bond between an atom A lying in the plane of the page and an atom B
located behind it is represented by a dashed wedge in a similar way (**1-29**).
Here though the thicker part of the wedge lies further from the viewer.

1-29   B �(llll)... A

Some common geometries found for small molecules are shown below using these
conventions (**1-30**).

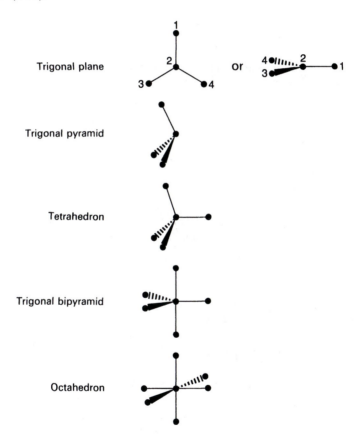

Trigonal plane

Trigonal pyramid

Tetrahedron

Trigonal bipyramid

Octahedron

1-30

1.4.2. VSEPR theory

In this theory, developed by Sidgwick, Powell, Nyholm and Gillespie (and sometimes
called the Nyholm–Gillespie theory) the shape of a molecule may be predicted by
counting the number of bond pairs and lone pairs around each atom, from whence

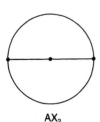

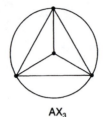

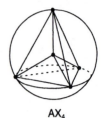

AX₂   AX₃   AX₄

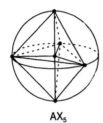

AX₅   AX₆

Figure 1.1. Minimum energy arrangements of n points on a sphere which simulates the arrangement of electron pairs around a central atom.

its abbreviation derives: valence shell electron pair repulsion. The rule is as follows: *The valence electron pairs associated with an atom, both bond pairs and lone pairs, arrange themselves so as to minimize the repulsions between them.* Figure 1.1 shows the lowest energy geometries for pairs of electrons on this model. The results are simple to understand since the problem is akin to that of the arrangement of charged particles on a sphere.

Using this rule we only need to know the Lewis structure for a molecule to be able to predict its geometry. We will apply the theory to a range of molecules starting with those which are highly symmetric.

(a) AX_n molecules

We first of all consider those AX_n molecules in which the atom A does not carry a lone pair. The case of $n = 2$ is the simplest to analyze. In order to minimize the repulsion between the two bond pairs around atom A we need to place one directly opposite the other. A linear geometry with an angle of 180° between the two A—X bonds results as in $BeCl_2$. When $n = 3$ minimization of the repulsion between three pairs of electrons leads to a trigonal planar geometry where the bond angles are 120° as in BF_3. For $n = 4$ the optimal geometry is a tetrahedron, for $n = 5$ a trigonal bipyramid and for $n = 6$ an octahedron. Table 1.3 summarizes the geometries found and **1-31** depicts the structures for the examples we have chosen.

Note that the last two examples are of molecules which do not obey the octet rule (they have 10 and 12 electrons respectively around phosphorus and sulfur). As we have noted these hypervalent molecules are typical of compounds from the third row of the periodic table and their heavier analogs. The first two examples do not satisfy the rule either, having too few electrons.

Table 1.3: Geometries of AX_n molecules

Formula	n	Geometry	X—A—X angle	Example
AX_2	2	linear	180°	$BeCl_2$
AX_3	3	trigonal plane	120°	BF_3
AX_4	4	tetrahedron	109.5°	CH_4
AX_5	5	trigonal bipyramid	120° and 90°	PF_5
AX_6	6	octahedron	90°	SF_6

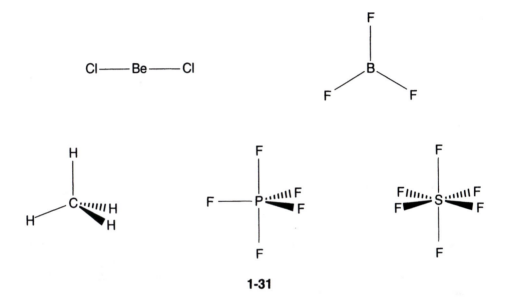

1-31

(b) Molecules containing lone pairs on A (AX_nE_m)

When the central atom contains lone pairs of electrons, which we label as E, the molecular geometry is obtained using the same rules. The geometry does, however, depend upon the number of lone pairs. Take, for example, the set of molecules where the atom A is surrounded by four pairs (**1-32**); methane, CH_4, is tetrahedral (AX_4), ammonia, NH_3, is pyramidal (AX_3E) and water, H_2O, is bent (AX_2E_2). All of these molecules are based upon a tetrahedral disposition of bond and lone pairs and the

1-32

AX₄	AX₃E	AX₂E₂
Tetrahedral	Pyramidal	Bent

Table 1.4: Geometries of molecules containing lone pairs

Formula	$n + m$	Geometry	Example
AX_2E	3	bent	SiH_2
AX_3E	4	pyramidal	NH_3
AX_2E_2	4	bent	OH_2

geometry of the atomic framework reflects this (**1-33**). Tables 1.3 and 1.4 summarize the various geometries found as a function of the number and nature of the electron pairs around the central atom for $n + m = 2, 3, 4$.

1-33

(c) Molecules with multiple bonds

The extra electron pairs associated with double or triple bonds occupy the same region of space between the two bonded atoms as the single bond. At the simplest level their presence does not influence the molecular geometry. So, for example in the CO_2 molecule, whose dominant Lewis structure is $\bar{O}\!=\!C\!=\!\bar{O}$, we only need to consider the four electrons around the central carbon atom involved in forming the single bonds when using the VSEPR rules. These four electrons (two bond pairs) thus lead to a linear AX_2 structure analogous to that of $BeCl_2$ (Table 1.3). The carbonate ion CO_3^{2-} (**1-19**) thus has three pairs of electrons around the carbon atom for VSEPR considerations and, as an AX_3 molecule, is planar. In practice therefore, one treats multiple bonds as a single pair for the determination of the angles between the bonds in a molecule.

(d) Asymetrically substituted molecules

In general the central atom A is attached to atoms of different types. However we can obtain a geometrical description close to reality by using the rules described above without distinguishing between the nature of the atoms bound to A. Accordingly the H—C—Cl angle in methyl chloride is expected to be close to $109.5°$ with reference to the symmetric structure of CH_4 and CCl_4. Experimentally this angle is $108.9°$. Similarly in ethylene, $H_2C\!=\!CH_2$ the H—C—C angles are $122°$ which again is very close to the ideal angle of $120°$. We can therefore get a very good idea of the geometry of a complex molecule by considering that of a symmetric reference geometry without taking into account the nature of the substituents. X—A—X' molecules, such as HCN, have to be linear. The same considerations apply to molecules which contain lone pairs. In molecules of the AX_nE_m type with $n + m = 4$ a tetrahedral geometry with angles of 109.5° is expected. In fact the angles between the bonds are different, but only slightly, from this ideal value when A comes from the second row. In ammonia the HNH angles are 107.3° and in water the HOH angle is 104.5°. When A belongs to the third row larger deviations are found; 93.3° and 92.1° in PH_3 and H_2S respectively. Generally the bond angles in AX_nE_m molecules are smaller than those in the symmetrical AX_{n+m} analogs, implying that the repulsions involving lone pairs are somewhat larger than those involving bond pairs.

1.4.3. Extensions of VSEPR theory

(a) Radicals

The VSEPR theory is based on the idea of repulsions between pairs of electrons, so we have to be careful in extending it to radicals, species which by definition have unpaired electrons. Take, for example, the $SiH_3 \cdot$ radical. Its geometry can be compared with that of its anion, SiH_3^-, and cation, SiH_3^+. The latter is an ion of the AX_3 type where repulsion between the three bond pairs leads to a trigonal planar structure. In SiH_3^- the presence of an extra lone pair at the central atom leads to a pyramidal geometry. In the $SiH_3 \cdot$ radical the situation lies between the two since here there is a single non-bonding electron on silicon. This electron exerts a repulsion on the bond pairs which is absent in the cation and less important than in the anion. We would expect perhaps a geometry intermediate between anion and cation. Experimentally the structure is found (**1-34**) to be slightly pyramidal with a bond

1-34

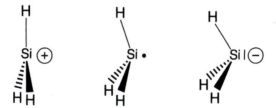

angle (112.5°) which lies between that found for the anion (94.5°) and that found for the cation (120°). This result is however not a general one. The $CH_3 \cdot$ radical which has the same number of valence electrons as $SiH_3 \cdot$ is flat, showing that in this case the repulsion exercised by the unpaired electron is too weak to force the molecule into a pyramidal structure. However, the interconversion of planar and pyramidal geometries requires little energy.

(b) Importance of resonance structures

If a molecule is described by a collection of resonance forms then it is possible that the geometries of all of them may not be the same. Take the aniline molecule for example. As we have already shown, this molecule is principally described by the two Kékulé structures, **1-35a**, with smaller contributions from the resonance

1-35a   **1-35b**   **1-35c**   **1-35d**

structures **1-35b, c, d**. Recall their importance in determining the C—N distance (Section 1.3.3). Here we show how they influence the bond angles at nitrogen. In the Kékulé structures, **a**, the nitrogen atom is surrounded by a lone pair and three bond pairs (AX$_3$E) which should lead to a tetrahedral environment of electron pairs and a pyramidal arrangement of the bonds. The pyramidal angle, ϕ, should then be close to 60° (**1-36a**). The geometry expected for the forms **b, c, d** is however different. Nitrogen is doubly bonded to carbon and now does not have a lone pair. This corresponds to an AX$_3$ system with a trigonal planar geometry and $\phi = 0°$. The experimental value of ϕ of 39° (**1-36**) which lies between the two limiting values of 0° and 60°, indicates the importance of these other structures. Not only is there a geometrical consequence of the participation of these resonance structures in the observed C—N distance, and the flattening of the amino group, but there is an energetic one too. The energy needed to convert the pyramidal to planar structure (called the inversion barrier) is only 8 kJ mol^{-1} to be compared with the 24 kJ mol^{-1} found for NH$_3$.

$\phi \approx 60°$ $\phi \approx 39°$ $\phi = 0°$

1-36   **a**   experimental geometry   **b, c,** or **d**

1.4.4. Dipole moments of polyatomic molecules

In the average polyatomic molecule the total dipole moment comes from the vector sum of the different dipole moments associated with the bonds in the molecules and contributions from the lone pairs. Consider as an example the AlCl$_3$ molecule, where each Al—Cl bond is polar as a result of aluminum having a smaller electronegativity than chlorine. The total dipole moment of this molecule is thus the sum of the three moments associated with the three Al—Cl bonds. Since the molecule is of the AX$_3$ type, the three vectors representing the bond dipole moments make angles of 120° with each other. If μ_0 is the magnitude of each vector then it is clear to see that the projection along both x and y axes, $0 + \mu_0 \cos 30° - \mu_0 \cos 30°$ and

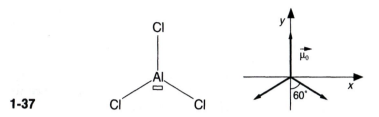

1-37

$\mu_0 - \mu_0 \cos 60° - \mu_0 \cos 60°$ respectively, is identically zero (**1-37**). Thus in this molecule the total dipole moment is zero even through the bonds are polar. This is a direct consequence of the very symmetrical structure adopted by the molecule. Conversely the experimental observation of a zero dipole moment for $AlCl_3$ is proof of a trigonal planar geometry for the molecule. If the molecule were a trigonal pyramid the three bond vectors would not sum to zero.

In the case of water, each OH bond is polar and there are two lone pairs on oxygen. The two bond moments do not cancel each other here since the molecule is bent (**1-38**), and a dipole moment results along the z-axis. In addition the lone pairs on the oxygen atom are located in a plane perpendicular to the molecular plane and also lead to a contribution to the total dipole moment along z. The total dipole moment for the molecule therefore comes from both the polarity of the O—H bonds and the presence of the lone pairs.

1-38

1.5. Conclusion

As we have just shown in the course of this first chapter, classical theories of electronic structure allow a basic understanding of the chemical reality of molecules. They have the advantage of being simple and sufficiently easy to apply that they are useful for chemists of widely differing persuasions. They also provide a rough picture of molecular structure which may be used before moving on to a more precise theory. Nevertheless their application runs into several difficulties which highlight their limitations.

The first difficulty lies in the fact that for many molecules it is impossible to find just one Lewis structure which is satisfactory. A whole set of such structures may have to be used. Individually though they are not chemically significant, and since we really don't know in general how to weigh them, the problem remains. This suggests that the basic assumption of the method, that electron pairs are either localized between two atoms or lie on a single atom, can be at fault. Other problems present themselves, even in the simplest molecules. We give some examples.

    (i) In the C_2 species the Lewis structure which obeys the octet rule and appears to give the largest number of bonds without the generation of formal charges is the quadruply bonded structure C≡≡C. There are many reasons to believe that this formula, which satisfies all of the rules enunciated above, well represents the structure of the molecule. However the CC distance in C_2 is 124 pm, a value intermediate between those for double and triple bonds. We should expect a much shorter value for a quadruple bond. Also the energy needed to break the bond in the C_2 molecule is smaller than that needed to destroy a triple bond. This suggests that the most important Lewis structures for C_2 should contain a double bond ($|C\!=\!C|$) and a triple bond ($\overset{+}{C}\!\!\equiv\!\!\bar{C}$ and $\bar{C}\!\!\equiv\!\!\overset{+}{C}$). So, the octet rule doesn't hold in these structures and some of them contain formal charges. How then can they be favored over C≡C?

(ii) In the O_2 molecule the Lewis structure $\bar{O}\!=\!\bar{O}$ fulfills all of the requisite conditions to properly describe the molecule. However it can be readily demonstrated experimentally that there are two unpaired electrons in this molecule. Why doesn't this property show itself in the electronic description of the molecule?

(iii) In the ethane molecule (C_2H_6) the two carbon atoms are singly bonded and the barrier for rotation around the C—C bond is rather small (around 12 kJ mol^{-1}). Contrast this with the situation in ethylene. Here the carbon atoms are doubly bonded and rotation around this bond is very high (around 260 kJ mol^{-1}). The second C—C bond is therefore of a very different nature than the first. A related point is that VSEPR is unable to explain the planar nature of ethylene. One can quite easily imagine two planar CH_2 groups arranged so that their molecular planes are perpendicular, but the molecule is quite decidedly flat.

(iv) The water molecule contains two oxygen lone pairs. Nothing distinguishes one from the other so that one would think that the energy associated with the removal of an electron from them would be the same. The experimental reality is quite different; the energies concerned are different, by 2.1 eV in fact. This shows conclusively that the two lone pairs in water are not equivalent. Why is this?

(v) Finally we mention a geometrical problem. The ·BH_2 radical is bent by virtue of the repulsion between the unpaired electron and the bond pairs. If the radical is excited (we use a star to denote this) the geometry becomes linear (1-39). Yet it always behaves chemically like a species with two B—H bonds and an unpaired electron on boron.

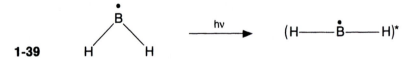

1-39

The explanation of these experimental results (and of many others) has to come from a more exact model for the electronic structure of molecules, and a model which uses results from quantum mechanics. The theory of molecular orbitals, outlined in Chapters 3 to 12 is one of the 'modern' models of molecular electronic structure. Before developing these ideas it is, however, necessary to study (in Chapter 2) the essential characteristics of the electronic structure of atoms, since these are the building blocks of molecules.

EXERCISES

1.1 Give the Lewis structure and the spatial geometry of the following molecules, propane ($CH_3CH_2CH_3$), methylamine (CH_3NH_2), ethanol (C_2H_5OH), dimethyl-ether (CH_3OCH_3), acetic acid (CH_3COOH), acetone (CH_3COCH_3).

1.2 (i) Write down the Lewis structure which obeys the octet rule for the molecules H_2CO and HCN.

(ii) Work out their geometries.

(iii) Predict the orientation of their dipole moment given that the moments associated with the C—H bonds are negligible. (From Table 1.2 the electronegativities of carbon and hydrogen are similar).

(iv) Give two Lewis structures for an isomer of HCN. (This is to say a molecule with an arrangement of bonds different from that used in (i) above.)

(v) With the knowledge that this isomer is linear, predict the most important Lewis structure.

1.3 Show that the two isomers of 1,2 dichloroethylene, *cis* and *trans*, have different dipole moments.

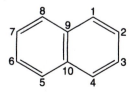

cis   *trans*

1.4 Considering the ammonium ion, NH_4^+, and the dimethylsulfoxide molecule (DMSO), $(CH_3)_2SO$.

(i) Show that the Lewis structure for each contains formal charges.

(ii) Give the geometry of each molecule.

(iii) What is the value of the dipole moment of NH_4^+?

1.5 A Lewis structure for naphthalene is shown below.

(i) Find two resonance forms which have no charge separation which are also important in the description of the molecule.

(ii) With the help of these resonance structures provide a rationale for the differences in bond lengths measured experimentally for this molecule.

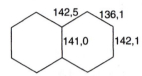

1.6 (i) Give a Lewis structure for cyanamine H_2NCN which does not place formal charges on the atoms. What is the spatial geometry for this structure?

(ii) Generate another Lewis structure by delocalizing the lone pair on the amino group. What is the geometry associated with this structure?

(iii) Experimentally it is found that the amino group is slightly pyramidal and the inversion barrier very small. Explain this result.

1.7 Collect together all the molecules with the formula H_4C_2O in which the octet rule is satisfied at each atom.

(i) Give the Lewis structure and the geometries of the two molecules containing the C—C—O unit.

(ii) Do the same for a cyclic molecule.

(iii) Give a structure for the molecule containing the C—O—C unit. Show that one cannot be written without charge separation. Complete the description of the molecule knowing that the two CO bonds are equal.

1.8 (i) Show that there exists for the carbon monoxide molecule (CO); (a) A Lewis structure where the octet rule is obeyed at each atom but which contains a separation of charge, and (b) a Lewis structure which does not have a separation of charge but does not obey the octet rule.

(ii) Consider the bond between M and CO in an MCO species where M is a Lewis acid, i.e., contains a vacant site (M for example could be a transition metal). Using the resonance forms found above write down the different bonding schemes for the M—CO bond in the two following cases; (a) M does not contain a lone pair, (b) M has a lone pair able to be delocalized in addition to a vacant site.

2 Properties of atoms

A number of experimental observations have shown that the movement of micro-scopic particles cannot be correctly described within the framework of classical or Newtonian mechanics. Since the development of quantum mechanics in the 1920s, this tool has become indispensable in understanding phenomena at the microscopic scale, from those associated with the atomic nucleus to those associated with molecules. The cornerstones of this approach require a substantial body of mathe-matical background but we will limit ourselves to just enough needed for an elementary description of atomic structure. We shall unashamedly sidestep many proofs and concepts which are too complex for this book and refer the interested reader to more specialist books in the bibliography.

2.1. Elements of quantum mechanics

Just as in classical mechanics, quantum mechanics possesses its fundamental equa-tion, the Schrödinger equation from which, *in principle*, most properties may be derived. We will illustrate some of its general characteristics by using the hydrogen atom as an example. This, with its single electron moving around a nucleus comprising a single proton, is surely the simplest system which one can study.

2.1.1. Schrödinger's equation

Within the language of quantum mechanics the electron is not described as a point mass associated with a trajectory in space, but rather as a wave represented by the mathematical function Ψ whose value depends upon the x, y, z coordinates of the space which it occupies. Ψ is usually referred to as the *wavefunction*. This very non-classical idea makes contact with the notion that atoms consist of a collection of electrons surrounding a nucleus in the following way. The square of this function $\Psi^2(x, y, z)$ represents the probability density of finding the electron at this point. (Sometimes Ψ will turn out to be a complex function containing real and imaginary parts. In this case we need to use $\Psi\Psi^*$ where Ψ^* is the complex conjugate of Ψ). In other words the probability dP of finding the electron in an infinitesimal volume $d\tau$ centered around the point given by the coordinates x_0, y_0, z_0 is given by the expression

$$dP = \Psi^2(x_0, y_0, z_0)\, d\tau \tag{1}$$

The probability of finding the electron somewhere in space has to be unity and so the function Ψ has to satisfy the relationship (2), where the integration takes place over all space. We say that Ψ is *normalized*.

$$\int_{space} \Psi^2(x, y, z)\, d\tau = 1 \tag{2}$$

The function Ψ is determined by solution of the Schrödinger equation (3). Here $\mathscr{H}$ is an operator, called the *Hamiltonian*

$$\mathscr{H}\Psi = E\Psi \tag{3}$$

which operates on the function Ψ and transforms it into another function. E is a scalar whose value is equal (in the cases we shall study) to the energy of the electron. Solution of equation (3) consists, therefore, of finding those functions, which after being transformed by the operator $\mathscr{H}$, may be written as a product of the original function multiplied by a scalar. Such solutions are called the *eigenfunctions* of the operator $\mathscr{H}$ and the scalars with which they are associated, the corresponding *eigenvalues*. We finally note that from all the possible solutions we only retain those which make physical sense, namely those which are normalized as in equation (2). *Solution of the Schrödinger equation amounts to searching for the collection of pairs* (Ψ_i, E_i) *which satisfy the relationships* (2) *and* (3).

The principles behind the solution of the Schrödinger equation are difficult to illustrate in a simple fashion, but we can give a simple analog from the field of vector geometry. Let us consider an operator $(\mathbf{R})$ which behaves like a planar mirror (2.1) and search for the eigenfunctions (v) and eigenvalues (λ) obtained by solution

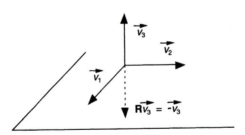

2-1

of the equation $\mathbf{R}v = \lambda v$. Obviously there are two types of functions. Those (for example v_1, v_2) which lie completely in this plane are transformed by the mirror plane operation into themselves and are therefore associated with an eigenvalue of $+1$ (i.e., $\mathbf{R}v_1 = (+1)v_1$). Those which lie perpendicular to the plane are transformed into vectors pointing in the opposite direction and therefore have an eigenvalue of -1. (i.e., $\mathbf{R}v_3 = (-1)v_3$.) Three solutions of $\mathbf{R}v = \lambda v$ are therefore $(v_1, +1)$, $(v_2, +1)$ and $(v_3, -1)$.

Among the various solutions of the Schrödinger equation, that which corresponds to the lowest energy of the system (Ψ_1, E_1) has a special importance. This is the state of the system, the *ground state*, which is the most stable. All the other solutions (Ψ_i, E_i) correspond to *excited states*.

2.1.2. Some important properties of the eigenfunctions

Some of the properties of the eigenfunctions of the Hamiltonian operator will be useful to us in what follows in subsequent chapters. To establish these we need first to state that the operator $\mathscr{H}$ is linear, that is to say it satisfies the relationships

$$\mathscr{H}(\lambda\Psi) = \lambda\mathscr{H}(\Psi) \tag{4}$$

$$\mathscr{H}(\Psi_i + \Psi_j) = \mathscr{H}(\Psi_i) + \mathscr{H}(\Psi_j) \tag{5}$$

(a) The sign of ψ

Let us suppose that Ψ_i is a normalized solution of the Schrödinger equation associated with an eigenvalue E_i such that

$$\mathscr{H}\Psi_i = E_i\Psi_i \tag{6}$$

with

$$\int_{space} \Psi_i^2 \, d\tau = 1 \tag{7}$$

Let us consider for the moment the function $\lambda\Psi_i$, where λ is a scalar. Combining the relationships (4) and (6) we get

$$\mathscr{H}(\lambda\Psi_i) = \lambda\mathscr{H}(\Psi_i) = \lambda E_i\Psi_i = E_i(\lambda\Psi_i) \tag{8}$$

Thus the function $\lambda\Psi_i$ is also a solution of the Schrödinger equation and is associated with the same eigenvalue (E_i) as Ψ_i itself. This function has to be normalized such that

$$\int_{space} (\lambda\Psi_i)^2 \, d\tau = 1 \quad \text{i.e.} \quad \lambda^2 \int_{space} \Psi_i^2 \, d\tau = 1 \tag{9}$$

so that $\lambda^2 = 1$ or $\lambda = \pm 1$. There are therefore two solutions for λ, the first corresponding to the initial function Ψ_i ($\lambda = 1$) and the second to its negative, $-\Psi_i$ ($\lambda = -1$). Thus if Ψ_i is a normalized solution of the Schrödinger equation, so is its negative, $-\Psi_i$, and thus corresponds to the same electronic description as a result. *A physical sense may not then be attributed to the sign of the wavefunction. The function Ψ_i may be used as equally well as its negative $-\Psi_i$.*

(b) Overlap and the orthogonality of eigenfunctions

The overlap integral of two functions Ψ_i and Ψ_j is the integral over all space of their product. We use a useful notation, due to Dirac, to write this in a condensed fashion

$$\langle \Psi_i \mid \Psi_j \rangle = \int_{\text{space}} \Psi_i \Psi_j \, d\tau \qquad (10)$$

where $d\tau$ is the infinitesimal volume element. (Equation (10) describes the situation for real functions Ψ. For complex functions we need to use the expression $\int_{\text{space}} \Psi_i^* \Psi_j \, d\tau$ where Ψ_i^* is the complex conjugate of Ψ_i.) One of the properties of the set of eigenfunctions Ψ is that the overlap integral is zero (equation (11)) if they are different but of course equal to unity (equation (12)) if they are normalized according to equation (2).

$$\langle \Psi_i \mid \Psi_j \rangle = 0 \text{ if } i \neq j \qquad (11)$$

$$\langle \Psi_i \mid \Psi_i \rangle = 1 \qquad (12)$$

Thus the collection of eigenfunctions of the Hamiltonian operator form a set of orthonormal functions.

An analog might be the set of orthogonal vectors i, j, k normalized to be of unit length which are orthogonal and define the x, y and z directions of three-dimensional space.

(c) Degenerate solutions

Suppose for the present that we have two different solutions Ψ_i and Ψ_j which are associated with the same eigenvalue E, i.e.,

$$\mathscr{H} \Psi_i = E \Psi_i \qquad (13)$$

$$\mathscr{H} \Psi_j = E \Psi_j \qquad (14)$$

We say that these two solutions are degenerate. Applying the Hamiltonian operator to an arbitrary linear combination of them ($\lambda \Psi_i + \mu \Psi_j$) and using the relationships of equations (4), (5), (13) and (14), we get

$$\mathscr{H} (\lambda \Psi_i + \mu \Psi_j) = \lambda \mathscr{H} (\Psi_i) + \mu \mathscr{H} (\Psi_j) = \lambda E \Psi_i + \mu E \Psi_j = E(\lambda \Psi_i + \mu \Psi_j) \qquad (15)$$

Thus all linear combinations of two degenerate eigenfunctions are themselves eigenfunctions of the Hamiltonian, associated with the same eigenvalue, E.

2.2. The hydrogen atom

2.2.1. Solutions of the Schrödinger equation

Solution of the Schrödinger equation for the case of the movement of a single electron under the influence of the nucleus leads to an infinite set of (Ψ_i, E_i). The two particles, nucleus and electron, interact via a Coulombic, electrostatic type of interaction and, if we were going to solve the equation algebraically, we would insert the kinetic and potential energy into the Hamiltonian to produce (as it turns out) a differential equation which is readily soluble.

(a) Allowed values of the energies (E_i) and atomic spectra

This book is not the place to detail the mathematical solution of the Schrödinger equation, but highlight the results. First, the eigenvalues of the Hamiltonian, E_i are all negative and second they are inversely proportional to n^2, where n is a positive integer called the principal quantum number.

$$E_n = -Ry/n^2 \qquad n = 1, 2, 3, \ldots \tag{16}$$

Ry is a constant with units of energy and equal to 13.6 eV*. The allowed values of the energy are therefore

$$E_1 = -Ry \qquad (n = 1) \qquad \text{ground state}$$
$$E_2 = -Ry/4 \qquad (n = 2) \qquad \text{first excited state}$$
$$E_3 = -Ry/9 \qquad (n = 3) \qquad \text{second excited state}$$

etc.

We will always use the convention that the lowest energy state is the one with the most negative value of the energy. So the electron in the hydrogen atom may not have an arbitrary energy, but one of the possibilities given by equation (16). We say that the energy is quantized. It only depends upon the value of the principal quantum number, n.

Suppose that a hydrogen atom finds itself in an excited state ($n > 1$). The return of the electron to the electronic ground state with $n = 1$ is accompanied by the liberation of energy ΔE equal to the energy difference between the ground state and the excited state initially populated.

$$\Delta E = -Ry/n^2 - (-Ry/1^2) = Ry(1 - 1/n^2) \tag{17}$$

This change in energy is accomplished via the *emission* of a photon with an energy equal to ΔE. From the Planck–Einstein relationship the frequency (v) of the emitted photon is given by

$$\Delta E = hv \tag{18}$$

---

* An electron volt (eV) is the energy acquired by an electron on moving through a potential difference of one volt. Although it is not an SI unit, it is frequently used to measure energies on the atomic scale. In numerical calculations its SI equivalent, 1 eV = 1.602×10^{-19} J should be used (see exercise 2.1).

where h is Planck's constant ($h = 6.62 \times 10^{-34}$ J s). Thus the frequency of the photon emitted in the present case is given by

$$h\nu = Ry(1 - 1/n^2)$$

i.e.,

$$\nu = Ry/h(1 - 1/n^2) \tag{19}$$

When the excited state is sufficiently high in energy ($n > 2$) the electron doesn't necessarily have to return to the ground state ($n = 1$), but can instead move to another excited state (labelled by n') of lower energy than the initial one. This implies $n' < n$. In this case the frequency of the emitted photon is given by the relationship

$$\nu = Ry/n(1/n'^2 - 1/n^2) \tag{20}$$

As a general result of such processes one obtains for each value of n' a series of spectral lines which together constitute the emission spectrum of the atom. They may be grouped according to the value of n' and are traditionally named after the physicists who discovered them (Figure 2.1). The Lyman series is a special one in that it corresponds to the return of excited electrons to the ground state with $n' = 1$.

The inverse of emission is *absorption*. A hydrogen atom in its electronic ground state can absorb a photon, using its energy to move to an excited state. Such an absorption process can only occur if the photon energy corresponds exactly to the energy difference between the excited state and that of the ground state, namely

$$h\nu = Ry(1 - 1/n^2) \tag{21}$$

The larger the value of n and therefore the higher in energy the excited state, the higher the frequency of the photon needed for excitation. When $n \to \infty$ the energy of the excited state tends to zero, giving rise to a situation where the electron and nucleus are completely separated from each other. This corresponds to *ionization* of the hydrogen atom, $H \to H^+ + e^-$, and the energy involved, ΔE_∞ is simply given by

$$\Delta E_\infty = Ry = 13.6 \text{ eV} \tag{22}$$

Thus the quantity Ry is the energy needed to ionize the hydrogen atom in its electronic ground state (the ionization potential) and is found experimentally to be just this, 13.6 eV. It is the smallest amount of energy needed to detach an electron from the atom in its ground state. The energy needed to ionize the atom in an excited state can be obtained in the same way. In general the rules which define the allowed frequencies of light for photon absorption are just the same as the ones we have described above for photon emission.

(b) Nomenclature for the eigenfunctions, Ψ_i

The principal quantum number n is sufficient to characterize the allowed values of the energy, but the situation is more complex when it comes to a description of

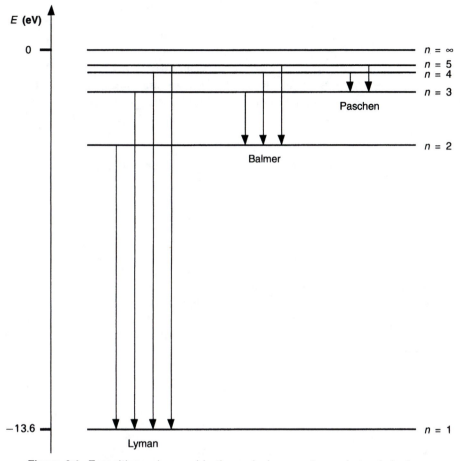

Figure 2.1. Transitions observed in the emission spectrum of atomic hydrogen.

the corresponding eigenfunctions. We need three quantum numbers to classify them.

(i) The principal quantum number, n, which as we have seen is a positive integer $(n = 1, 2, 3, \ldots)$.

(ii) A secondary quantum number, l often called the angular momentum, or azimuthal quantum number. It is an integer, positive or zero and is always less than n.

$$0 \leq l < n \tag{23}$$

(iii) The magnetic quantum number, m. This is an integer, positive, negative or zero, lying between the values $\pm l$

$$-l \leq m \leq +l \tag{24}$$

Each wavefunction Ψ_i is characterized by a set of three quantum numbers n, l, m.

The simplest case is that for n equal to 1. Since the quantum number l has to be less than n, it must be equal to zero. As a consequence m is also equal to zero. There is then just a single collection of numbers ($n = 1$, $l = 0$, $m = 0$) which are possible and these describe a single wavefunction, or solution of the Schrödinger equation.

When n is equal to 2, l may take on the two values 0 and 1. As before if l is zero then m must be so too. If however l is equal to 1, there are three possibilities for m; -1, 0 and $+1$. There are therefore four possible solutions (n, l, m) corresponding to the set of quantum numbers $(2, 0, 0)$, $(2, 1, -1)$, $(2, 1, 0)$ and $(2, 1, +1)$. These four functions are degenerate since they correspond to the same value of the principal quantum number, $n = 2$, with an energy of $E = -Ry/4$.

We can simply label these different functions by using a shorthand of the form nX_m where X is a letter used to represent a given value of l in the following way

l	0	1	2	3
X	s	p	d	f

There is nothing magical about these labels. They are the first letters of some descriptive terms early spectroscopists used to characterize atomic spectra (namely sharp, principal, diffuse, and fundamental) and have been adopted for use in the present context. By convention when l is equal to zero (the ns functions) we don't need to specify the value of m, since it is always equal to zero. The names of the different solutions for the hydrogen atom for n up to 3 are given below.

$n = 1$	$l = 0$	$m = 0$	$1s$
$n = 2$	$l = 0$	$m = 0$	$2s$
	$l = 1$	$m = +1$	$2p_{+1}$
		$m = 0$	$2p_0$
		$m = -1$	$2p_{-1}$
$n = 3$	$l = 0$	$m = 0$	$3s$
	$l = 1$	$m = +1$	$3p_{+1}$
		$m = 0$	$3p_0$
		$m = -1$	$3p_{-1}$
	$l = 2$	$m = +2$	$3d_{+2}$
		$m = +1$	$3d_{+1}$
		$m = 0$	$3d_0$
		$m = -1$	$3d_{-1}$
		$m = -2$	$3d_{-2}$

There are a total of n^2 solutions with the same value of the principal quantum number n, and therefore with the same energy, $-Ry/n^2$.

2.2.2. Description of the eigenfunctions

The various eigenfunctions which describe these electronic situations depend upon the spatial coordinates i.e. $\Psi_{n,l,m}(x, y, z)$. Relying on the spherical symmetry of the hydrogen atom it is advantageous to express these functions in terms of the spherical coordinates, r, θ, ϕ shown in **2-2**.

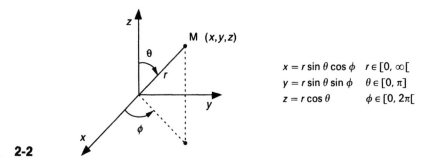

$$x = r \sin \theta \cos \phi \quad r \in [0, \infty[$$
$$y = r \sin \theta \sin \phi \quad \theta \in [0, \pi]$$
$$z = r \cos \theta \quad \phi \in [0, 2\pi[$$

2-2

(a) Analytical form

Because of the form of the mathematics of the hydrogen atom problem it turns out that the wavefunctions may be written as a simple product of two functions, one radial in extent and the other containing all of the angular dependence.

$$\Psi_{n,l,m}(r, \theta, \phi) = R_{n,l}(r) Y_{l,m}(\theta, \phi) \tag{25}$$

The first term, the radial one ($R_{n,l}(r)$) only dependent on r, contains the principal quantum number n and the angular momentum quantum number l. The second term $Y_{l,m}(\theta, \phi)$, the angular part, only depends on the variables θ, ϕ. Both terms are normalized so that the total wavefunction itself is normalized. The volume element $d\tau$ may be written in terms of spherical coordinates as $d\tau = r^2 \sin \theta \, dr \, d\theta \, d\phi$. Equations (26) and (27) show this process formally.

$$\int_0^\infty R_{n,l}^2(r) r^2 \, dr = 1 \tag{26}$$

$$\int_0^\pi \int_0^{2\pi} Y_{l,m}^2(\theta, \phi) \sin \theta \, d\theta \, d\phi = 1 \tag{27}$$

As we will see it will prove very useful to use a pictorial representation for the wavefunctions. However, since they in general depend upon the three variables r, θ, and ϕ, it is impossible to rigorously represent their shape in two dimensions. The most satisfactory way is to draw a contour map of the function by drawing lines of constant Ψ in a plane of interest.

(b) The 1s function ($n = 1$, $l = 0$, $m = 0$)

The analytical expression for this function is

$$\Psi_{1s} = \left[\frac{2}{\sqrt{a_0^3}} \exp\left(-\frac{r}{a_0} \right) \right] \frac{1}{\sqrt{4\pi}} \tag{28}$$

The expression doesn't appear to contain the variables θ and ϕ. In fact the angular contribution is equal to $\sqrt{(1/4\pi)}$. The radial part varies via the term $\exp(-r/a_0)$ and the expression $2/a_0^{3/2}$ assures that this part of the wavefunction is normalized. a_0 is a universal constant of length, known as the *Bohr radius*, equal to 52.9 pm.

This wavefunction is said to be spherically symmetrical since its value at a point only depends on the distance of that point to the nucleus. It is thus easy to precisely describe the behavior of the 1s function using a plot of its dependence on r, the single variable on which it depends (2-3a). The amplitude of the wavefunction is largest at

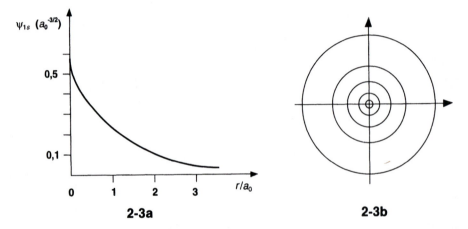

2-3a   2-3b

the nucleus and decreases exponentially via the term in $-r/a_0$. It is important to note that the wavefunction has the same sign (positive from equation (26)) for all values of r. The contour diagrams for Ψ_{1s} are easy to construct since the surfaces of constant Ψ are spheres (i.e. constant r) and their intersection with a plane containing the nucleus are circles (2-3b).

These two representations are not very convenient ones in a practical sense. A representation which captures the essence of the wavefunction in terms of spherical symmetry and a sign which doesn't change with r is shown in 2-4a. Ψ_{1s} is represented

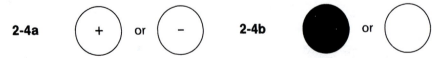

2-4a   +   or   −   2-4b   ●   or   ○

by a circle centered at the nucleus and containing a positive sign to show that the wavefunction is positive everywhere. An equally valid possibility is one which contains a negative sign, indicative of a wavefunction which is negative everywhere.

Recall that from Section 2.1.2a this gives a completely equivalent description of the electron. The convention we will use in this book is shown in **2-4b**. Hatching or shading is used to indicate where the wavefunction is positive, and the circle is left empty when it is negative.

(c) The 2s function ($n = 2$, $l = 0$, $m = 0$)

The analytic expression for this function is

$$\Psi_{2s} = \left[\frac{1}{\sqrt{8a_0^3}} \left(2 - \frac{r}{a_0} \right) \exp\left(-\frac{r}{2a_0} \right) \right] \frac{1}{\sqrt{4\pi}} \tag{29}$$

Just as for the $1s$ orbital the angular part of the $2s$ function is also constant and equal to $\sqrt{(1/4\pi)}$. Thus the wavefunction depends only on r, and as a consequence is spherically symmetrical. Its amplitude tends to zero as r tends to infinity via the exponential term, just as for the $1s$ function. What is new here however is that the term $2 - r/a_0$ goes to zero when $r = 2a_0$, and the probability of finding the electron at this distance from the nucleus, described by the surface of a sphere of radius $2a_0$, is identically zero. We say that the $2s$ function possesses a *spherical node*. As defined in equation (29) it is positive when r is less than and negative when r is greater than $2a_0$. *A surface where the wavefunction is zero everywhere on it is called a nodal surface. The wavefunction changes sign on moving from one side to the other.*

It is again possible to describe the behavior of Ψ_{2s} as a function of r in a simple way since the wavefunction does not explicitly contain the variables θ, ϕ (**2-5a**). It

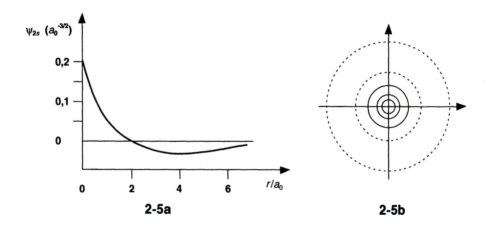

2-5a   2-5b

has a maximum at $r = 0$, changes sign at $r = 2a_0$, as indicated by the change from solid to dashed contour lines (**2-5b**) and then approaches zero as r becomes large. It is however more convenient to use the representation **2-6** which comprises a pair of concentric circles, containing plus and minus signs, or hatched and unhatched areas to describe the sign of the wavefunction as described earlier.

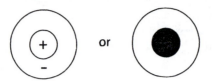

2-6

(d) The 2p functions ($n = 2$, $l = 1$, $m = +1, 0, -1$)

Although the function which describes $2p_0$ is real, it turns out that the functions describing $2p_{+1}$ and $2p_{-1}$ are complex conjugates of each other. In general the ϕ dependence of $Y_{l,m}(\theta, \phi)$ turns up in the form $e^{im\phi}$ leading to values of $e^{i\phi}$, 1 and $e^{-i\phi}$ for p_1, p_0 and p_{-1} respectively. Since $2p_{+1}$ and $2p_{-1}$ are degenerate, judiciously chosen linear combinations of the two are also valid wavefunctions, as described in Section 2.1.2c. Thus we could write for one combination $(e^{i\phi} + e^{-i\phi})/2 = \cos \phi$ and $(e^{i\phi} - e^{-i\phi})/2i = \sin \phi$ for the other. This leads to two new, real, orthonormal functions. Along with the function for $2p_0$ we now have the three functions

$$\Psi_{2p_x} = \left[\frac{1}{2\sqrt{6a_0^3}} \frac{r}{a_0} \exp\left(-\frac{r}{2a_0} \right) \right] \sqrt{\frac{3}{4\pi}} \sin \theta \cos \phi \tag{30-a}$$

$$\Psi_{2p_y} = \left[\frac{1}{2\sqrt{6a_0^3}} \frac{r}{a_0} \exp\left(-\frac{r}{2a_0} \right) \right] \sqrt{\frac{3}{4\pi}} \sin \theta \sin \phi \tag{30-b}$$

$$\Psi_{2p_z} = \left[\frac{1}{2\sqrt{6a_0^3}} \frac{r}{a_0} \exp\left(-\frac{r}{2a_0} \right) \right] \sqrt{\frac{3}{4\pi}} \cos \theta \tag{30-c}$$

These functions take on a simple analytic form by transformation back to the cartesian coordinates of **2-2**. The angular part of the three wavefunctions of equation (30) multiplied by r are just the cartesian functions x, y and z. Such a correspondence leads to the following description of the wavefunctions:

$$\Psi_{2p_x} = Nx \exp\left(-\frac{r}{2a_0} \right) \tag{31-a}$$

$$\Psi_{2p_y} = Ny \exp\left(-\frac{r}{2a_0} \right) \tag{31-b}$$

$$\Psi_{2p_z} = Nz \exp\left(-\frac{r}{2a_0} \right) \tag{31-c}$$

where

$$N = \frac{1}{2\sqrt{6a_0^5}} \sqrt{\frac{3}{4\pi}}$$

This new nomenclature emphasizes the similarities which exist between the three functions. Each possesses the same local geometric properties, they just point along a different goemetrical axis, x, y or z. It is sufficient to take a look at just one of them (p_z for example in equation (31-c)) to be able to understand them all.

For a given value of z the function $2p_z$ has the same value for all points located at the same distance, r, from the origin (**2-7a**). We say that this function is *cylindrically symmetrical* about the z-axis. On the other hand $2p_z$ is of opposite sign for two points related by the xy plane, namely $z_A = -z_B$ and $r_A = r_B$ (**2-7b**). This function is thus said to be *antisymmetric* with respect to the xy plane. Finally Ψ_{2p_z} is identically zero within the xy plane. This plane is consequently a nodal plane of p_z.

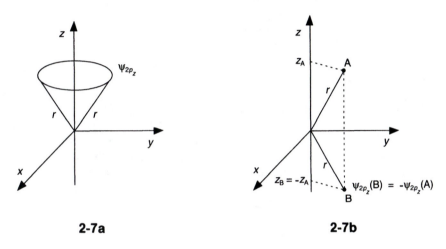

2-7a   2-7b

A contour description of the $2p_z$ function is given in **2-8a**. As before the change from solid to dashed contour lines indicates the change in sign of the wavefunction. Generally we prefer the representation shown in **2-8b**, one which contains all of the essential information about the orbital, namely cylindrical symmetry around the z-axis, the presence of a nodal xy plane and a function which is antisymmetric with respect to it. We say that this function has two *lobes*, one positive (shaded in **2-8b**) and one negative (unshaded).

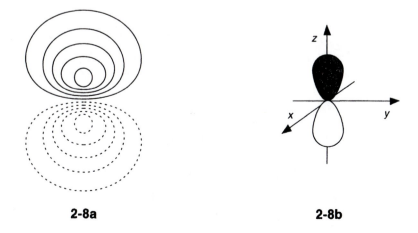

2-8a   2-8b

The representations we use for $2p_x$ and $2p_y$ can be obtained from **2-8b** by changing the axis involved, and are shown in **2-9**. Notice the special way we draw the $2p_x$ orbital to indicate that it is directed perpendicular to the page.

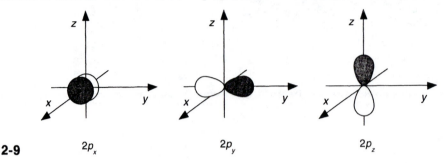

2-9 $2p_x$ $2p_y$ $2p_z$

(e) Radial probability density

The probability of finding an electron somewhere is an important concept but one which is sometimes difficult to portray. Another way which is frequently used to characterize the wavefunction is the radial probability density which is the probability of finding the electron in the volume enclosed by the two spheres of radii r and $r + dr$ (**2-10**)*. We need to integrate the square of the function over the angular coordinates,

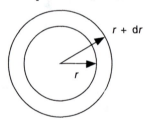

$r + dr$

r

2-10

θ and ϕ:

$$dS = \int_{\theta, \phi} R^2_{n,l}(r)\, Y^2_{l,m}(\theta, \phi) r^2 \sin \theta \, dr \, d\theta \, d\phi$$

$$= R^2_{n,l}(r) r^2 \, dr \int_{\theta, \phi} Y^2_{l,m}(\theta, \phi) \sin \theta \, d\theta \, d\phi$$

Taking into account the normalization of $Y_{l,m}(\theta, \phi)$ we obtain an expression of the form

$$dS = R^2_{n,l}(r) r^2 \, dr \tag{32}$$

---

* The form of the volume element as defined here is equal to $4\pi r^2 \, dr$, which immediately leads to the possibility of calculating dS via the relationship $dS = \Psi^2 4\pi r^2 \, dr$. However this is only the case for a spherically symmetrical function (i.e., s functions) whose probability density, Ψ^2, is constant within the volume considered.

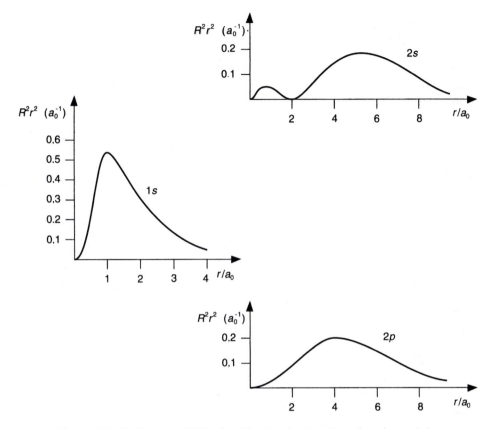

Figure 2.2. Radial probability densities for the functions ψ_{1s}, ψ_{2s} and ψ_{2p}.

The radial probability density is then written as

$$dS/dr = R^2_{n,l}(r)r^2 \tag{33}$$

This is shown for the 1s, 2s and 2p functions in Figure 2.2. In each case the density is zero at the nucleus and tends to zero as r tends to infinity. It is also zero for the case of the 2s function at $r = 2a_0$ where there is a spherical node. Each curve has a maximum value, located at $r = a_0$ for the 1s function and clearly at larger r, at $5.24a_0$ and $4a_0$ for the 2s and 2p functions respectively. (The 2s function has a smaller maximum at $r = 0.76a_0$). These maxima correspond to the most probable distance of finding the electron from the nucleus. The value of r to which they correspond characterizes the spatial extent of the wavefunction, or its '*radius*'. The radii of the 2s and 2p functions are clearly larger than that for the 1s function. This is a general result which may be extended to all the hydrogenic wavefunctions. The radius of the wavefunction (defined as that radius where there is a maximum in the radial probability density) increases with n (and is close to n^2a_0) but depends only slightly on the value of l.

2.2.3. Electron spin

The characterization of the electron by a wavefunction of the type we have described uses just three quantum numbers, n, l and m. A complete description of the electron in the hydrogen atom however, requires consideration of a further concept. In addition to the classical ideas of mass and charge, the electron possesses a permanent magnetic moment, μ. It is difficult to visualize this purely quantum mechanical concept in classical terms, but it too is quantized. The magnetic moment may take one of two values and is generally expressed in units of Bohr magnetons (μ_B) viz:

$$\mu = -2m_s\mu_B$$

Here m_s is called the spin quantum number which may only adopt one of two values $\pm\frac{1}{2}$. By convention an electron with $m_s = \frac{1}{2}$ is called an electron with α spin and one with $m_s = -\frac{1}{2}$ is called an electron with β spin. *It is necessary to use these four quantum numbers, n, l, m and m_s to completely characterize the electron in the hydrogen atom.*

The experimental demonstration of the existence of a magnetic moment associated with the spin of the electron is due to two physicists, Stern and Gerlach. If one sends a beam of silver atoms (an element containing a single unpaired electron) through an inhomogeneous magnetic field it splits into two as indicated by the development of two separate regions of metallic silver deposited on a plate downstream of the magnet. The existence of just two areas of metallic silver show that the magnetic moment can take just two values. The atoms with $m_s = \frac{1}{2}$ are deflected one way and those with $m_s = -\frac{1}{2}$ the other.

2.2.4. Hydrogen-like atoms

An exactly analogous treatment to the one we have just given for hydrogen is applicable to all hydrogen-like atoms, namely those that possess a single electron moving around a nucleus of charge $+Ze$. These are the ions He^+ ($Z = 2$), Li^{2+} ($Z = 3$) and Be^{3+} ($Z = 4$) etc. The results pertaining to these situations are qualitatively identical to those found for hydrogen itself in terms of the form of the wavefunctions Ψ_{1s}, Ψ_{2s}, Ψ_{2p} etc. Again the energy only depends on the principal quantum number, n. There are however several quantitative differences which appear.

(i) The eigenvalues are of the form $E_n = -RyZ^2/n^2$ leading to the important result that *the energy of the orbital is lowered as the nuclear charge increases.*

Thus whereas the energy of the 1s orbital is -13.6 eV($-Ry$) in hydrogen, it is -54.4 eV($-4Ry$) in the helium cation, He^+. This lowering of the energy arises from the stronger electrostatic interaction between electron and nucleus as the nuclear charge increases.

(ii) The analytical expression for the eigenfunctions may be generated from those obtained for hydrogen by replacing a_0 by a_0/Z in the relevant equations. For example the expression for the 2s function is

$$\Psi_{2s} = \left[\sqrt{\frac{Z^3}{8a_0^3}} \left(2 - \frac{Zr}{a_0} \right) \exp\left(-\frac{Zr}{2a_0} \right) \right] \frac{1}{\sqrt{4\pi}} \tag{34}$$

This last point is important for it shows that the wavefunctions in the hydrogen-like atoms are more contracted, i.e., of smaller spatial extent than in hydrogen itself. So, whereas the radial maximum in the radial probability function for the 2s orbital is about $5a_0$ (265 pm) in hydrogen it is about $5(a_0/Z) = 2.5a_0$ (132 pm) in He^+.

2.3. Many-electron atoms

In the case of hydrogen and hydrogen-like atoms an analytic solution of the Schrödinger wave equation is possible. The single interaction between electron and nucleus is simple to treat but the quantum mechanical problem for the situation where there are in addition interactions between electrons, the case in the many-electron atom, prevents such an analytical result. Since it is impossible to find the exact analytical wavefunctions for the Hamiltonian operator applicable to the many-electron atom, it is necessary to make some approximations in order to determine the wavefunctions which best approximate the state of affairs.

2.3.1. The orbital approximation

Let us call e_i the set of three spatial coordinates (x_i, y_i, z_i) appropriate to the electron i. The wavefunction which describes the collection of electrons depends on the coordinates of all the electrons and may be written as a many-electron function as

$$\Psi(e_1, e_2, \ldots, e_i, \ldots, e_z) \tag{35}$$

This function is a solution of the Schrödinger wave equation:

$$\mathscr{H}\Psi(e_1, e_2, \ldots, e_i, \ldots, e_z) = E\Psi(e_1, e_2, \ldots, e_i, \ldots, e_z) \tag{36}$$

In the orbital approximation one looks for solutions which approach that of the many-electron function by writing a product of single-electron functions, χ_i which only depend upon the coordinates of a single electron.

$$\Psi(e_1, e_2, \ldots, e_i, \ldots, e_z) = \chi_1(e_1)\chi_2(e_2) \ldots \chi_i(e_i) \ldots \chi_z(e_z) \tag{37}$$

The single-electron functions are called the *atomic orbitals* (AOs) of the atom. They are themselves solutions of an equation (38) which although having the same form as the Schrödinger equation, is considerably simpler since it only contains the coordinates of a single electron.

$$\hat{h}\chi_i(e_i) = \varepsilon_i \chi_i(e_i) \tag{38}$$

There are an infinite set of solutions χ_i associated with the eigenvalues ε_i, the energies of the atomic orbitals, χ_i. The wavefunction Ψ, describing the many-electron atom is thus the product of a given set of the χ_i, the choice of which is determined by a set of rules which we will describe later (Section 2.3.4b).

2.3.2. Mathematical description and nomenclature of atomic orbitals

By analogy with the eigenfunctions for the hydrogen-like atoms, each orbital χ is written as a product of radial and angular functions

$$\chi_{n,l,m}(r, \theta, \varphi) = R_{n,l}(r) Y_{l,m}(\theta, \varphi) \tag{39}$$

The angular part of this expression, $Y_{l,m}(\theta, \varphi)$ is identical to that found for the hydrogen atom and is determined by the values of the two quantum numbers l and m. The radial part, set by the values of the two quantum numbers n and l, is determined by both the nuclear charge and the presence of the other electrons (see Section 2.5.2). Its functional form is thus expected to be rather different from the $R_{n,l}(r)$ of the hydrogen atom. As before, each orbital is characterized by the three quantum numbers n, l and m appropriate for the one-electron atom. The rules which determine the allowed values of these quantum numbers are also the same as those for the hydrogen-like atoms; $n = 1, 2, 3, \ldots, 0 \le l < n$ and $-l \le m \le +l$. Ortho-normal AOs, described as before, with the labels $1s$, $2s$, $2p$ etc., result.

The close similarities between the AOs for the many-electron atom and the wavefunctions for the hydrogen atom and the hydrogen-like atoms means that, although it is not strictly correct we use the same language to describe the orbitals in one- and many-electron atoms. Irrespective of whether the functions are determined via the orbital approximation or as exact solutions of the wave equation for one-electron atoms, we shall use the term atomic orbital (AO) for both.

2.3.3. Atomic orbital energies

For the hydrogen-like atoms we saw that the energy of the atomic orbital (ε_n), in this case the same as the eigenvalue E_n only depends upon the principal quantum number n ($\varepsilon_n = E_n = -RyZ^2/n^2$). Classification of the AOs in terms of increasing energy doesn't pose any particular problem. The energy of the orbital increases with increasing n and orbitals with the same n are degenerate, $2s$ and $2p$ for example. The situation is more complex in many-electron atoms. *In many-electron atoms, the energy of an atomic orbital depends on the two quantum numbers n and l.*

An immediate consequence of this result is that AOs with the same n and l remain degenerate. For example the three $2p$ orbitals are strictly degenerate, as are the five $3d$ orbitals. Such a group of orbitals with the same energy is called a *sub-shell*, the term *shell* itself is reserved for all the orbitals with the same n. In this way the 2 shell ($n = 2$) is made up of the two sub-shells 2s and 2p.

There are two general rules which allow, at least in part, the energetic ordering of the orbitals

(i) For the same value of l the orbital energy increases with increasing n. Thus

$$\varepsilon_{1s} < \varepsilon_{2s} < \varepsilon_{3s} < \cdots$$

$$\varepsilon_{2p} < \varepsilon_{3p} < \varepsilon_{4p} < \cdots$$

(ii) For the same value of n the orbital energy increases with increasing l. So

$$\varepsilon_{2s}\,(l = 0) < \varepsilon_{2p}\,(l = 1)$$

$$\varepsilon_{3s}\,(l = 0) < \varepsilon_{3p}\,(l = 1) < \varepsilon_{3d}\,(l = 2)$$

These two rules are not, however, sufficient to completely fix the energetic ordering. For example they do not allow the placement of 2p relative to 3s. However calculations show that it is the value of the principal quantum number which dominates here, i.e., $\varepsilon_{2p} < \varepsilon_{3s}$. In fact for all atoms the lowest five orbitals are found in the following order

$$\varepsilon_{1s} < \varepsilon_{2s} < \varepsilon_{2p} < \varepsilon_{3s} < \varepsilon_{3p}$$

Beyond this group the situation gets complicated. For example it is not possible to predict using these rules the energetic ordering of the 4s and 3d AOs since the actual state of affairs depends on the atom being considered, and sometimes its oxidation state.

2.3.4. The electronic configuration of atoms

The electronic configuration of an atom is the assignment of the available electrons to the different sub-shells open to them. The number of electrons in a sub-shell is generally indicated by an exponent. For example, $1s^2$ signifies the presence of two electrons in the 1s orbital. We speak then of two electrons 'occupying' the 1s orbital. Two rules limit the configurations which are possible.

(a) The Pauli exclusion principle

One way of stating this is that *in an atom no two electrons may have the same values for the four quantum numbers n, l, m, m_s.* This principle has two important consequences.

(i) If two electrons have the same spin (that is to say the same value of the spin quantum number) they must occupy two different orbitals. (i.e., at least one of the n, l and m quantum numbers must be different for the two electrons.)

(ii) If two electrons do occupy the same orbital their spins must be different. A corollary to this is that there may be a maximum of only two electrons per orbital and these with opposed spins ($m_s = +\frac{1}{2}$ and $m_s = -\frac{1}{2}$). By way of examples an s sub-shell (one orbital) may hold up to two electrons, a p sub-shell (three orbitals) up to six electrons and a d sub-shell (five orbitals) up to ten electrons.

(b) The Aufbau process—Klechkowsky's rule

To determine the electronic configuration of the atom we usually look for that filling of the sub-shells which leads to the lowest energy of the system. This gives rise to the electronic ground state. Such a filling process is often called the *Aufbau* or *building up* process. Other configurations which are possible, but higher in energy, give rise to excited states. For the ground state, electrons fill the sub-shells in the order given by *Klechkowsky's rule*.

In many-electron atoms the sub-shells are filled in the order of increasing $(n + l)$. If two sub-shells have the same value of $n + l$ then the one with the smaller n is filled first. Thus we get the following filling sequence

$$1s < 2s < 2p < 3s < 3p < 4s < 3d < 4p \text{ etc}$$

$$n + l \quad 1 \quad 2 \quad 3 \quad 3 \quad 4 \quad 4 \quad 5 \quad 5$$

For the first five sub-shells this order is the same as that presented earlier. $2p$ $(n + l = 3)$ is filled before $3s$ $(n + l = 3)$ because the principal quantum number of $2p$ is smaller. For the same reason $3p$ is filled before $4s$. The orbital filling pattern generated by Klechkowsky's rule is usually shown in the form of a chart (**2-11**) where the arrows indicate the order in which the sub-shells are filled. We can obtain in this way, with some exceptions which we will return to in Section 2.4.1, the electronic configurations of all the atoms. By way of example, the electronic configuration of sodium (Na, $Z = 11$) is: $1s^2 2s^2 2p^6 3s^1$, that of calcium (Ca, $Z = 20$) $1s^2 2s^2 2p^6 3s^2 3p^6 4s^2$, and that of titanium (Ti, $Z = 22$) $1s^2 2s^2 2p^6 3s^2 3p^6 4s^2 3d^2$.

2.3.5. Hund's rule

(a) Formulation

When several electrons need to be placed in a degenerate set of orbitals without their number being sufficient to completely fill them (i.e. two electrons in each), there can be several different ways of arranging them. For example in the carbon atom $(1s^2 2s^2 2p^2)$ the two electrons in the $2p$ sub-shell may both occupy the same atomic orbital or be located in different ones. Hund's rule tells us which is the most stable arrangement of electrons for the lowest energy configuration. *When several electrons occupy degenerate AOs the most stable configuration is the one containing the largest number of parallel electron spins.*

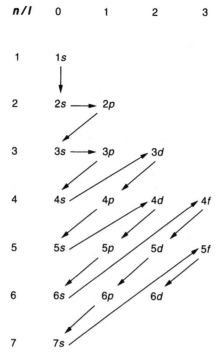

2-11

The three $2p$ orbitals of the carbon atom are equivalent, and three ways of arranging the electrons are shown in **2-12**. In these diagrams the upward pointing arrow represents an electron with α spin and a downward pointing one an electron with β spin. Using Hund's rule it is the arrangement 3 which is predicted to have the lowest energy. It is relatively easy to understand why arrangement 1 is the least favored. Since the two electrons occupy the same orbital, they can come close to each other in space and experience a strong electrostatic repulsion. This energetically unfavorable interaction is reduced when the two electrons occupy different regions of space as is the case when they occupy different orbitals. Thus arrangements 2 and 3 are more stable than 1. The stabilization of configuration 3 over 2 however does not have a classical explanation at all, but one which only comes from quantum mechanics. A pair of electrons with parallel spins (i.e., the same value of m_s) are lowered in energy relative to the same arrangement with antiparallel spins by a purely quantum mechanical process called the *exchange interaction*.

2-12

(b) Diamagnetism and paramagnetism

Let us consider the magnesium atom ($Z = 12$) with the ground state configuration $1s^2 2s^2 2p^6 3s^2$. We can represent its atomic structure by the scheme **2-13** in which

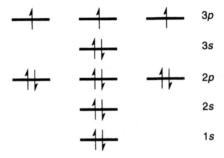

2-13

the orbitals are arranged in order of increasing energy. In the magnesium atom all of the electrons are paired. In other words each occupied orbital contains one electron with α and one with β spin. The spin magnetic moments of each of the electrons in such a pair cancel each other out so that overall the magnesium atom does not have an intrinsic spin magnetic moment. Systems with such a property are called *diamagnetic*.

The ground state configuration of phosphorous with 15 electrons may be written $1s^2 2s^2 2p^6 3s^2 3p^3$ (**2-14**). The three $3p$ electrons each occupy a different orbital with

2-14

identical spins to satisfy Hund's rule. These electrons are not paired, their spin magnetic moments do not cancel and the total magnetic moment for phosphorus is non-zero. This atom is thus *paramagnetic*.

2.3.6. Core and valence electrons

The electronic configuration of the atom assigns locations to all of its electrons but we will find it convenient to divide the electrons into two groups. The first are the *core electrons* which occupy deep-lying AOs and the second are the *valence electrons* which occupy the higher lying ones. Most often the valence electrons are regarded as all those in orbitals associated with the largest value of the principal quantum number. The remainder make up the core. Thus carbon $(1s^2 2s^2 2p^2)$ has two core electrons $(1s^2)$ and four valence electrons $(2s^2 2p^2)$. Similarly magnesium $(1s^2 2s^2 2p^6 3s^2)$ has ten core electrons $(1s^2 2s^2 2p^6)$ and two valence electrons $(3s^2)$. There is one difficulty with this nomenclature for the elements with a partially filled d sub-shell. (The transition elements are discussed in Section 2.4.1.) In effect since the nd and $(n + 1)s$ orbitals are close in energy all of the electrons occupying them are considered to be valence electrons. Thus titanium $(1s^2 2s^2 2p^6 3s^2 3p^6 4s^2 3d^2)$ contains four valence electrons $(4s^2 3d^2)$ not two. This distinction between core and valence electrons is an important one, since it is the valence electrons (those in the outermost shell) which determine the chemical properties of the elements.

As a shorthand notation we often replace the set of core electrons by the chemical symbol of the noble gas which contains this number of electrons when writing down the electronic configuration. So the electronic configuration for phosphorus ($1s^2 2s^2 2p^6 3s^2 3p^3$) may be written more compactly as $\{Ne\}3s^2 3p^3$. The part of the electronic configuration actually written out in such an expression is then the *valence configuration* of the atom ($3s^2 3p^3$ for phosphorus).

2.4. The periodic classification of the elements

2.4.1. Organization by rows

The periodic classification of the elements is shown in Figure 2.3. The labels used to describe the columns (1–18) are those recommended by IUPAC at the end of the 1980s. The way the table is assembled follows directly through the filling of the sub-shells and shells of atomic orbitals using Klechkowsky's scheme described earlier. Each period begins by the filling of an ns sub-shell and is completed by the filling of an np sub-shell, except of course for $n = 1$ where there is no p sub-shell. The principal quantum number n increases by one on moving from the end of one period to the beginning of the next.

The first period only contains two elements since it corresponds to the filling of the first shell which only contains a single s orbital ($1s^1$ for H and $1s^2$ for He).

The second period begins with the filling of the $2s$ orbital and continues with the filling of the three $2p$ orbitals. Eight elements result from lithium (Li) with the valence configuration $2s^1$ to neon (Ne) with the configuration $2s^2 2p^6$.

The situation is identical for the third period in that the $3s$ orbital (Na:$3s^1$; Mg:$3s^2$) and then the $3p$ orbitals (from Al:$3s^2 3p^1$ to Ar:$3s^2 3p^6$) are successively filled. An important point to note, and one we will return to, is that the $3d$ orbitals, with the same value of the principal quantum number as the $3s$ and $3p$ orbitals are not filled in this period. Klechkowsky's rule (Section 2.3.4b) prevents their filling until after that of the $4s$ orbital.

The fourth period begins with the filling of the $4s$ sub-shell (K:$4s^1$; Ca:$4s^2$) and ends with the filling of the $4p$ sub-shell (Ga:$4s^2 4p^1$ to Kr:$4s^2 4p^6$). Between these two groups lies a *transitional series*, so named because one set of orbitals, of principal quantum number n, are filled after one of the orbitals of principal quantum number $n + 1$ is filled. The collection of elements here are known as the *first transition metal series*, which correspond to the progressive occupation of the $3d$ orbitals. This series contains ten elements since there are a total of five $3d$ orbitals which can hold two electrons each. Therefore the valence electronic configuration of scandium (Sc) is $4s^2 3d^1$ and that of zinc (Zn) $4s^2 3d^{10}$. In total the fourth period contains eighteen elements ($2 + 10 + 6$). We must note, however, two exceptions to Klechkowsky's rule among this transition metal series. The lowest energy configuration for chromium (Cr) is $4s^1 3d^5$, the expected arrangement ($4s^2 3d^4$) being less stable. In the same way the lowest energy configuration of copper (Cu) is $4s^1 3d^{10}$ and not $4s^2 3d^9$. In order to understand these exceptions we must first recognize that the $4s$ and $3d$ orbitals are close in energy for these elements, so that the promotion energy $4s \rightarrow 3d$ is very small. Thus the energies of the configurations, obtained by summing just the orbital

1	2	3	4	5	6	7	8	9	10	11	12	13	14	15	16	17	18
H $1s^1$																	**He** $1s^2$
Li $1s^2 2s^1$	**Be** $1s^2 2s^2$											**B** $1s^2 2s^2 2p^1$	**C** $1s^2 2s^2 2p^2$	**N** $1s^2 2s^2 2p^3$	**O** $1s^2 2s^2 2p^4$	**F** $1s^2 2s^2 2p^5$	**Ne** $1s^2 2s^2 2p^6$
Na $(Ne)3s^1$	**Mg** $(Ne)3s^2$											**Al** $(Ne)3s^2 3p^1$	**Si** $(Ne)3s^2 3p^2$	**P** $(Ne)3s^2 3p^3$	**S** $(Ne)3s^2 3p^4$	**Cl** $(Ne)3s^2 3p^5$	**Ar** $(Ne)3s^2 3p^6$
K $(Ar)4s^1$	**Ca** $(Ar)4s^2$	**Sc** $(Ar)3d^1 4s^2$	**Ti** $(Ar)3d^2 4s^2$	**V** $(Ar)3d^3 4s^2$	**Cr** $(Ar)3d^5 4s^1$	**Mn** $(Ar)3d^5 4s^2$	**Fe** $(Ar)3d^6 4s^2$	**Co** $(Ar)3d^7 4s^2$	**Ni** $(Ar)3d^8 4s^2$	**Cu** $(Ar)3d^{10} 4s^1$	**Zn** $(Ar)3d^{10} 4s^2$	**Ga** $(Ar)3d^{10}4s^2 4p^1$	**Ge** $(Ar)3d^{10}4s^2 4p^2$	**As** $(Ar)3d^{10}4s^2 4p^3$	**Se** $(Ar)3d^{10}4s^2 4p^4$	**Br** $(Ar)3d^{10}4s^2 4p^5$	**Kr** $(Ar)3d^{10}4s^2 4p^6$
Rb $(Kr)5s^1$	**Sr** $(Kr)5s^2$	**Y** $(Kr)4d^1 5s^2$	**Zr** $(Kr)4d^2 5s^2$	**Nb** $(Kr)4d^4 5s^1$	**Mo** $(Kr)4d^5 5s^1$	**Tc** $(Kr)4d^5 5s^2$	**Ru** $(Kr)4d^7 5s^1$	**Rh** $(Kr)4d^8 5s^1$	**Pd** $(Kr)4d^{10}5s^0$	**Ag** $(Kr)4d^{10}5s^1$	**Cd** $(Kr)4d^{10}5s^2$	**In** $(Kr)4d^{10}5s^2 5p^1$	**Sn** $(Kr)4d^{10}5s^2 5p^2$	**Sb** $(Kr)4d^{10}5s^2 5p^3$	**Te** $(Kr)4d^{10}5s^2 5p^4$	**I** $(Kr)4d^{10}5s^2 5p^5$	**Xe** $(Kr)4d^{10}5s^2 5p^6$
Cs $(Xe)6s^1$	**Ba** $(Xe)6s^2$	**La** $(Xe)5d^1 6s^2$	**Hf** $(Xe)4f^{14}5d^2 6s^2$	**Ta** $(Xe)4f^{14}5d^3 6s^2$	**W** $(Xe)4f^{14}5d^4 6s^2$	**Re** $(Xe)4f^{14}5d^5 6s^2$	**Os** $(Xe)4f^{14}5d^6 6s^2$	**Ir** $(Xe)4f^{14}5d^7 6s^2$	**Pt** $(Xe)4f^{14}5d^9 6s^1$	**Au** $(Xe)4f^{14}5d^{10}6s^1$	**Hg** $(Xe)4f^{14}5d^{10}6s^2$	**Tl** $(Xe)4f^{14}5d^{10}6s^2 6p^1$	**Pb** $(Xe)4f^{14}5d^{10}6s^2 6p^2$	**Bi** $(Xe)4f^{14}5d^{10}6s^2 6p^3$	**Po** $(Xe)4f^{14}5d^{10}6s^2 6p^4$	**At** $(Xe)4f^{14}5d^{10}6s^2 6p^5$	**Rn** $(Xe)4f^{14}5d^{10}6s^2 6p^6$
Fr $(Rn)7s^1$	**Ra** $(Rn)7s^2$	**Ac** $(Rn)6d^1 7s^2$															

LANTHANIDE SERIES

58	59	60	61	62	63	64	65	66	67	68	69	70	71
Ce $(Xe)4f^1 5d^1 6s^2$	**Pr** $(Xe)4f^3 6s^2$	**Nd** $(Xe)4f^4 6s^2$	**Pm** $(Xe)4f^5 6s^2$	**Sm** $(Xe)4f^6 6s^2$	**Eu** $(Xe)4f^7 6s^2$	**Gd** $(Xe)4f^7 5d^1 6s^2$	**Tb** $(Xe)4f^9 6s^2$	**Dy** $(Xe)4f^{10}6s^2$	**Ho** $(Xe)4f^{11}6s^2$	**Er** $(Xe)4f^{12}6s^2$	**Tm** $(Xe)4f^{13}6s^2$	**Yb** $(Xe)4f^{14}6s^2$	**Lu** $(Xe)4f^{14}5d^1 6s^2$

ACTINIDE SERIES

90	91	92	93	94	95	96	97	98	99	100	101	102	103
Th $(Rn)6d^2 7s^2$	**Pa** $(Rn)5f^2 6d^1 7s^2$	**U** $(Rn)5f^3 6d^1 7s^2$	**Np** $(Rn)5f^4 6d^1 7s^2$	**Pu** $(Rn)5f^6 7s^2$	**Am** $(Rn)5f^7 7s^2$	**Cm** $(Rn)5f^7 6d^1 7s^2$	**Bk** $(Rn)5f^9 7s^2$	**Cf** $(Rn)5f^{10}7s^2$	**Es** $(Rn)5f^{11}7s^2$	**Fm** $(Rn)5f^{12}7s^2$	**Md** $(Rn)5f^{13}7s^2$	**No** $(Rn)5f^{14}7s^2$	**Lr** $(Rn)5f^{14}6d^1 7s^2$

Figure 2.3. The Periodic Table. The electronic configuration is given for each element.

energies, $4s^1 3d^5$ and $4s^2 3d^4$ for chromium, and $4s^1 3d^{10}$ and $4s^2 3d^9$ for copper, are very close. The inversion observed for chromium comes about because of the particular stability of the half-filled shell of electrons whose spins are parallel, another manifestation of Hund's rule. The exchange energy which results (see Section 2.3.5a) is sufficient to compensate for the $4s \rightarrow 3d$ promotion energy. A similar effect is responsible for the completion of the filled sub-shell in the case of copper leading to the $4s^1 3d^{10}$ configuration lying lowest in energy.

The structure of the fifth period is identical to that of the fourth. It begins with the filling of the $4s$ sub-shell (Rb, Sr) and ends with that of the $5p$ sub-shell (from In to Xe). Between these two sets of elements lie the ten elements corresponding to the successive filling of the $4d$ sub-shell, giving rise to the *second transition metal series*. The exceptions to Klechkowsky's rule are numerous in this series, namely Nb $(5s^1 4d^4)$, Mo $(5s^1 4d^5)$, Ru $(5s^1, 4d^7)$, Rh $(5s^1 4d^8)$, Pd $(5s^0 4d^{10})$ and finally Ag $(5s^1 4d^{10})$. Here the $5s$ and $4d$ levels are very close in energy. Although it is the case that the energetics controlling these observations are indeed a balance between those of the one-electron AO energies and the energies of the electron–electron interactions (the exchange energy for example) it is difficult to give a simple explanation of the origin of all of these inversions.

The sixth period encompasses two transitional series, the *third transition metal series* and the set of elements called the *lanthanides*. The filling of the $6s$ sub-shell (Cs and Ba) is followed by that of the $4f$ sub-shell (leading to the lanthanides) in turn followed by that of the $5d$ sub-shell (leading to the third transition metal series), before the $6p$ sub-shell (from Tl to Rn) is filled to complete the period. The situation is complicated by the fact that the two transitional series overlap in the sense that for lanthanum an electron occupies a $5d$ orbital $(6s^2 5d^1)$ but in the following series of fourteen elements, it is the group of seven $4f$ orbitals which are being filled. Following this set of elements the filling of the $5d$ orbitals continues with the nine elements from hafnium (Hf) to mercury (Hg). In this period which contains 32 elements there are four exceptions to Klechkowsky's rule; La $(6s^2 5d^1 4f^0)$, Gd $(6s^2 5d^1 4f^7)$, Pt $(6s^1 5d^9 4f^{14})$ and Au $(6s^1 5d^{10} 4f^{14})$.

Finally the seventh and last period is not complete since not all of the possible elements are actually known in practice. In this period where the $7s$, $6d$ and $5f$ orbitals (filling of the last giving rise to the actinides) are close in energy, exceptions to Klechkowsky's rule are frequent.

2.4.2. Organization by column: chemical families

The structure of the periodic classification clearly shows how elements with the same valence electron configuration fall neatly into columns. (Some exceptions are found for elements in the transitional series.) Since the valence electrons are responsible for the chemical properties of the elements we can understand why they are often so similar for the elements from the same column. Such a group is often called a *chemical family*. Examples from Figure 2.3 include the alkali metals, the halogens and the noble gases.

(i) The last column (18) contains the set of elements whose valence configuration is of the form $ns^2 np^6$ (except for helium where it is $1s^2$). This family is called the inert, rare or noble gases (He, Ne, Ar, Kr, Xe). The fact that the ns and np sub-shells are completely filled confers a special stability on these elements. They are monatomic gases under ambient conditions and are almost completely chemically inert.

(ii) The penultimate column (17) contains the halogens with valence electron configurations ns^2np^5 (F, Cl, Br, I). The elements are stable as the dimers F_2, Cl_2, Br_2, I_2. They form simple compounds with hydrogen, the hydrogen halides HF, HCl, HBr, and HI which all have acidic properties in aqueous solution. On the other hand they can readily capture an electron to give the anions F^-, Cl^-, Br^- and I^-, which are isoelectronic with the adjacent noble gas which follows them in the periodic classification. So, for example, Cl^- is isoelectronic with Ar, both having the configuration $3s^23p^6$. Such ions, by analogy with the noble gases have a special electronic stability.

(iii) The first column (1) of the periodic table contains the alkali metals, Li, Na, K, Rb, and Cs, with the configuration ns^1. (As we indicated in Chapter 1, hydrogen is not considered an alkali metal since its behavior is often different from that of the other elements of the group.) The alkali metals form water-soluble salts with the halogens, NaCl and KBr for example. The outermost electron of the atom may readily be removed which leads to the generation of a cation, isoelectronic with the adjacent noble gas which precedes it in the periodic classification. Thus K^+ is isoelectronic with Ar, $\{Ne\}3s^23p^6$.

We can put together these last two results to understand the formation of NaCl. For the alkali metals there is a tendency to readily lose an electron, and for the halogens a tendency to accept an electron. Such an ionic salt, written Na^+Cl^-, comes about via electron transfer from alkali to halogen.

2.5. Electronic parameters of many-electron atoms

Starting off from the ground electronic configuration of an atom it is possible to develop several concepts which will be useful in the rest of the book. For this we have to introduce further approximations.

2.5.1. Screening

In many-electron atoms the electrons occupy different AOs which, just as in the one-electron atom, have a radius which increases with the value of the principal quantum number (see Section 2.2.2e). The deepest lying AOs are thus the most contracted or, alternatively, the electron density associated with them lies closest to the nucleus. The higher-lying AOs are contrarily more diffuse. The outer electrons which occupy the higher-energy orbitals not only lie further from the nucleus than the electrons in deep-lying orbitals but see a lower effective nuclear charge since these deep-lying electrons are of the opposite charge to the nucleus. Thus the inner shells of electrons tend to 'screen' the nucleus from the outermost electrons. Effectively the outer electrons 'see' a smaller *effective nuclear charge Z^**, related to the real charge

Z by the screening constant σ. This parameter σ represents the mean effect exercised by the inner electrons. Thus we may write

$$Z^* = Z - \sigma \tag{40}$$

The constant, σ, depends strongly on the orbital occupied by the electron under consideration. Obviously the $1s$ electron cannot be screened by the electrons in the $2s$ or $2p$ orbitals since it lies closer to the nucleus. On the contrary an electron occupying a $3s$ orbital will be strongly screened by the internal electrons occupying the $1s$, $2s$ and $2p$ orbitals.

2.5.2. The effective charge: Slater's rules

Slater proposed an empirical method for the calculation of the screening constants. It consists first of sorting the AOs into different groups.

$$1s/2s, \; 2p/3s, \; 3p/3d/4s, \; 4p/4d/4f/5s, \; 5p/\text{etc}\ldots$$

The value of σ associated with an electron occupying a given AO is determined by the screening contribution by the other electrons occupying orbitals in the same group (weak screening) and by the electrons occupying orbitals in lower groups. The latter are closer to the nucleus and thus exercise a stronger screening. In practice σ is calculated by adding up the different contributions (σ_i) using the following rules.

   (i) For an electron in a $1s$ orbital the screening from another $1s$ electron is equal to 0.30.

   (ii) For an electron occupying an ns or np orbital the screening resulting from an electron in an orbital with principal quantum number n' is

$$\begin{array}{lll} \sigma_i = 1 & \text{if } n' < n - 1 & \text{complete screening} \\ \sigma_i = 0.85 & \text{if } n' = n - 1 & \text{strong screening} \\ \sigma_i = 0.35 & \text{if } n' = n & \text{weak screening} \\ \sigma_i = 0 & \text{if } n' > n & \text{no screening} \end{array}$$

  (iii) For an electron in an nd or nf orbital the screening constant is 0.35 for an electron in the same group and 1 for all the others.

The rules are summarized in Table 2.1. As an example consider the phosphorus atom ($1s^2 2s^2 2p^6 3s^2 3p^3$). According to the grouping scheme proposed by Slater there are three groups of electrons to consider; two $1s$ electrons, eight electrons in $2s$ and $2p$ and five electrons in $3s$ and $3p$. Each group has a different screening constant.

   (i) A $1s$ electron is screened by one other

$$\sigma_{1s} = 1 \times 0.30 = 0.30$$

Table 2.1: Slater's rules for the calculation of the screening constant.

	$n' < n - 1$	$n' = n - 1$	$n' = n$	$n' > n$
$1s$	–	–	0.30	0
ns, np	1	0.85	0.35	0
nd, nf	1	1	0.35	0

(ii) A $2s$ or $2p$ electron is screened by two $1s$ electrons and by seven other $2s$ or $2p$ electrons

$$\sigma_{2s, 2p} = (2 \times 0.85) + (7 \times 0.35) = 4.15$$

(iii) A $3s$ or $3p$ electron is screened by two $1s$ electrons, eight $2s$ or $2p$ electrons and by four $3s$ or $3p$ electrons.

$$\sigma_{3s, 3p} = (2 \times 1) + (8 \times 0.85) + (4 \times 0.35) = 10.2$$

Thus the effective charges felt by the different electrons are

$$Z^*_{1s} = 15 - 0.3 = 14.7$$
$$Z^*_{2s, 2p} = 15 - 4.15 = 10.85$$
$$Z^*_{3s, 3p} = 15 - 10.2 = 4.8$$

These values fall between $Z^* = 15$ (no screening) and $Z^* = 1$ (complete screening). They vary considerably depending upon the orbital under consideration. In particular we note that *the effective charge felt by the valence electrons is much smaller than that felt by the core electrons.*

2.5.3. Orbital radii and atomic size

For the hydrogen-like atoms we noted earlier that the radius of an orbital, defined as the point where a maximum is found in the radial probability density, was given approximately by $n^2 a_0/Z$. An approximate value (see exercise 2.8) for the radius (ρ) of an AO in a many-electron atom is given by

$$\rho = n^2 a_0/Z^* \tag{41}$$

Thus the radius depends both on the effective charge Z^* and the principal quantum number associated with the AO. The orbital is most contracted (i.e., ρ is smallest) when Z^* is large and n small. Since as n increases, Z^* decreases, the inner orbitals are expected to be much more contracted than the outer ones. So for phosphorus

$$\rho_{1s} = 3.6 \text{ pm}, \ \rho_{2s} = \rho_{2p} = 19.6 \text{ pm}, \ \rho_{3s} = \rho_{3p} = 99.5 \text{ pm}$$

This example underscores the general result that *in an atom the radii of the valence orbitals are much larger than those of the core orbitals.*

The valence electrons are therefore the electrons which are furthest from the nucleus and those which experience the weakest effective nuclear charge. They are then the electrons which are the most weakly bound to the nucleus since the electrostatic energy between the two is roughly proportional to $-Z^*/\rho$. This simple result allows us to readily understand how the valence electrons are the ones which are most sensitive to outside perturbations, such as the approach of another atom, and are therefore primarily responsible for the chemical properties of the elements. We finally note that the radius of the valence orbitals give an estimate of the 'size' of an atom, since it represents the most probable distance from the nucleus of the outermost electrons. This parameter is called the atomic radius. In the case of phosphorus the atomic radius is about 100 pm.

2.6. Evolution of atomic properties

It is useful to study the way certain properties of the atoms depend upon the row and column of the periodic table. In what follows we will consider the elements of the first five rows with the exception of the transition metal series which we will not discuss in this book.

2.6.1. Atomic orbital parameters

(a) Effective charge

Use of Slater's rules (Section 2.5.2) allows the computation of the effective nuclear charge seen by the valence electrons for the elements of the table. These are shown in Table 2.2 and will be considered in order of increasing Z. On moving from the left to the right along a given row of the table the ns and np orbitals are progressively filled. On moving from one element to the next, the actual nuclear charge increases by unity. This is only partially compensated by the increase (of 0.35) in the screening constant. As a result, for a given row of the periodic table the effective charge Z^* seen by a valence electron increases with Z. For example, Z^* increases from 1.30 (Li) to 5.85 (Ne) for the second row and from 2.20 (Na) to 6.75 (Ar) in the third. An important characteristic is the abrupt discontinuity observed on moving from one row to the next, as a result of the increase (by unity) in the value of the principal quantum number, n. So Z^*, equal to 5.85 for neon, drops sharply to only 2.2 for sodium. In this case the change in the screening is much larger than the variation in Z which is always equal to one. All of the other electrons in sodium lie in earlier groups than the single valence electron and thus are highly effective in screening. (σ_i is 1 for the $1s$ electrons and 0.85 for the $2s$ and $2p$ electrons.) In neon the outermost electron is one of a group which screen it much less effectively. The result is a sudden

Table 2.2: Effective charge (Z^*) felt by the valence electrons for the elements of the first five rows.

H							He
1.0							1.70
Li	Be	B	C	N	O	F	Ne
1.30	1.95	2.60	3.25	3.90	4.55	5.20	5.85
Na	Mg	Al	Si	P	S	Cl	Ar
2.20	2.85	3.50	4.15	4.80	5.45	6.10	6.75
K	Ca	Ga	Ge	As	Se	Br	Kr
2.20	2.85	5.00	5.65	6.30	6.95	7.60	8.25
Rb	Sr	In	Sn	Sb	Te	I	Xe
2.20	2.85	5.00	5.65	6.30	6.95	7.60	8.25

jump in screening on moving from neon (4.15) to sodium (8.8). Thus the effective charge felt by the valence electrons drops abruptly on moving from one row to the next in the periodic table.

(b) Atomic radius

Recall that the atomic radius is equal to the distance from the nucleus where there is a maximum in the radial probability density associated with the outermost electrons. It is given (equation (41)) in terms of n and Z^* as

$$\rho = n^2 a_0 / Z^* \text{ where } a_0 = 52.9 \text{ pm}$$

There are once again two important characteristics to note when examining the variation in ρ as a function of Z (Table 2.3). On moving from the left to the right of the periodic table along a given row although n is constant Z^* increases (Table 2.2) resulting in a diminution of ρ. These variations can be considerable. Using this measure of size an atom of neon is five times 'smaller' than one of lithium.

On moving from one row to another a sharp discontinuity appears; the atomic radius increases by a factor of six for example between neon ($Z = 10$) and sodium ($Z = 11$). Both the increase in the value of the principal quantum number (from 2 to 3) and the decrease in Z^* (from 5.85 to 2.2) contribute to this effect. *Thus the atomic size decreases on moving from left to right across a row of the periodic table (e.g., Li to Ne) and the atomic radius increases abruptly on moving from the end of one row of the periodic table to the beginning of the next.*

We finally note that the atomic radius gives information too about the atomic polarizability, a measure of the deformation experienced by the electron cloud when under the influence of an external electric field. When the atomic radius is large, the

Table 2.3: Atomic radii (in pm) calculated using equation (41).

H							He
53							31
Li	Be	B	C	N	O	F	Ne
163	109	82	65	55	47	41	36
Na	Mg	Al	Si	P	S	Cl	Ar
217	168	137	115	100	88	78	71
K	Ca	Ga	Ge	As	Se	Br	Kr
332	256	146	129	116	105	96	88
Rb	Sr	In	Sn	Sb	Te	I	Xe
386	300	171	151	135	122	112	103

peripheral electron density interacts relatively weakly with the nucleus and is easily distorted under these conditions. Atoms with smaller radii, containing more tightly bound electrons are less susceptible to deformation. Simple arguments suggest that the polarizability increases as ρ^3. From Table 2.3 the most polarizable atom is therefore rubidium (Rb) and the least is helium.

(c) Orbital energies

On moving from one atom to another the energy associated with a given orbital changes. From a purely qualitative point of view we can give two factors which contribute to this effect, namely the variation in the radius of the orbital under consideration and the effective charge seen by the electron in it. The energy of the orbital is lowered when the interaction between the nucleus and electron is strong, that is to say the effective charge is large and the orbital radius small. In Table 2.4 are listed the orbital energies given by Slater* for the occupied valence orbitals of the elements of the first four rows of the periodic table. (Parenthetically we note that although the ns and np orbitals are characterized by the same values of Z^* and therefore have the same orbital radii, their energies are different ($\varepsilon_{2s} < \varepsilon_{2p}$) and often significantly. This shows that the energy of an orbital doesn't only depend upon the parameters n and Z^* of our simple theory but on the quantum number l too). On moving from left to right across a row of the table, the energy of a given orbital drops considerably. For example the $2s$ orbital drops from -5.4 eV for lithium to -19.4 eV for carbon and then to -48.4 eV for neon. In the same way the $3p$ orbital drops in energy from -6.0 eV for aluminum to -15.8 eV for argon. This stabilization is easy to understand. The effective charge seen by the electron increases (see Table 2.2) and the radius of the orbital correspondingly decreases (see Table 2.3). Both

---

* J. C. Slater, *Quantum Theory of Atomic Structure*, Vol 1, McGraw-Hill (1960), p. 206.

Table 2.4: Valence orbital energies expressed in eV.

	H							He
$1s$	−13.6							−24.6
	Li	Be	B	C	N	O	F	Ne
$2s$	−5.4	−9.4	−14.7	−19.4	−25.6	−32.4	−40.1	−48.4
$2p$	—	—	−5.7	−10.7	−12.9	−15.9	−18.6	−21.6
	Na	Mg	Al	Si	P	S	Cl	Ar
$3s$	−5.2	−7.6	−11.3	−15.0	−18.4	−20.9	−25.3	−29.2
$3p$	—	—	−6.0	−7.8	−9.8	−11.7	−13.7	−15.8
	K	Ca	Ga	Ge	As	Se	Br	Kr
$4s$	−4.4	−6.1	−12.6	−15.6	−17.7	−20.9	−24.5	−27.2
$4p$	—	—	−6.0	−7.5	−9.2	−10.9	−12.6	−14.0

of these related effects contribute to the lowering of the energy of the orbital. Exactly the same relationship in terms of the changes in effective charge and radius lead to the abrupt change in the energies of orbitals of the same type (l quantum number) on moving from one row to the next. So the energy of the $2s$ orbital in neon lies at −48.4 eV but the $3s$ orbital of sodium at only −5.2 eV.

The energetic changes associated with the valence orbitals of the elements within a column are more difficult to understand. On descending a column the effective charge increases at first and then remains constant (Table 2.2) but the orbital radius continues to increase from top to bottom (Table 2.3). These two effects tend to work in opposite directions when it comes to determining the orbital energy. In general the increase in radius is the dominant factor, i.e., the valence energy levels drop in energy on moving from the bottom to the top of a column. In general, the energies of the valence s and p orbitals drop on moving from the left to the right along a row of the periodic table and on moving from the bottom to the top of a column.

2.6.2. Relationship with measurable properties

(a) Ionization potential

The *ionization potential* (IP) of an atom is the smallest amount of energy needed to remove an electron.

$$A \rightarrow A^+ + e^- \qquad \Delta H = IP$$

It corresponds therefore to the loss of an electron from the outermost sub-shell. (We will note later the exceptions provided by the transition metals.) It is clear to see that there is a direct correspondence between the IP and the energy of the orbital occupied by the ejected electron. Turning first of all to the simplest case of the

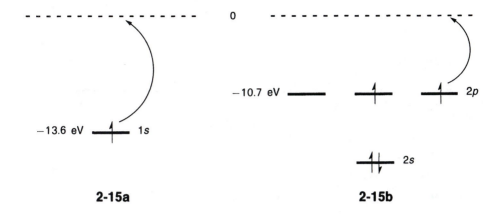

2-15a   **2-15b**

hydrogen atom which has only a single electron (**2-15a**), its IP is 13.6 eV, i.e., exactly the energy of the $1s$ orbital with a change of sign. In many-electron atoms the situation is more complex (of course) and the IP depends upon two factors.

(i) The energy of the orbital which contains the electron to be ejected. If this were the only factor then the IP would be equal to the orbital energy with a change in sign. This result is referred to as *Koopman's theorem.*

(ii) The reorganization energy associated with the cation which is formed. This is a new factor, absent in the hydrogen atom case since the cation is but a proton. The reorganization of the remaining charge density around the nucleus in the cation comes in part from a reduction in the screening contant as the result of the loss of an electron and from a change in the interactions between the electrons.

Invariably the second factor has a relatively small proportional effect on the IP. For example, carbon has an IP of 11.3 eV, not widely different from that expected by consideration of the orbital energy alone (Table 2.4) of 10.7 eV (**2-15b**). *Generally the orbital energy of the outermost electron gives quite a good approximation of the ionization potential.*

The ionization potentials of the elements of the first five rows of the periodic table are given in Table 2.5. One can compare these experimental values with those for the orbital energies of Table 2.4. The general trends found for the IPs mimic those observed for the orbital energies. *The IP increases on moving from the left to the right along a row of the periodic table and on moving from the bottom to the top. The atoms with the smallest IPs are the alkali metal family of the first column.* From Table 2.5 the IP is largest for helium (24.6 eV) and smallest for rubidium (4.2 eV).

However, we must note two exceptions to these trends. The first concerns the drop (rather than an increase) in the IP which occurs on moving from column 2 to column 13. This is simple to understand, there is a change in the nature of the orbital holding the outermost electron, from s to p. The IP drops since ε_{2s} (Be) $< \varepsilon_{2p}$(B). The second exception is an interesting one, since it vividly demonstrates the importance of electron–electron interactions. This is the drop in IP on moving from column 15 to 16. For nitrogen and oxygen this corresponds to a drop from 14.5 eV (for nitrogen) to 13.6 eV (for oxygen). This trend is opposite to that expected on the basis (Table 2.4)

Table 2.5: Ionization potentials expressed in eV.

1							18
H 13.6	2	13	14	15	16	17	**He** 24.6
Li 5.4	**Be** 9.3	**B** 8.3	**C** 11.3	**N** 14.5	**O** 13.6	**F** 17.4	**Ne** 21.6
Na 5.1	**Mg** 7.6	**Al** 6.0	**Si** 8.2	**P** 10.5	**S** 10.4	**Cl** 13.0	**Ar** 15.8
K 4.3	**Ca** 6.1	**Ga** 6.0	**Ge** 7.9	**As** 9.8	**Se** 9.8	**Br** 11.8	**Kr** 14.0
Rb 4.2	**Sr** 5.7	**In** 5.8	**Sn** 7.3	**Sb** 8.6	**Te** 9.0	**I** 10.5	**Xe** 12.1

of the orbital energies; -15.9 eV ($\varepsilon_{2p}(O)$) and -12.9 eV ($\varepsilon_{2p}(N)$). **2-16** and **2-17** show pictorially the ionization processes for the two atoms. **2-18a** shows the trends in the ionization potentials for the second-row atoms. The valence orbital energies of Table 2.4 increase smoothly across the series.

2-16

2-17

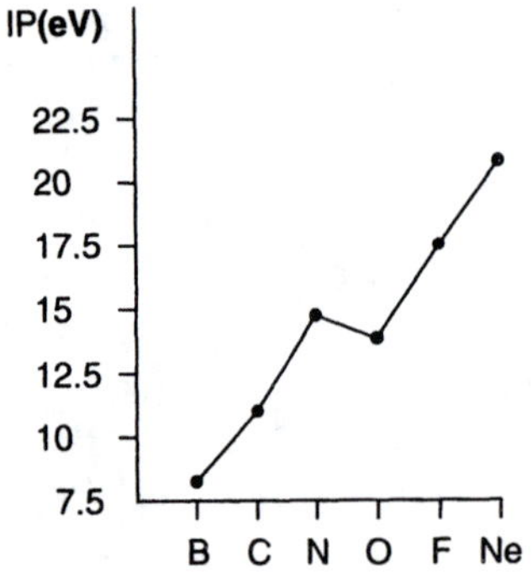

2-18a

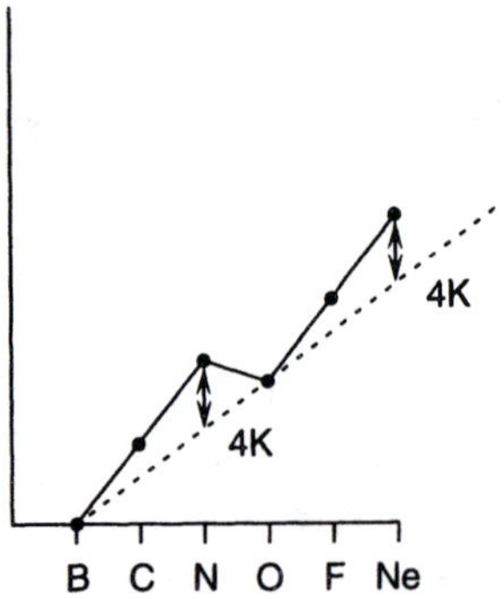

2-18b

Table 2.6: Exchange energy for the p^1–p^6 configurations

	Exchange energy		
	neutral (or anion)	cation (or neutral)	Δ
B	0K	0K	0
C (B⁻)	2K	0K	2K
N (C⁻)	6K	2K	4K
O (N⁻)	6K	6K	0
F (O⁻)	8K	6K	2K
Ne (F⁻)	12K	8K	4K

The obvious difference between the two is that in the case of oxygen the single electron with down spin is ionized but for nitrogen, one of three electrons with the same spin is removed. So for nitrogen, on ionization there is a loss of exchange energy while for oxygen there is no such loss. We may use a very simple model to put this idea on a semiquantitative basis. Table 2.6 shows the exchange energy for each of the neutral atoms with p^n valence configurations and their corresponding cations. These are obtained in the following way. We associate an energy of 2K with each pair of electrons with the same spin relative to the pair of electrons with spins paired. So for the p^1 configuration where there are no pairs of electrons the exchange energy is zero. For p^2 there is just one pair (exchange energy $=2K$). For p^3 there are now three pairs of electrons with the same spin (total exchange energy $=6K$). For p^4–p^6 the pattern repeats itself for the electrons of down spin. In the last column of Table 2.6 is the difference (Δ) in exchange energy between neutral atom and cation, obtained by simple subtraction. **2-18b** shows then how we might envisage the variation in IP as the $2p$ sub-shell is filled. First there is a sloping background (dashed line) which accommodates the changes in Z^* with electron configuration. Superimposed is the change in exchange energy from Table 2.6 to give the total ionization energy expected (solid line). Notice the resemblance of this plot to the actual saw-tooth behavior of the experimentally determined IPs of **2-18a**. The values of the AO energies of Table 2.4 average out these two effects. A similar saw-tooth plot is obtained if we take into account the coulombic repulsion arising from pairs of electrons in the same orbital, rather than the exchange energy. Both effects are of course important, and highlight the influence of electron–electron interactions on the ionization potential variation.

One can equally well eject electrons other than the outermost one. The energy change involved (ionization energy) is directly correlated with how deep lying is the orbital concerned. For example the three different ionization energies for oxygen are

$$1s \text{ electron } E = 543.1 \text{ eV}$$

$$2s \text{ electron } E = 41.6 \text{ eV}$$

$$2p \text{ electron } E = 13.6 \text{ eV (IP)}$$

Finally one can define a second, IP_2, (or third, IP_3, etc.,) ionization potential which represents the smallest amount of energy needed to detach an electron from the singly charged cation (or doubly charged cation etc.) produced after the first (or second etc.) ionization.

$$A^+ \rightarrow A^{2+} + e^- \qquad \Delta H = IP_2$$

$$A^{2+} \rightarrow A^{3+} + e^- \qquad \Delta H = IP_3$$

etc.

The second IP is always larger than the first, in spite of the fact that the electron often comes from the same sub-shell, since the outermost electron in the ion is less effectively shielded than in the neutral atom. For example in the carbon atom the effective charge seen by a $2p$ electron is equal to 3.25 in the neutral species but 3.60 in the singly charged cation. The six ionization potentials for carbon are:

$IP_1 = 11.3$ eV	$IP_3 = 47.9$ eV	$IP_5 = 392.1$ eV
$IP_2 = 24.4$ eV	$IP_4 = 64.5$ eV	$IP_6 = 490.0$ eV
$2p$ electrons	$2s$ electrons	$1s$ electrons

(b) Electron affinity

The *electron affinity* (EA) measures the capacity of an atom to accept an extra electron. The production of a stable anion is an exothermic process ($\Delta H < 0$), but conventionally, values of the electron affinity are reported after reversing the sign. The larger the value of the EA the greater the stability of the anion A^- relative to a neutral atom A plus an electron

$$A + e^- \rightarrow A^- \qquad \Delta H = -EA$$

By analogy with our analysis of the variations found for the IPs we can correlate the values of the EAs with the properties of the orbital that receives the extra electron. So for the case of chlorine, with a valence electron configuration $3s^2 3p^5$, this electron has to occupy the last free place in the set of $3p$ orbitals whose orbital energy is -13.7 eV. On the most naïve level the gain in energy associated with the electron capture process should therefore be close to this value, i.e., the EA should then be close to $+13.7$ eV. However the value measured experimentally (3.6 eV) is far from this figure. This is a very general result as may be seen by a comparison of the EA values given in Table 2.7 with the orbital energy values of Table 2.4. A part of the solution to this problem is immediately apparent once we realize that the electron affinity of an atom is identical to the ionization potential of its singly charged negative ion.

$$A^- \rightarrow A + e^- \qquad \Delta H = EA(\text{atom A}) = IP(\text{ion } A^-)$$

Table 2.7: Electron affinities expressed in eV.

1	2	13	14	15	16	17	18
H 0.75							**He** 0.0
Li 0.62	**Be** 0.0	**B** 0.28	**C** 1.26	**N** 0.0	**O** 1.46	**F** 3.40	**Ne** 0.0
Na 0.55	**Mg** 0.0	**Al** 0.44	**Si** 1.39	**P** 0.75	**S** 2.08	**Cl** 3.62	**Ar** 0.0
K 0.50	**Ca** 0.0	**Ga** 0.30	**Ge** 1.23	**As** 0.81	**Se** 2.02	**Br** 3.37	**Kr** 0.0
Rb 0.49	**Sr** 0.0	**In** 0.30	**Sn** 1.11	**Sb** 1.07	**Te** 1.97	**I** 3.06	**Xe** 0.0

Thus our arguments above concerning the difference in first and second ionization potentials of neutral atoms are applicable to this case too. For carbon, for example the three ionization potentials to consider are:

$$\text{IP of } C^- \;(=\text{EA of } C) = 1.26 \text{ eV}$$
$$\text{IP of } C \;(\text{IP}_1) = 11.3 \text{ eV}$$
$$\text{IP of } C^+ \;(\text{IP}_2) = 24.4 \text{ eV}$$

Each time the number of valence electrons changes by one, the corresponding ionization potential varies by around 10 eV for carbon. Z^* for the three species are 2.90 (C^-), 3.25 (C) and 3.60 (C^+).

A second observation concerning the figures of Table 2.7 are the zero values of the EA for some atoms. Thus for Be, N and Ne no stable anion exists. For Be and Ne this result is easy to understand. The added electron would have to occupy the next highest sub-shell since both of these species have full sub-shells. This costs energy and the overall process of electron attachment is unfavorable. As a result all of the elements of groups 2 (valence configuration ns^2) and 18 (valence configuration ns^2np^6) have zero electron affinities.

The realization that the EA of A is just the IP of A^- enables a further correlation with our earlier discussion. **2-19a** shows a plot of the observed EAs for the $2p^n$ series. Notice that it has the same saw-tooth behavior as that seen in **2-18a** for the ionization potentials of the neutral atoms. The EA drops between carbon and nitrogen in just the same way that the IP drops between nitrogen and oxygen. From Table 2.6 we can construct a qualitative prediction of the variation across the p^n series by adding to a sloping background, increasing with increasing Z, the effect of the difference in exchange energy between anion and neutral atom (**2-19b**). The critical electron count associated with the saw-tooth discontinuity is different by one electron from that

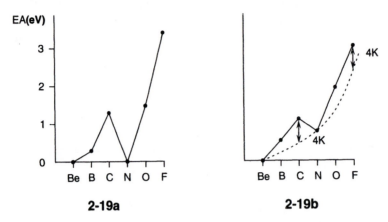

2-19a   **2-19b**

found in **2-18** for the IPs since there we were concerned with the difference in exchange energy between neutral atom and cation. In general, given the exceptions associated with the ns^2, ns^2np^3 and ns^2np^6 valence configurations we have described how EA tends to increase on going from left to right across the table.

The general ideas of this chapter would predict an increase in the electron affinities on moving from the bottom to the top of a column of the periodic table mirroring such predictions found experimentally for the ionization potentials. This is in general true but with the exceptions that the elements of the second row of the table (B–F) have smaller values than those of the third row (Al–Cl). To summarize, *the electron affinity has a tendency to increase on moving from left to right across a row of the periodic table from column 13 to column 17 and on moving from bottom to top with a reversal of this order between second row and third rows.* The halogens have the largest electron affinities.

2.6.3. Electronegativity scales

As we indicated in the first chapter, it is important to have an idea, albeit qualitative, of the way an atom polarizes the electron cloud within a molecule, that is to say its capacity to attract electron density. The quantity which we use to characterize this property is the electronegativity. It is not feasible to measure experimentally the polarization created by the presence of given atoms in molecules, but it is very useful to employ an electronegativity scale which allows a qualitative feeling for the concept.

There are three scales commonly in use. The first two are defined using some of the atomic properties which we have just described.

(i) Mulliken's scale. The electronegativity, $\chi(A)$ of an element A is proportional to the sum of the ionization potential (IP) and the electron affinity (EA).

$$\chi(A) = K[IP(A) + EA(A) + C]$$

where K and C are constants.

(ii) The Allred–Rochow scale. Here the electronegativity is proportional to the attractive force exerted on the outermost electron by the nucleus, via the following relationship

$$\chi(A) = KZ^*/r^2 + C$$

Here K and C are constants (different from those in (i)) and r is defined as the covalent radius of the atom. The values of this parameter for atoms of interest come from experimental data on the bond lengths in molecules containing the atom concerned. The covalent radius for chlorine, for example, is just half the Cl—Cl distance in the Cl_2 molecule. In practice the values for r are close to those found for ρ, the atomic radius we defined earlier (Section 2.5.3).

(iii) Pauling's scale. The third scheme is defined using some molecular properties. Pauling noticed that the bond dissociation energy, D_{AB}, of a heteronuclear diatomic molecule AB is generally larger than that of each of the AA and BB bond dissociation energies. For example the dissociation energy of HF is 570 kJ mole^{-1} but those of H_2 and F_2 are 436 and 159 kJ mol^{-1} respectively.

Table 2.8: Three electronegativity scales, (a) Mulliken, (b) Allred-Rochow, and (c) Pauling.

a

H						
3.1						

Li	Be	B	C	N	O	F
1.3	2.0	1.8	2.7	3.1	3.2	4.4

Na	Mg	Al	Si	P	S	Cl
1.2	1.6	1.4	2.0	2.4	2.7	3.5

b

H						
2.2						

Li	Be	B	C	N	O	F
1.0	1.5	2.0	2.5	3.1	3.5	4.1

Na	Mg	Al	Si	P	S	Cl
1.0	1.2	1.5	1.7	2.1	2.4	2.8

c

H						
2.2						

Li	Be	B	C	N	O	F
1.0	1.6	2.0	2.6	3.0	3.4	4.0

Na	Mg	Al	Si	P	S	Cl
0.9	1.3	1.6	1.9	2.2	2.6	3.2

In Pauling's scale the difference in electronegativity between A and B is determined by the values of the diatomic dissociation energies.

$$[\chi(A) - \chi(B)]^2 = K(D_{AB} - \sqrt{D_{AA} \times D_{BB}})$$

Pauling set the value for fluorine at 4.0.

The variations in electronegativity across the periodic table are very similar and independent of the scale which is used. Table 2.8 shows values for the first three rows. We notice that the electronegativity increases on moving from left to right across the periodic table and from bottom to top.

In Table 2.8 the most electronegative element is fluorine and the least electronegative is sodium. For the first two scales the variation in χ across the periodic table is easy to understand given our earlier discussion of the concepts of IP, EA, Z^* and ρ. Pauling's scale though, is perhaps the one most commonly used. We note though that there is no value of the electronegativity for the noble gases (column 18) since there are no diatomic molecules for these atoms. The Mulliken and Allred–Rochow scales do not suffer from this problem.

2.6.4. Electronegativity, orbital energy and orbital radius

A result which we will often use in the rest of this book relies on the comparison between Tables 2.4 and 2.8. This is that *the valence s and p orbital energies vary in a way which is paralleled by the variation in atomic electronegativity.* Thus the steady increase in electronegativity in the series boron (2.0), carbon (2.6), nitrogen (3.0), oxygen (3.4) and fluorine (4.0), corresponds with a steady lowering of the 2p orbital energy; -5.7 (B), -10.7 (C), -12.9 (N), -15.9 (O) and -18.6 eV (F). In the same way, the more electronegative the atom, the more contracted the orbitals (Table 2.3).

EXERCISES

Single-electron atoms

2.1 Consider the hydrogen atom in the $3p$ excited state

(i) Which electronic transitions are possible with emission of energy?

(ii) Calculate the wavelengths of light associated with these transitions. (In SI units, $Ry = 2.18 \times 10^{-18}$ J; $h = 6.62 \times 10^{-34}$ J s; $c = 3 \times 10^8$ m s^{-1}.) Recall that wavelength and frequency are related via the expression $\lambda v = c$.

(iii) Determine the ionization potential for hydrogen in this excited state in eV and in kJ mol^{-1}. Recall N_A (Avogadro's number) $= 6.02 \times 10^{23}$.

2.2 (i) Calculate (in eV) the IPs of the ions He$^+$ and C^{5+} in their ground electronic state. $(Ry = 13.6$ eV.)

(ii) Consider the Li^{2+} cation in its second excited state. What is the degeneracy of the wavefunctions describing this state? What is the ionization potential (in eV) of the ion in this state?

Many-electron atoms

2.3 Consider the sulfur atom $(Z = 16)$

(i) Give the electronic configuration of the lowest energy state.

(ii) Calculate the radii of the different occupied AOs for this atom (equation (41)).

2.4 Calculate the radii of the atoms of fluorine $(Z = 9)$, chlorine $(Z = 17)$ and bromine $(Z = 35)$. Which is the most, and which the least, polarizable?

2.5 Give the electronic configuration for platinum (Pt, $Z = 78$) which is in accord with Klechkowsky's rule. Knowing that the $6s$ and $5d$ levels are very close in energy, suggest two other configurations which should be close in energy to this.

2.6 Consider all the atoms with Z less than or equal to 20. In their electronic ground state

(i) Which of them are diamagnetic?

(ii) Which of them have a single unpaired electron?

(iii) Which of them have two unpaired electrons?

Analytic calculation of the orbital radius

2.7 Recalling that the radial probability density dS/dr of an atomic orbital is defined as: $dS/dr = R_{n,l}^2(r)r^2$ calculate the position of the maximum in this density for the 1s, 2s and 2p orbitals of the hydrogen atom. The analytical expressions for the different functions are given by equations (28)–(30).

2.8 An approximate expression for the radial part of the orbital wavefunctions of many-electron atoms was proposed by Slater.

$$R_{n,l}(r) = N(r/a_0)^{n-1} \exp(-Z^*r/na_0)$$

where N is a normalization constant. Calculate the radius of an orbital of this type. You should recover equation (41).

Orthonormalization of the hydrogen wavefunctions

2.9 Show that the 1s function (equation (28)) of the hydrogen atom is normalized, given that the volume element $d\tau$ expressed in spherical polar coordinates is $d\tau = r^2 \sin\theta \, dr \, d\theta \, d\phi$ (r goes from 0 to infinity, θ from 0 to π and ϕ from 0 to 2π). Show that the radial and angular parts of the wavefunction are normalized independently of each other. Note that

$$\int_0^\infty x^2 \exp(-ax) \, dx = 2/a^3$$

2.10 (i) Show that the 1s and 2s functions are both orthogonal to the three 2p functions. This result will be able to be established using the angular parts only of the respective analytic functions given by equations (28)–(30).

(ii) Show that the 1s and 2s functions are also orthogonal. Use the relationships

$$\int_0^\infty x^2 \exp(-ax) \, dx = 2/a^3 \qquad \int_0^\infty x^3 \exp(-ax) \, dx = 6/a^4.$$

Part II
Building up molecular orbitals and electronic structure

3 Interaction of two atomic orbitals on different centers

The attempt to use quantum mechanics to provide an electronic description of molecules raises a number of difficulties, some of which have already been encountered in the treatment of the isolated atom. Others are directly connected with the polyatomic nature of the molecule. The Schrödinger equation, which in principle allows the calculation of the molecular wavefunction, is much more difficult to solve than in the case of the many-electron atom. (The exception to this statement concerns molecules with a single electron such as H_2^+.) The most usual resolution of this problem makes use of a number of simplifying assumptions. The picture which results provides an excellent approximation to the real state of affairs.

3.1. Basic approximations

Three approximations are frequently used to calculate the molecular wavefunction.

3.1.1. The Born–Oppenheimer approximation

In a molecule the nuclei move together as a block during the displacements associated with translation and rotation, and move relative to each other during vibrations. All of these motions carry a contribution to the total energy of the molecule and have to be taken into account in the determination of the total wavefunction of the system since it depends both upon the nuclear coordinates (R) and the electronic coordinates (r).

The *Born–Oppenheimer approximation* writes the total wavefunction $\Psi(r, R)$ as a product of two parts; one $\Xi(R)$ which describes the nuclei and the other $\Phi_{el}(r)$ the electrons

$$\Psi(r, R) = \Phi_{el}(r)\Xi(R) \tag{1}$$

The justification for this is that since the nuclei are so much heavier than the electrons, they move much more slowly. Thus equation (1) represents a description of a set of mobile electrons moving in the field of frozen nuclei. Within the framework of this approximation the Schrödinger equation as a function of the coordinates of all the particles, nuclei and electrons, is replaced by one which still contains the electron coordinates as variables but uses a fixed geometry for the nuclear ones. The problem is thus reduced to solution of $\Phi_{el}(r)$ at a given molecular (i.e., nuclear) geometry. In this way the energy of the system may be calculated as a function of geometry by performing a series of computations at different values of the set of coordinates, R. *In what follows we will restrict ourselves to the study of the molecular electronic wavefunction $\Phi_{el}(r)$.*

3.1.2. The orbital approximation

We described the use of this approximation in Chapter 2 when studying the electronic situation in many-electron atoms. It is used both for atoms of this type and also for molecules since it is not possible to find an exact analytic solution for the electronic wavefunction for systems containing more than a single electron. An approximate solution is found by writing the many-electron wavefunction as a product of one-electron functions

$$\Phi_{el}(e_1, e_2, \ldots, e_n) = \phi_1(e_1)\phi_2(e_2) \ldots \phi_n(e_n) \tag{2}$$

The one-electron functions ϕ_i are called the *molecular orbitals* (MOs) of the system under consideration.

The molecular case poses another problem however. Remember that for atoms the one-electron functions are the atomic orbitals χ_i, whose mathematical form is derived from those found for the hydrogen atom. These, of course, are known exactly. It is not at all clear though how to choose the form of the molecular orbitals, $\phi_1, \phi_2, \ldots, \phi_n$, but one approach is almost universally used in chemistry.

3.1.3. The form of the MOs: the LCAO approximation

The simplest form used for describing molecular orbitals is the *linear combination of atomic orbitals (LCAO) approximation.* We write the molecular orbital (ϕ_i) as a linear sum of contributions from the atomic orbitals of the molecule χ_j

$$\phi_i = \sum_j c_{ij}\chi_j \tag{3}$$

In this expression the c_{ij} are the coefficients or the weights of the AO χ_j in the molecular orbital ϕ_i. A simple justification for the validity of the approach, is that in the interior of the molecule an atom does not completely lose its identity and retains many of the characteristics of the isolated atom. For example the elements having a large electron affinity and a large ionization potential, are the most electronegative elements, and those which have a tendency to attract electrons in the molecule. In equation (3) the AOs, χ_j are supposedly known and we just have to determine the coefficients c_{ij}. In principle there are an infinite number of functions χ_j (1s, 2s, 2p, 3s, 3p, etc.) for each atom, but in order for the problem to be manageable we need to truncate the summation in expression (3) by selecting a reasonable number of AOs. Two simplifications narrow this choice.

(i) Orbitals describing core electrons are ignored. This is understandable in part from the contraction of these orbitals. Their amplitude is only large close to the nucleus and they play a negligible role in the formation of chemical bonds. Although it is not a quantum mechanical approach, we recall here too that Lewis' theory only considered the valence electrons when viewing bond formation (Chapter 1).

(ii) Included for each atom in the sum of equation (3) are all the occupied valence orbitals and also those orbitals, which might be empty in the isolated atom but have the same value of the principal quantum number. For example for carbon ($1s^2 2s^2 2p^2$) only the 2s and 2p orbitals are included, for hydrogen ($1s^2$) the 1s orbital and for helium ($1s^2$) the 1s orbital. For lithium ($1s^2 2s^1$) we use both 2s and 2p orbitals. (Some variations on this are used for transition metal systems.)

The LCAO theory is not an approximation in itself. If it were possible, the calculation using an infinite set of AOs should lead to an exact solution for the MO. It is the restriction that the set of AOs included is limited to a small number which introduces the approximation. In fact it works well for much of the periodic table but it is often necessary to increase the 'basis set' to be able to calculate the MOs and properties of several molecules.

Determination of the form of the MOs, ϕ_i, reduces to finding the set of coefficients c_{ij} which characterize it. In practice these may be calculated using a result from the variation theorem. Essentially this theorem states that if we have a wavefunction such as that of equation (3), then the best approximation to the energy can be achieved by minimizing the analytical expression for the energy with respect to all of the c_{ij}. This leads to a direct determination of these coefficients. When the coefficients are known, the calculation of the energy of the orbital (ε_i) follows via a relationship of the type used for the many-electron atom in Chapter 2 (equation (38)). However in this book we will follow a different path. We shall show that the orbital problem may be analyzed in a very qualitative way and one which relies on a number of simple rules, based on symmetry, electronegativity and on the relative energies of the AOs of the isolated atoms. In this chapter we will treat the simplest cases one could imagine, namely MOs which are linear combinations of only two atomic orbitals.

3.2. Construction of MOs

Let us consider two atoms A and B each carrying a single atomic orbital, χ_1 and χ_2 respectively (**3-1**). We shall see that this in fact is not the great restriction it appears

$$\chi_1 \qquad\qquad \chi_2$$

3-1 $$A \text{------} B$$

at first sight. The MOs of the molecule AB are simply the linear combinations of the AOs χ_1 and χ_2. In current parlance we say that the AOs χ_1 and χ_2 *interact* to give the MOs of the AB molecule. The simplest situation of this type corresponds to real systems such as H_2 or H_2^+ ($\chi_1 = \chi_2 = 1s_H$), He_2^+ or He_2^{2+} ($\chi_1 = \chi_2 = 1s_{He}$) and

HeH$^+$ ($\chi_1 = 1s_{He}$, $\chi_2 = 1s_H$). Although an analysis of this problem will be quite sufficient to understand the structure of these small molecules, the principal motivation is the generation of a set of general rules which may be applied to the more complex systems which we will discuss later.

3.2.1. Interaction of two identical AOs

In the homonuclear diatomics, such as H_2, $H_2{}^+$, $He_2{}^+$ or $He_2{}^{2+}$, the two atoms, A and B, are the same. We will call them A_1 and A_2. The two AOs χ_1 and χ_2 are identical with the same form and the same energy, but they are centered at two different points in space, χ_1 on A_1 and χ_2 on A_2. The formation of the MOs in these systems therefore results from the interaction of *two degenerate atomic orbitals*.

(a) Study of the electron density

Within the framework of the LCAO approximation, each MO may be written as a simple linear combination of the two AOs

$$\phi = c_1\chi_1 + c_2\chi_2 \tag{4}$$

So for each MO there are two coefficients to calculate, c_1 and c_2. We can see how various restrictions on their relative values arise by study of the probability density for an electron located in a given MO. As described earlier (Section 2.1.1) the probability density is given by the square of the wavefunction:

$$\phi^2 = (c_1\chi_1 + c_2\chi_2)^2 = c_1^2\chi_1^2 + c_2^2\chi_2^2 + 2c_1c_2\chi_1\chi_2 \tag{5}$$

This expression may be divided into three parts. The first, $c_1^2\chi_1^2$, is only important in those regions of space where χ_1 itself is large, namely in the vicinity of atom A_1. The integral $\int c_1^2\chi_1^2 \, d\tau$ may be written as $c_1^2\langle\chi_1 | \chi_1\rangle$ using the notation introduced in Chapter 2. It represents, approximately, the probability of finding the electron close to atom A_1. In the same way, the integral of the second term $c_2^2\langle\chi_2 | \chi_2\rangle$ gives the probability of finding the electron close to atom A_2. The last term $2c_1c_2\langle\chi_1 | \chi_2\rangle$ is only important when *both* χ_1 and χ_2 are non-negligible, namely in the region between the two atoms A_1 and A_2.

(b) Some mathematical descriptions of the MOs

Some very important results may be derived from the symmetry properties of the A_2 diatomics. Since the molecule is symmetrical about the A—A bond, the two nuclei are completely equivalent. Put another way, the electrons in the molecule have the same probability of being close to atom A_1 as they do of being close to atom A_2. This means that the two terms describing the electron probability from the quantum mechanical description must be equal i.e.

$$c_1^2\langle\chi_1 | \chi_1\rangle = c_2^2\langle\chi_2 | \chi_2\rangle \tag{6}$$

Since χ_1 and χ_2 are normalized ($\langle \chi_1 \mid \chi_1 \rangle = \langle \chi_2 \mid \chi_2 \rangle = 1$) the result puts strong restrictions on the values of the coefficients, namely;

$$c_1^2 = c_2^2 \qquad \therefore \ c_1 = \pm c_2 \tag{7}$$

This is a very important result. It means that starting from two AOs, χ_1, χ_2 one can construct two MOs which in fact only contain one unknown, the normalization constant N_+, N_-.

$$\phi_+ = N_+(\chi_1 + \chi_2) \qquad \phi_- = N_-(\chi_1 - \chi_2) \tag{8}$$

To obtain $N_\pm$ we must normalize these functions. For ϕ_+

$$\langle \phi_+ \mid \phi_+ \rangle = 1$$
$$N_+^2(\langle \chi_1 \mid \chi_1 \rangle + \langle \chi_2 \mid \chi_2 \rangle + 2\langle \chi_1 \mid \chi_2 \rangle) = 1 \tag{9}$$

$\langle \chi_1 \mid \chi_2 \rangle$ is the overlap integral, S, between the two AOs χ_1 and χ_2 (Section 2.1.2b). The orbitals are normalized so that

$$N_+^2(1 + 1 + 2S) = 1$$

$$N_+ = 1/[2(1 + S)]^{1/2} \tag{10}$$

An analogous calculation for the orbital ϕ_- leads to

$$N_- = 1/[2(1 - S)]^{1/2} \tag{11}$$

The two MOs for the homonuclear diatomic molecule thus become

$$\phi_+ = \frac{1}{[2(1 + S)]^{1/2}} (\chi_1 + \chi_2) \qquad \phi_- = \frac{1}{[2(1 - S)]^{1/2}} (\chi_1 - \chi_2) \tag{12}$$

So, starting off from two AOs, χ_1 and χ_2, we have obtained two MOs ϕ_+ and ϕ_-. This relationship between the number of AOs and the number of MOs they generate is a general one, even in more complex systems; n AOs give rise to n MOs. If we calculate the overlap between the MOs we get

$$\langle \phi_+ \mid \phi_- \rangle = \frac{1}{2(1 - S^2)^{1/2}} (\langle \chi_1 \mid \chi_1 \rangle + \langle \chi_1 \mid \chi_2 \rangle - \langle \chi_1 \mid \chi_2 \rangle - \langle \chi_2 \mid \chi_2 \rangle)$$

$$= \frac{1}{2(1 - S^2)^{1/2}} (1 + S - S - 1)$$

$$= 0 \tag{13}$$

Thus, just as atomic orbitals on the same atom are orthogonal, so molecular orbitals are orthogonal to each other.

(c) The form of the MOs

In the orbital ϕ_+ the atomic orbital coefficients have the same sign, and thus they add together, a condition which we describe as being in phase. If we consider a point outside the region of space between the two nuclei, close to A_1 and therefore far from A_2, the form of the wavefunction shows that the amplitude of χ_2 is small but that of χ_1 large. Thus in this region the MO ϕ_+ has practically the same form as the AO χ_1. Correspondingly, in the region close to A_2 and far from A_1, ϕ_+ resembles the AO χ_2. It is in the internuclear region where ϕ_+ is distinctly different from either χ_1 or χ_2 alone. In fact χ_1 and χ_2 have similar contributions in the internuclear region and neither is negligible. The in-phase addition of χ_1 and χ_2 in this region leads to an amplitude of ϕ_+ which is clearly larger than that from χ_1 or χ_2 alone. This characteristic appears in the schematic contour diagram for the ϕ_+ function shown in **3-2**. Thus *the function ϕ_+ is characterized by a large amplitude in the internuclear region*.

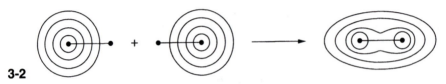

3-2

In contrast the amplitudes of the two functions χ_1 and χ_2 are subtracted in the ϕ_- orbital where the coefficients of χ_1 and χ_2 are opposite in sign, or out of phase. As before, close to the nuclei the wavefunction strongly resembles χ_1 or χ_2 but in the internuclear region, the subtraction of the two functions leads to a small amplitude of ϕ_-. If we consider a point, M, equidistant between the two nuclei, then by symmetry at this point the contributions from χ_1 and χ_2 are exactly equal but opposite in sign. The result is simple

$$\phi_-(M) = \chi_1(M) - \chi_2(M) = 0 \qquad (14)$$

The plane which bisects the A—A axis is thus a *nodal plane*. A contour diagram is shown in **3-3**, and just as for the atomic orbital case the change in sign of the function is shown by the use of dashed lines. Thus the MO ϕ_- is characterized by a small amplitude in the inernuclear region.

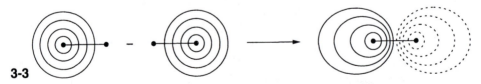

3-3

Conventionally we describe MOs in terms of the contributions from the AOs from which they are derived. The signs of the AOs are indicated by the conventions adopted in Chapter 2 (**2-4**). When the function is positive the orbital is hatched and when the function is negative it is unhatched. Additionally the size of the AO used reflects the magnitude of its coefficient in the LCAO expansions; the larger the coefficient, the larger the orbital size. In ϕ_+ (**3-4**) the AO coefficients are equal in absolute size (circles of the same radius) and are of the same sign (circles which are either both hatched or both unhatched since ϕ_+ and $-\phi_+$ have the same physical significance.) In ϕ_- (**3-5**) the coefficients are again equal in absolute value (same radius) but now different in sign (opposite hatching characteristics). *The schematic representation of a molecular orbital gives the relative signs and relative weights of the coefficients.*

3-4 φ_+ ⬤—⬤   or   ◯—◯

3-5 φ_- ⬤—◯   or   ◯—⬤

(d) Bonding and antibonding orbitals

Let us return to the probability density functions associated with the two MOs ϕ_+ and ϕ_-.

$$\phi_+^2 = \frac{1}{2(1+S)}(\chi_1^2 + \chi_2^2 + 2\chi_1\chi_2) \tag{15}$$

$$\phi_-^2 = \frac{1}{2(1-S)}(\chi_1^2 + \chi_2^2 - 2\chi_1\chi_2) \tag{16}$$

Ignoring the normalization term the two density functions differ only in the sign of the cross term $2\chi_1\chi_2$, which we recall, is only important in the region between the nuclei. In ϕ_+ this density adds to that already associated with the nuclei ($\chi_1^2 + \chi_2^2$) and an electron occupying this orbital has an increased probability, relative to the atoms of finite separation, of being between the nuclei. An electron in this orbital lies at a lower energy than one in an isolated AO. Although the detailed breakdown of the energy is beyond the scope of our approach, both the kinetic and potential energy of the electron are reduced. We can see that the electron between the two nuclei is attracted to both, rather than just a single nucleus, and in this orbital, where the internuclear probability is enhanced via the term $2\chi_1\chi_2$, this is important. Because of this energetic stabilization this orbital is called a *bonding orbital*.

The converse is true in ϕ_-. Here the electron density $2\chi_1\chi_2$ is subtracted from the terms associated with the density around the nuclei ($\chi_1^2 + \chi_2^2$). There is now a reduced probability of finding the electron between the nuclei, and the stabilization described above for the bonding orbital is absent. In fact the electron is less stable in this orbital than in an isolated orbital. Such an orbital is called an *antibonding orbital*.

The term $2\chi_1\chi_2$ clearly plays an important role in determining the energetics here.

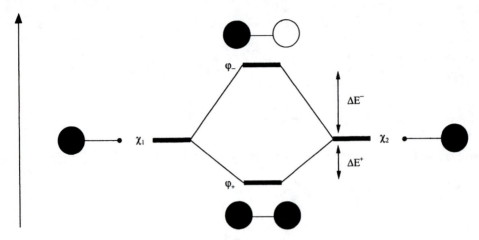

Figure 3.1. Molecular orbital diagram showing the interaction between two identical atomic orbitals in a homonuclear molecule.

Its integral over all space $2\langle \chi_1 \mid \chi_2 \rangle$ is twice the overlap integral between χ_1 and χ_2. This overlap controls the way molecular orbitals are constructed and is a crucial ingredient in the formation of a bond between two atoms.

(e) Molecular orbital energies: interaction diagrams

The energies of the two MOs ϕ_+ and ϕ_-, in contrast to the energies of the AOs from which they are derived, are not equal. Summarizing the discussion above, *the bonding orbital ϕ_+ is stabilized and the antibonding orbital ϕ_- destabilized with respect to the energies of the starting AOs.*

Let us call ΔE^+ the stabilization energy of the bonding orbital ($\Delta E^+ = \varepsilon(\chi_1) - \varepsilon(\phi_+)$) and ΔE^- the destabilization of the antibonding MO ($\Delta E^- = \varepsilon(\phi_-) - \varepsilon(\chi_1)$). These two quantities are thus, by definition, positive, and a general result connecting the two is that *the stabilization of the bonding orbital (ΔE^+) is smaller than the destabilization of the antibonding orbital (ΔE^-).*

It is now possible to construct an interaction diagram between the two atomic orbitals χ_1 and χ_2 in Figure 3.1. The atomic and molecular levels are represented by horizontal lines arranged in order of increasing energy. Since the AOs χ_1 and χ_2 are degenerate they lie at the same level on the diagram. Reading the diagram is made easier by including the graphical description of the orbitals. (In fact we have simplified the diagram a little here. Since the normalization constants, equations (10) and (11) include S (>0) the coefficients for the bonding orbitals are smaller than those for the antibonding ones.)

The strength of the interaction between the AOs is measured by the stabilization ΔE^+, or the destabilization ΔE^-, of the bonding or antibonding orbital respectively. As we have noted these depend upon the size of the overlap integral (S) between the two AOs. The larger the overlap the larger those energies ΔE^+ and ΔE^-, as shown in **3-6**. *The two quantities ΔE^+ and ΔE^- are proportional to the overlap integral S.*

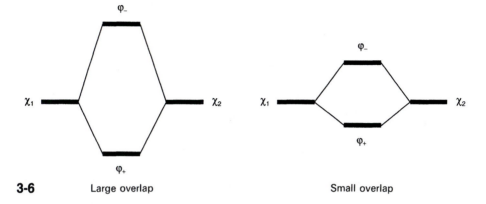

3-6   Large overlap   Small overlap

3.2.2. Interaction of two different AOs

Let us consider the case of a diatomic heteronuclear molecule AB, in which the two atoms are not the same. We assume once again that each atom only has a single AO (χ_1 on A and χ_2 on B) and that their energies are different (the energy of χ_2 will be chosen to lie deeper than that of χ_1). An example of a molecule of this type is the ion HHe^+ where $A = H$, $B = He$, $\chi_1 = 1s_H$, $\chi_2 = 1s_{He}$, $\varepsilon_1 = -13.6\,eV$ and $\varepsilon_2 = -24.6\,eV$.

(a) Expression for the MOs

The determination of the details of the molecular orbitals is more complex than in the homonuclear case. In effect the symmetry arguments used before which allowed us to find expressions for ϕ_+ and ϕ_- do not hold here. The two AOs, χ_1 and χ_2 are different and the electron densities associated with A and B need not be equal. In practice we need to perform a calculation to uncover the nature of the orbitals, but there are two rules which apply in general.

(i) The interaction of two AOs (χ_1 and χ_2) which are non-degenerate leads to a bonding orbital (ϕ_+) and an antibonding orbital (ϕ_-).

(ii) In the bonding orbital the larger coefficient is associated with the AO which started off at lower energy and in the antibonding orbital the opposite is true, the larger coefficient is associated with the AO which started off at higher energy.

In other words when $\varepsilon_2 < \varepsilon_1$ *the bonding orbital is largely located on* χ_2 *and the antibonding orbital largely localized on* χ_1. Using our pictorial scheme the two orbitals look as in 3-7. The antibonding orbital, ϕ_-, in which the coefficients are of different signs, has a nodal surface. Since the absolute values of the atomic orbital coefficients are not equal this does not lie midway between A and B as in the symmetric case. It lies closer to atom B.

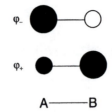

3-7

(b) The interaction diagram

Just as in the degenerate case, the important electron density lies between the nuclei in the bonding orbital ϕ_+ where the two AOs are in phase. Contrariwise this density is extruded from this region in the antibonding orbital ϕ_-. As a consequence the bonding orbital is stabilized below the initial level of the deeper-lying AO and the antibonding orbital is destabilized above the higher energy AO. We can profitably define $\Delta E^+ \ (=\varepsilon(\chi_2) - \varepsilon(\phi_+))$ and $\Delta E^- \ (=\varepsilon(\phi_-) - \varepsilon(\chi_1))$ just as in the symmetric case. Just as before $\Delta E^+ < \Delta E^-$. This collection of results is shown in the orbital diagram of Figure 3.2.

As before the interaction between the two AOs depends upon the overlap integral S. The larger S the larger the interaction. For the non-degenerate case we can show that it is proportional to the square of the overlap integral. Another factor which is important here is the energy separation between the two orbitals. The largest interaction between the two orbitals is found for small values of $\Delta\varepsilon = \varepsilon_1 - \varepsilon_2$. *In general the two quantities ΔE^+ and ΔE^- are proportional to the function $S^2/\Delta\varepsilon$.* This formula is only applicable for the case where $\Delta\varepsilon$ is large enough. For $\Delta\varepsilon = 0$ or close to zero then, as described earlier the interaction energy is proportional to S alone.

Now it is possible to qualitatively justify the approximation which involved neglect of the core orbitals in the construction of the molecular orbital diagrams.

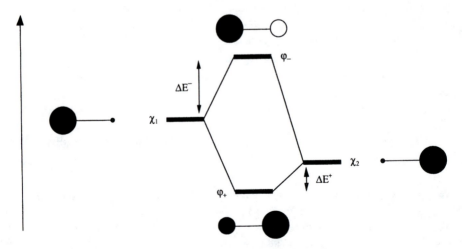

Figure 3.2. Molecular orbital diagram showing the interaction between two different atomic orbitals in a heteronuclear molecule.

These orbitals lie so deep in energy that there is a large energy separation between them and the valence orbitals (large $\Delta\varepsilon$). In addition they are strongly contracted, have small orbital radii and thus have extremely small overlap with orbitals (of any type) on adjacent atoms (small S). These two factors both contribute to negligibly small interactions between these two sets of orbitals and are usually ignored.

3.2.3. Orbitals with zero overlap

Recall that when two orbitals interact, the stabilization of the bonding MO which is formed is proportional to S if the starting orbitals have the same energy and to $S^2/\Delta\varepsilon$ if they don't. In both cases *if the overlap integral between two orbitals centered on different atoms is zero (i.e., orthogonal orbitals) then there will be no interaction between them.* This apparently trivial result is, in fact, quite important. Since formation of a bonding molecular orbital results in deformation of the electronic cloud and allows an electron simultaneously to interact with both nuclei, if $S = 0$ then there will be no such deformation and no resulting stabilization.

3.3. Application to some simple diatomic molecules

We will limit these examples to those ions and molecules formed from those atoms with a single valence orbital, namely H and He. In what follows we will describe the electronic energy in a molecule E_e as a simple sum of the energies of the individual electrons; $E_e = \sum_i n_i \varepsilon_i$ where n_i is the number of electrons occupying the MO ϕ_i with energy ε_i.

3.3.1. Level filling rules

Once the MOs have been constructed from the AOs we must now put electrons into them to generate the electronic structure for the molecule. The orbital filling rules are just the same as for AOs.

(i) The MOs are filled in order of increasing energy.
(ii) Each MO cannot accommodate more than two electrons. These two have opposite spins.

In some systems, more complex than those envisaged here, fewer electrons will be present than needed to fill a given set of degenerate orbitals and we will need to place these electrons in different orbitals with their spins parallel according to Hund's rule. In molecules, contrary to the success of the rule in atoms, Hund's rule sometimes breaks down. However the exceptions are infrequent and will not be discussed here.

3.3.2. Systems with two or four electrons

The general result, derived above, is that when two orbitals, degenerate or not, interact a bonding orbital is formed which is lower in energy than either of the AOs, and an antibonding orbital formed, which is higher in energy than either of the AOs.

If the molecule possesses two valence electrons in their lowest energy state they have to occupy the bonding orbital with their spins paired. With respect to the situation where the atoms are separated, each electron is thus stabilized by ΔE^+ (**3-8**). This gives rise to an electronic stabilization energy $2\Delta E^+$. The simplest

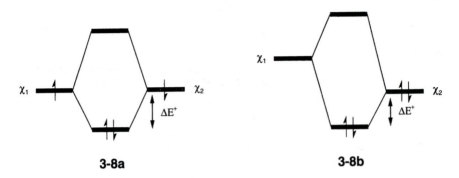

3-8a   **3-8b**

example is of course the molecule H_2 where each hydrogen atom carries a $1s$ orbital containing a single electron (**3-8a**). Such an energetic stabilization is very important. It is responsible for the existence of hydrogen as a diatomic molecule under normal conditions. An H_2 molecule is more stable than 2H atoms by $2\Delta E^+$. The bond energy ($2\Delta E^+$) is worth 440 kJ mol^{-1} for hydrogen, the energy needed to split the molecule into two separated hydrogen atoms. Another example for consideration with two electrons is the molecular ion HHe^+. This time the interacting orbitals have different energies (**3-8b**). The dissociation energy needed to break HHe^+ into H^+ ($1s$ empty) and He ($1s$ doubly occupied) is 178 kJ mol^{-1}.

Now consider the case where the set of interacting orbitals contains four electrons. Two electrons go into the bonding orbital ϕ_+ where each is stabilized by ΔE^+, and two go into the antibonding orbital ϕ_- where each is destabilized by ΔE^- (**3-9**). The

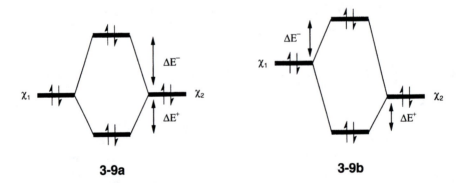

3-9a   **3-9b**

total energy of the system relative to the isolated atoms is thus $2(-\Delta E^+ + \Delta E^-)$. Since the antibonding orbital is destabilized more than the bonding orbital is stabilized (i.e. $\Delta E^- > \Delta E^+$) then the overall energy change on forming the molecule is unfavorable and the two atoms will be more stable lying far apart than they will together in a molecule. If the two atoms are placed close together they will repel each other and move away from each other. Experimentally, this is just what is found.

The two species He_2 (**3-9a**) and HHe^- (**3-9b**) are not known as stable molecules but as separated atoms.

The strikingly different behavior of the two-electron (H_2) and four-electron case (2He) is simple to understand. In these simple systems one can immediately see the link between the Lewis and molecular orbital pictures of bonding. *The presence of two electrons in the bonding orbital and the absence of electrons in the antibonding orbital gives rise to a chemical bond.* The two atoms are more stable together in the molecule than they are separated. However, with two more electrons, which have to occupy the antibonding orbital, this energetic stabilization is lost. *In general terms an interaction between two orbitals is always stabilizing for the case of two electrons but destabilizing for the case of four electrons.* This is an important result and is used extensively in later chapters.

3.3.3. Total energy of the molecule: the Morse curve for H_2

So far we have focussed just on the stabilization (or not) of the electrons due to the formation of molecular orbitals. This is not the only factor to be considered in tracing the energetics of bond formation. The nuclei interact too, on bringing the atoms together. This is always repulsive since they are both positively charged. This energetic contribution needs to be taken into account to properly account for the existence of two hydrogen atoms separated by 74 pm in the H_2 molecule.

The electronic stabilization energy ($2\Delta E^+$) is proportional to the overlap integral between the two $1s$ orbitals concerned. As we will see in the next section this gradually increases from zero (at long distances) to a maximum of 1.0 at an internuclear separation (R) of zero. If the geometry was determined by the variation in the total electronic stabilization energy of the molecule, then the optimal distance would be $R = 0$. The two hydrogen atom nuclei would then coalesce into a helium atom nucleus. This is clearly not the case. The correct description is obtained by considering both the coulombic repulsion between the two charged nuclei of the same sign (varying as $1/R$) and the stabilization afforded electronically. The coulombic repulsion (E_n) becomes fierce at small values of the internuclear separation and overwhelms the electronic stabilization energy (E_e) as shown in Figure 3.3. The addition of the two curves leads to an energy minimum in the total energy, E_t. At shorter distances the system experiences a strong nuclear repulsion and at longer distances the molecule loses some of its electronic stabilization energy. A curve of this type is called a *Morse curve*, although Morse actually specified an algebraical form for it.

To summarize, the repulsion between the nuclei offsets the electronic stabilization afforded by AO overlap and the formation of MOs. The equilibrium distance is a compromise between the two. In general we can see that the stronger the electronic stabilization the shorter the bond length. For the case of four electrons it is clear that the coulombic repulsion of the nuclei is accentuated by the electronic repulsion of the electrons.

3.3.4. Systems with one or three electrons

When only one electron occupies the bonding orbital, the electronic stabilization from **3-8** is just half as large as that obtained for the case of two electrons. As a result,

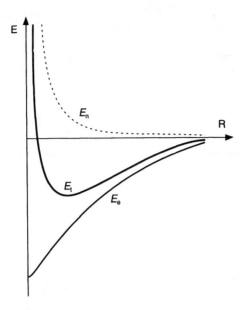

Figure 3.3. Variation in the electronic energy (E_e), the nuclear repulsion (E_n) and the total energy (E_t, Morse curve) for the H_2 molecule as a function of the distance R between the two atoms.

systems which only have one bonding electron are less stable than those with two. Thus the molecular ion H_2^+, although it is indeed stable with respect to H and H^+ has a smaller dissociation energy (263 kJ mol^{-1}) than found for dihydrogen (440 kJ mol^{-1}). There is a geometrical consequence of this difference; the H—H distance is longer in H_2^+ (106 pm) than in H_2 (74 pm). Another example of a one-electron system might be the dication HHe^{2+}, but this is not a stable species. Here the electronic stabilization energy, expected to be weaker than in the degenerate H_2^+ case because of the large difference in energy between the 1s levels of H and He, is insufficient to compensate for the strong nuclear repulsion between H^+ and He^+.

For three electron systems it is not possible to predict *a priori* if the electronic interaction is a stabilizing or destabilizing one. Two electrons occupy the bonding orbital and one the antibonding orbital. The change in electronic energy is therefore equal to

$$\Delta E = -2\Delta E^+ + \Delta E^- \tag{17}$$

We only have the relationship $\Delta E^+ < \Delta E^-$ to work with, and this does not allow a firm conclusion concerning the sign of ΔE in equation (17). So it is generally not possible to decide using simple orbital ideas whether a particular system might be stable. Experimentally the HeH molecule dissociates spontaneously into He and H, whereas the cation He_2^+ is stable. In the second example the presence of an antibonding electron in ϕ_- reduces the attraction between the nuclei compared to He_2^{2+}, resulting in an increasing bond length on going from He_2^{2+} (74 pm) to He_2^+

(108 pm). The calculated equilibrium distance in He_2^{2+} is the same as that found for H_2 (74 pm).

3.4. Overlap and symmetry

It should be clear by now that within the orbital picture, overlap is an important factor in determining the form of the molecular orbital diagram. In general it depends upon the internuclear separation, the nature of the orbitals involved (*s* or *p*) and their relative orientation.

3.4.1. 1*s*/1*s* overlap

Consider two AOs $1s_a$ and $1s_b$, localized on the two atoms A and B. The two AOs $1s_a$ and $1s_b$ are spherically symmetric and thus their overlap depends only on the distance, R, between the two nuclei. Recall that the *s* orbital wavefunction decreases as $\exp(-\lambda r)$ where r is the distance from the nucleus (see Chapter 2). The amplitude of the 1*s* orbital rapidly tends to zero as r increases. As a consequence the overlap between the two 1*s* orbitals at infinite distance is zero and increases as R decreases, i.e., as the nuclei become closer. If the two AOs are identical functions (for example two hydrogen 1*s* orbitals), at $R = 0$ they are obviously located at the same point and thus their overlap integral has to be equal to 1. If they are not identical the overlap integral will be less than 1. Thus the plot of $\langle 1s_a \mid 1s_b \rangle$ as a function of R is a simple curve with a value of 1 (for identical orbitals) at $R = 0$ decreasing to zero at infinity (Figure 3.4). Since in a molecule such as H_2 the overlap integral curve is of this form, the interaction energy between the two atomic orbitals and hence the energy difference between bonding (ϕ_+) and antibonding (ϕ_-) orbitals ($\Delta E^+ + \Delta E^-$), increases as the distance decreases as in **3-6**. We usually call overlap of this type, *σ-type overlap*. This is a term with several meanings (unfortunately) which will become apparent later.

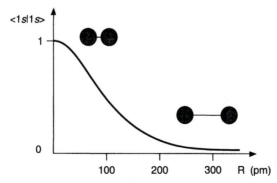

Figure 3.4. Variation in overlap integral between two hydrogen 1*s* orbitals as a function of the internuclear separation, R.

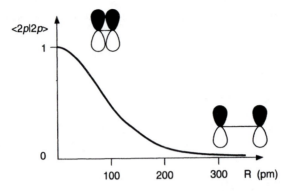

Figure 3.5. Variation in overlap integral between two carbon $2p$ orbitals as a function of the internuclear separation, R. The axes of the p orbitals lie perpendicular to the internuclear axis.

3.4.2. Overlap between 'parallel' 2p orbitals (π-type overlap)

If we consider two $2p$ orbitals whose axes lie parallel to each other and perpendicular to the internuclear separation then their overlap behavior is very similar to that found for the s/s overlap. There are two interactions of this type. If we choose the internuclear axis as z then there are the two pairs of interactions $[p_a^x, p_b^x]$ and $[p_a^y, p_b^y]$. The overlap tends to zero for large internuclear separations and is equal to 1 at $R = 0$ if the two p orbitals are identical (Figure 3.5). These two points are connected by a smooth curve which never changes sign. Such lateral or sideways overlap is labeled π-type. It is clearly quite different from the σ-type overlap of Figure 3.4, but the two overlap versus distance curves do look qualitatively similar.

3.4.3. 1s/2p overlap

The overlap integrals involving atomic orbitals which are not spherically symmetric depend upon the angular geometry in addition to the distance apart of the two atoms. The overlap integral between a $1s$ orbital and a $2p$ orbital depends upon R and the angle θ shown in **3-10**. For the case where θ is fixed at 90° (**3-11**) the overlap integral

3-10

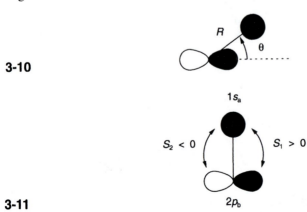

3-11

is identically zero. This comes about in the following way. We can imagine breaking the overlap integral down into two halves, the overlap of the s_a orbital with each lobe of the p orbital. Since the two lobes are completely equivalent each contribution must have the same absolute value $|S_1| = |S_2|$. However since the lobes have different signs the two contributions have opposite signs ($S_1 = -S_2$) so that the total overlap integral $\langle 1s_a \mid 2p_b \rangle$ is identically zero for $\theta = 90°$ irrespective of the value of the internuclear separation, R.

When θ is fixed at $0°$ the overlap integral $\langle 1s_a \mid 2p_b \rangle$ only depends on R. As before it is zero for infinite R. As R decreases we may regard the overlap integral being composed of two contributions, S_1 and S_2, which are not equivalent (3-12a). There is the contribution via the overlap of $1s_a$ with the 'near' lobe of $2p_b$ (S_1), which is always positive, and the contribution from the 'far' lobe of $2p_b$ (S_2), which is always negative. Generally $|S_1| > |S_2|$ and thus the overlap integral is always positive where this is true, but for $R = 0$ (3-12b) S_1 and S_2 are equal in magnitude leading to a zero

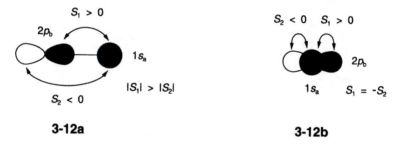

3-12a   **3-12b**

value of the overlap integral. Between these two points, infinite R and $R = 0$, the overlap integral must pass through a maximum, and this is shown in Figure 3.6. The maximum in the overlap integral for a $2p$ and $1s$ orbital varies with the identity of the atoms but occurs approximately at the internuclear separation found for e.g. CH, NH and OH bonds (i.e., at about 100 pm).

In the general case (3-10) we can show that the overlap integral is proportional to $\cos \theta$. If $S_R(\theta)$ is the overlap between the two orbitals for a fixed value of R

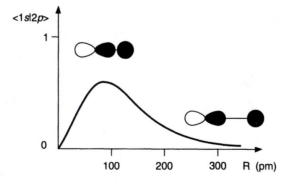

Figure 3.6. Variation in overlap integral between a hydrogen $1s$ orbital and a carbon $2p$ orbital, whose axis lies along the internuclear axis, as a function of the internuclear distance, R.

then the general expression for the geometrical dependence of the overlap integral is

$$S_R(\theta) = S_R(0) \cos \theta \qquad (18)$$

3.4.4. Symmetry ideas

(a) Overlap integrals zero by symmetry

We saw in the preceding section how a zero overlap develops between s and p orbitals when the axis of revolution of the p orbital is perpendicular to the internuclear axis (**3-11**). This result may also be obtained by recourse to simple symmetry arguments (**3-13**). Consider the xy plane, the function s_a is symmetric with respect to this plane,

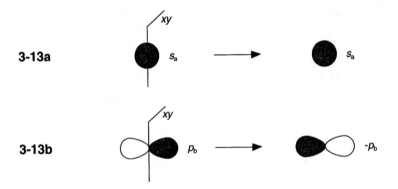

3-13a

3-13b

that is to say it remains unchanged during the 'operation' which reflects all of it lying on the left-hand side of the plane to the right-hand side and vice-versa (**3-13a**). Contrarily the p_b function is antisymmetric, since although on reflection the form of the function remains unchanged there is a change of sign (**3-13b**). The product $s_a p_b$ is thus an antisymmetric function too with respect to this plane. As a result its integral over all space, the overlap integral $S = \langle s_a \mid p_b \rangle$, is identically zero. **3-14** shows how

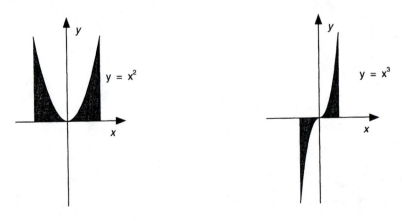

3-14

the symmetry of a function controls its integral properties by recourse to some simple mathematics. Consider the two functions $y = x^2$ and $y = x^3$. The first is symmetric with respect to the operation $x \to -x$ and the second is antisymmetric. It is clear to see that the integral of the antisymmetric function is exactly zero, but that of the symmetric function non-zero. This leads to a very general result. *If one AO, χ_i, is symmetric with respect to an element of symmetry and another χ_j, antisymmetric, the overlap integral between them $\langle \chi_i | \chi_j \rangle$ is identically zero.*

By way of another example consider the overlap between two p orbitals whose axes of revolution are perpendicular (**3-15**). The overlap integral $\langle p_a^x | p_b^z \rangle$ is zero, since p_b^z is symmetric but p_a^x is antisymmetric with respect to the yz plane (**3-15a**). In the same way if one considers the xz plane as in **3-15b**, since p_a^y is antisymmetric but

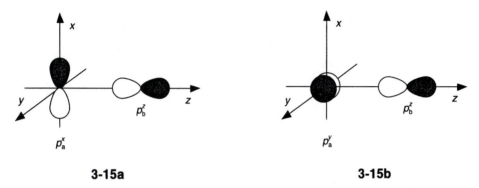

3-15a   **3-15b**

p_b^z is symmetric with respect to this symmetry element. Once again the overlap integral between them is zero.

(b) Symmetry elements

Symmetry properties are extremely important for the construction of molecular orbitals since their use provides a way to readily determine which AOs have zero overlap integrals and which thus do not interact. In order to characterize the 'symmetry' of a molecule we need to collect together all of the symmetry operations which, although they may exchange symmetry-equivalent atoms, leave the structure of the molecule intact. These elements of symmetry may be associated with axes, planes and points. Five types of symmetry operation exist in molecules.

   (i) The operation which leaves all of the atoms fixed in space. This is the identity or null operation, E.
  (ii) Rotation of $2\pi/n$ around an axis (say z), labeled as C_n^z.
 (iii) Reflection of a molecule in a plane, labeled as σ.
  (iv) Inversion of the coordinates of the atoms of a molecule through a point, labeled as i.
   (v) A combination of a rotation of $2\pi/n$ followed by reflection in a plane perpendicular to this axis. This operation is labeled as S_n.

These five have been listed here for completeness. The identity operation, apparently a trivial one has to be included when the treatment of symmetry passes beyond our simple usage here and passes into the area of group theory. Elements of type (v) are sometimes difficult to visualize. We will in fact use the symmetry elements (ii)–(iv) in the construction of orbital diagrams.

As an example consider the water molecule, H_2O. It is a bent triatomic molecule and we set up the coordinate system so that the z axis bisects the H—O—H angle (3-16). The symmetry operations for this molecule are

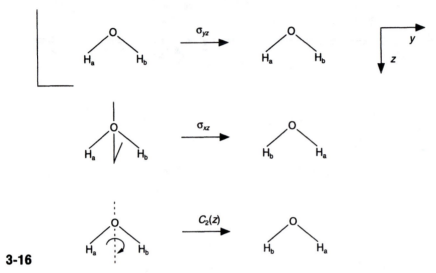

3-16

(i) The mirror plane in the plane of the molecule (σ_{yz}) which leaves all three atoms in their place.

(ii) The mirror plane perpendicular to the plane of the molecule (σ_{xz}) which exchanges the two (symmetry equivalent) hydrogen atoms.

(iii) Rotation by 180° around the z axis (C_2^z) which similarly exchanges the two hydrogen atoms.

(c) Symmetry properties of the MOs of H_2

The hydrogen molecule has many elements of symmetry, rotation axes, mirror planes and an inversion center. We will just concentrate on two which will be sufficient to characterize its molecular orbitals.

(i) Rotation of any angle, $2\pi/\alpha$ around the internuclear axis z (3-17) conserves the positions of both atoms. We write this symmetry operation as C_α.

(ii) A center of inversion lies at the mid-point between the two atoms H_a and H_b. With reference to 3-17 the inversion center lies at the origin.

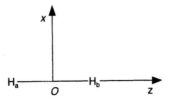

3-17

Now we can establish the symmetry properties of the two MOs, ϕ_+ and ϕ_-, for H_2 by seeing how they behave as a result of these two operations. We notice that both of these orbitals remain unchanged after any rotation around the z axis (**3-18a, c**). We call such orbitals σ orbitals to denote this cylindrical symmetry. As

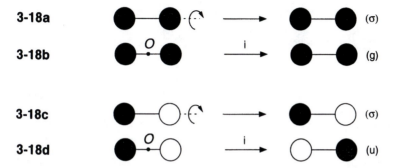

far as the inversion operation is concerned the orbital ϕ_+ remains unchanged (**3-18b**) but ϕ_- changes sign (**3-18d**) as a result of such an operation. The orbital ϕ_+ is thus symmetric with respect to the inversion center and receives a g label (from the German, gerade). The orbital ϕ_- though is antisymmetric with respect to this symmetry element and is labeled u (ungerade). The bonding and antibonding orbitals in H_2 are thus of σ_g and σ_u symmetry respectively. Another notation, often used in this area is to use σ_{H_2} in place of σ_g and $\sigma_{H_2}^*$ in place of σ_u. Sometimes too (but not in this book) one will find the antibonding orbital labeled with an asterisk to give σ_g and σ_u^* orbitals.

3.5. Application of symmetry ideas to some polyatomic molecules

Symmetry ideas and the notion of overlap find an important application in the study of planar and linear molecules.

3.5.1. σ/π separation

For a planar molecule the molecular plane is clearly a plane of symmetry for the molecule. All of the AOs carried by the atomic constituents of the molecule may be then divided into two groups, those that are symmetric with respect to this plane and those which are antisymmetric. All of the AOs of one group are orthogonal ($S = 0$) to those of the other group, and so a considerable simplification ensures now that one may just consider the interaction of AOs within each group. The symmetric AOs lead to a group of MOs also symmetric with respect to the molecular plane which forms the σ framework of the molecule. Similarly the interaction between the antisymmetric AOs leads to the construction of the π system. This σ–π separation between the two groups of orbitals is an extremely useful one. The π orbital structure is generally easier to determine than that of the σ orbitals. It also turns out to be probably the more important of the two when considering molecular reactivity.

3.5.2. The π MOs of ethylene

The ethylene molecule is planar with each carbon atom lying in an approximately trigonal environment (**3-19**).

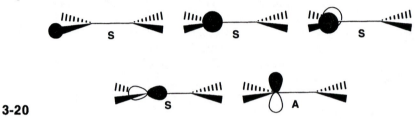

3-19

In the construction of the MOs, each hydrogen atom contributes its $1s$ orbital and each carbon atom its collection of $2s$ and $2p$ AOs. Let us analyze first of all the symmetry properties of these AOs with respect to the molecular plane, xz. This is easy. Since all of the atoms lie in the plane, all of the s orbitals are symmetric with respect to reflection, as are the AOs p_x and p_z on each carbon (**3-20**). Only the p_y

3-20

orbitals are antisymmetric. Thus the π system of ethylene is built from just these two orbitals, each centered on a carbon atom. They overlap with each other to give a non-zero overlap integral as shown in the previous section, and thus give rise to a bonding orbital (labeled π_{CC}) and an antibonding orbital (labeled π^*_{CC}) which contain equal contributions from the two carbon AOs since the interaction is of the degenerate type (**3-21**). In the ethylene electronic ground state there are two electrons

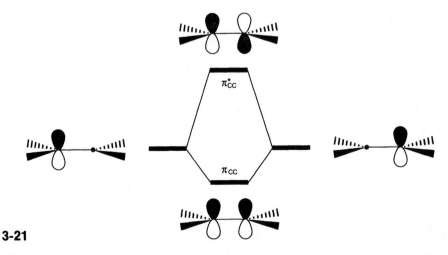

3-21

in the π system, located with their spins paired in the π_{CC} bonding orbital. Since the π^*_{CC} orbital is empty a π bond is formed between the two carbon atoms, one of the two bonds between the carbon atoms in the Lewis structure $H_2C{=}CH_2$. The other bond comes via the occupation of a σ bonding orbital the construction of which will be discussed in Chapter 8. The π bond in ethylene contributes a bond energy of about $270\ kJ\ mol^{-1}$. If the planes of the two CH_2 groups were arranged perpendicular to each other as in **3-22** then by symmetry the overlap between them would be zero,

3-22    $S = 0$

no π stabilization would exist and there would be no π bond between the carbon atoms. It is this π stabilization at the planar geometry which leads to the well-known planarity of the ethylene molecule, one of the observations noted in Chapter 1 which traditional theories cannot tackle.

An interesting effect exists in the first excited state of ethylene. Here one electron from the π_{CC} level is promoted to the π_{CC}^* level, a process which occurs under the influence of light. Now, as we saw before in Section 3.2.1 since the antibonding orbital is destabilized more than the bonding orbital is stabilized, such a state of affairs is less stable than two non-interacting p orbitals. It is just like the situation in He_2 except now there is only one electron (rather than two) in the bonding and antibonding orbitals. For He_2 the energy is lowered by moving He atoms apart, but for ethylene this energetically unfavorable situation is relieved by simple rotation around the C—C axis as in **3-22**. This is an example of a type of reaction called a *photochemical rearrangement process* where the excited state has a different geometry to the ground state.

3.5.3. π System of formaldehyde

The Lewis structure of this planar molecule is shown in **3-23**. Its π system is formed from the interaction of the p_y AOs on the carbon and oxygen atoms. The construction of the π molecular orbital diagram for formaldehyde is similar to that for ethylene. It differs from that system in that the interaction between the two AOs is now not a degenerate one. Since oxygen is more electronegative than carbon, its $2p$ orbitals lie deeper than those of carbon (see Chapter 2). Following the rules which we developed in Section 3.2.2 a bonding orbital (π_{CO}) largely localized on oxygen, and an antibonding orbital (π_{CO}^*) largely localized on carbon, results (**3-24**). In the ground state configuration of H_2CO there turn out to be two electrons to be located in the π system and these occupy the π_{CO} orbital with their spins paired. The π bond which results corresponds to one of the bonds between carbon and oxygen (C=O) in the Lewis structure of **3-23**.

3-23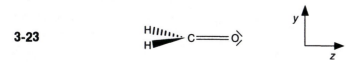

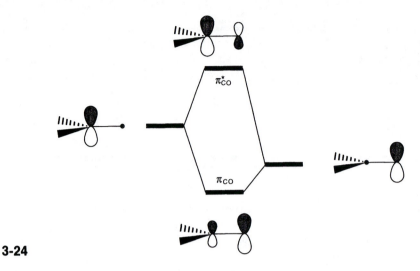

3-24

3.5.4. A comparison between ethylene and formaldehyde

If we limit ourselves to consideration of the π systems only, ethylene and formaldehyde are related by the substitution of a carbon $2p$ orbital by an oxygen $2p$ orbital. The consequences of this in terms of the character of the orbitals have just been discussed. The difference in orbital energies of the starting AOs does, however, have consequences for the energies of the MOs in the two molecules. Thus we find $E(\pi_{CO}) < E(\pi_{CC}) < E(\pi_{CO}^{*}) < E(\pi_{CC}^{*})$. We will come back to this point in the following chapter.

3.5.5. The π orbitals of acetylene

Acetylene is a linear molecule whose Lewis structure is H—C≡C—H. Each plane which contains the internuclear axis (z) is a symmetry plane of the molecule (3-25). In contrast to the situation in ethylene or formaldehyde it is not possible to define

3-25   H——C≡≡≡C——H

the π system in a unique way. We can get around this difficulty by arbitrarily deciding on two perpendicular planes which contain the z-axis. We choose the xz and yz planes. It is now possible to easily define two different π systems for the molecule by considering the AOs antisymmetric with respect to one or other of these two planes. So $p_{C_1}^{y}$ and $p_{C_2}^{y}$ constitute one π system, antisymmetric with respect to the xz plane and $p_{C_1}^{x}$ and $p_{C_2}^{x}$ constitute another, antisymmetric with respect to the yz plane. All of the other AOs are symmetric with respect to both of these planes and thus are orthogonal to both sets of π orbitals.

Since the p_x and p_y AOs on each carbon atom have the same energy, interaction within each π system reduces to interaction between two pairs of degenerate AOs. The overlap integrals $\langle p_{C_1}^x \mid p_{C_2}^x \rangle$ and $\langle p_{C_1}^y \mid p_{C_2}^y \rangle$ on the other hand are equal since the x and y axes are equivalent. This leads to identical interactions within each of the π systems with the result that a pair of *degenerate bonding* and a pair of *degenerate antibonding* orbitals result (**3-26**). In the electronic configuration appropriate for the

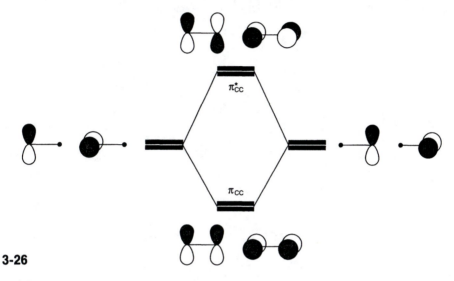

3-26

ground state of acetylene, there are four electrons located in the π systems. Both π_{CC} bonding orbitals are occupied, each with two spin-paired electrons. The two π bonds which result may be identified with two of the three C—C bonds of the Lewis structure H—C≡C—H. The third bond comes from the σ framework and we will describe how it comes about in Chapter 8.

3.6. Conclusions

It is useful to collect together the main results associated with the interaction of two atomic orbitals located on two centers.

   (i) Orbitals can only interact if their overlap integral is non-zero.
  (ii) The interaction of two AOs leads to the formation of two MOs, one bonding and one antibonding.
 (iii) The bonding MO is more stable than the lower energy orbital of the starting AOs.
 (iv) The antibonding MO is less stable than the higher energy orbital of the starting AOs.
  (v) The bonding orbital is stabilized less than the antibonding orbital is destabilized.
 (vi) If the AOs are degenerate, their interaction is proportional to their overlap integral, S.

(vii) If the AOs are non-degenerate, their interaction is proportional to $S^2/\Delta\varepsilon$, where $\Delta\varepsilon$ is the energy separation between the AOs. In this case the bonding MO is preferentially localized on the atom with the deeper lying AO, usually the more electronegative atom. The antibonding MO is preferentially localized on the atom which holds the higher energy AO.

Although we have chosen simple systems to illustrate these rules, they will find many applications throughout this book for molecules which are much more complex.

EXERCISES

3.1 Calculate the atomic orbital coefficients of the π and π^* orbitals of ethylene knowing that the overlap integral between the two p AOs is 0.27. What would the coefficients be if overlap is neglected in the normalization process? Show that the stabilization energy of the bonding orbital and the destabilization energy of the antibonding orbital are equal if overlap is neglected.

3.2 The calculated MOs for the HHe^+ ion are given by the expressions

$$1\sigma = 0.877(1s_{He}) + 0.202(1s_H)$$
$$2\sigma = 0.798(1s_{He}) - 1.168(1s_H)$$

(i) Justify the relative sizes of the coefficients in 1σ and 2σ.
(ii) Given that the MOs are normalized, calculate the overlap integral between the AOs, $1s_H$ and $1s_{He}$.
(iii) Verify that 1σ and 2σ are orthogonal.
(iv) The coefficient of $1s_H$ is larger than unity in 2σ. Is this acceptable?

3.3 Describe in a qualitative fashion how the overlap integral between two $2p$ orbitals located along the *same* axis varies with the distance R between the nuclei.

3.4 Consider the following molecules: CH_3Cl, NH_3, CH_2CO, *cis* and *trans* $ClHC{=}CHCl$. For each write down:

(i) Its Lewis structure.
(ii) Its geometry using the VSEPR model.
(iii) All of its elements of symmetry.

3.5 (i) Describe the π systems of ethylene (C_2H_4) and acetylene (C_2H_2).
(ii) Knowing that the C—C distance is 134 pm in ethylene and 120 pm in acetylene show how the energies of the π and π^* levels differ between the to molecules.

3.6 Consider the two geometries shown for X_2H_4 molecules. The first is planar and the second is obtained by rotation of one XH_2 group by 90°.

1   **2**

(i) Which is more stable for the ethylene molecule (X = C) and why?
(ii) How many extra electrons are there in hydrazine (X = N)?
(iii) Assuming that these electrons occupy the π manifold of orbitals which geometry (1 or 2) is more stable?
(iv) With the help of VSEPR suggest a more precise geometry for hydrazine.

4 The fragment orbital method; application to some model systems

In the preceding chapter we showed how the interaction of two atomic orbitals located on adjacent atoms leads to the formation of two molecular orbitals, one bonding and one antibonding between the two atoms. Since we are usually interested in molecules more complex than H_2 or its ions, the problem of constructing the molecular orbitals for such large systems becomes more difficult. It is necessary in general to consider more than just one atomic orbital per atom. For example, the valence orbitals of carbon which need to be considered are the one $2s$ and three $2p$ orbitals. In addition, the majority of molecules contain more than just two atoms. Thus there are many orbital interactions to be taken into account simultaneously. There do exist calculational methods of course which enable determination of the energies of the molecular orbitals and their atomic orbital description (the coefficients) for molecules of all types of complexity. Certainly one can use the numbers which come from such calculations as a starting point for the analysis of the electronic structure, but one can also search for an understanding of the form of the molecular orbitals, and their relative energies by using a qualitative approach analogous to the one developed earlier for the simple problem of two orbitals localized on two atomic centers.

One particularly fruitful method consists of imagining the complex molecule, whose electronic structure is not known, split into two smaller units whose molecular orbitals are known. The orbitals of the complete molecule are then derived via the interaction of the orbitals of these two sub-systems in exactly the same way as we constructed the molecular orbitals of the H_2 molecule by allowing the two $1s$ hydrogen atomic orbitals to interact with each other. This approach is known as the *fragment orbital method*. We will illustrate its utility by studying a number of model systems, H_n, which contain hydrogen atoms in different geometrical arrangements. The examples are discussed in order of increasing complexity, rather than by the value of n.

The question of the stability of these molecules will not be tackled in this chapter. We do note however that the molecule H_3^+ is known as a gas-phase species and its geometry has been determined to be equilateral triangular in shape. Many other molecules of general formula H_n^+ have been observed in a mass spectrometer. Although these molecules may seem bizarre ones to the reader they are useful in another way. The energy level patterns found for these H_n molecules do in fact apply to their conjugated hydrocarbon analogs. Thus the set of energy levels found for a cyclic H_6 look very similar indeed to the π levels of benzene, C_6H_6. Looking ahead to later chapters we will use the orbitals developed here for H_n in the generation of the level structure of more complex systems.

4.1. Molecular orbitals of some model systems, H_n

4.1.1. Square planar H_4

(a) The problem

Let us consider four hydrogen atoms situated at the vertices of a square. The most natural fragmentation of this problem is to consider H_4 as being made up of two H_2 entities, H_a—H_b and H_c—H_d (4-1). Each fragment is then characterized by a bonding

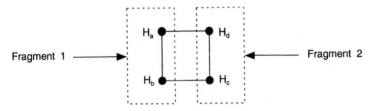

4-1

orbital (σ_{H_2}) and an antibonding orbital ($\sigma_{H_2}^*$). The molecular orbitals of H_4 are then obtained via the interaction of the orbitals of these two fragments. Note that we use for the geometries of the fragments the dimensions found in the more complex molecule rather than the geometry expected for the fragment alone. There are therefore initially four interactions to consider (4-2), two between degenerate orbitals (1 and 2) and two between non-degenerate orbitals (3 and 4).

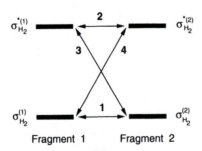

4-2   Fragment 1   Fragment 2

(b) Overlap between the fragment orbitals

First consider the overlap integrals associated with interaction 1 between $\sigma_{H_2}(1)$ and $\sigma_{H_2}(2)$ (4-3a). The overlap both between orbitals located on neighboring atoms (H_a, H_d and H_b, H_c) and on the atoms further away (H_a, H_c and H_b, H_d) are of the same sign. The total overlap integral, the sum of these four terms, is therefore different from zero, and so much so, that a strong interaction occurs between those two fragment orbitals. We can do the same analysis for the interaction 2 between $\sigma_{H_2}^*(1)$ and $\sigma_{H_2}^*(2)$ (4-3b). In this case the overlap integrals involving neighboring centers

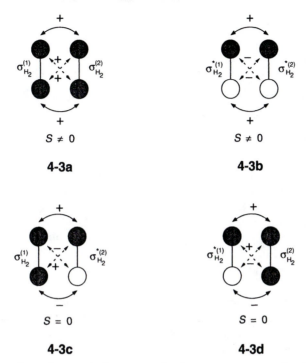

are positive but those between the atoms further away across the diagonal of the square are negative. However, since the overlap integral drops off with distance the latter are smaller in magnitude. The total overlap integral is therefore non-zero. Finally, consider the interaction 3 between $\sigma_{H_2}(1)$ and $\sigma^*_{H_2}(2)$ (**4-3c**). Notice that the positive overlap between $1s_a$ and $1s_d$ is exactly cancelled by the negative overlap between $1s_b$ and $1s_c$, since the distances between these overlapping pairs of atomic orbitals are the same. Likewise the overlaps between the pairs of centers further apart have opposite signs and therefore cancel. That between $1s_d$ and $1s_b$ is positive. That between $1s_a$ and $1s_c$ is negative. The total overlap between $\sigma_{H_2}(1)$ and $\sigma^*_{H_2}(2)$ is therefore rigorously equal to zero, and, directly as a consequence, the interaction 3 is zero. It is clear that the same conclusion may be drawn by considering the interaction 4 between $\sigma^*_{H_2}(1)$ and $\sigma_{H_2}(2)$ (**4-3d**).

(c) Introduction of symmetry

The detailed analysis of the overlap integrals between the fragment orbitals allows us to see how the problem of constructing the MO diagram for square planar H_4 in fact comes down to just two interactions (1 and 2) between degenerate orbitals. We can draw the same conclusion by using the symmetry properties of the fragment orbitals by choosing a symmetry element which is common to the two fragments. In our example here we use the plane P, which passes through the middle of the fragments of H_a—H_b and H_c—H_d and perpendicular to the molecular plane (**4-4**).

With reference to this plane P, the σ_{H_2} orbitals are symmetric (**S**) and the orbitals $\sigma^*_{H_2}$ antisymmetric (**A**). Now we can apply the rule which we have already used for the case of the interaction of two atomic orbitals located on different centers. There

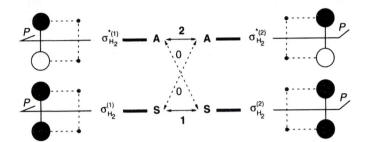

4-4

cannot be any interaction between two orbitals unless they are of the same symmetry (S or A, 4-4). This gives directly the same result established in the preceding section. We will often use such symmetry arguments to simplify the assembly of molecular orbital diagrams from their fragment orbitals. Interactions, *identically zero by symmetry* are readily eliminated in this way. We may comment that a rigorous discussion of the symmetry properties of the orbitals necessitates consideration of *all* of the symmetry elements of the system. It requires the use of *group theory*. When in a simplified approach we just retain a single element, its choice is crucial. We have already said that this element must be common to the two fragments. However this is not a sufficient condition. It must also be the case that the orbitals separate into two groups, symmetric and antisymmetric with respect to the symmetry element, so as to be able to eliminate the interactions between orbitals of different symmetries. If, for example, we had retained the molecular plane containing the four hydrogen atoms, we would have concluded, with just cause, that the four fragment orbitals were symmetric with respect to this plane. This would have led, incorrectly, to the conclusion that none of the interactions was zero by symmetry. This 'bad choice' of symmetry element in fact masks the symmetry properties of σ_{H_2} and $\sigma^*_{H_2}$ which we need in the analysis. We will return to this point in the next chapter when studying the construction of the molecular orbitals of trigonal AH_3 molecules.

(d) MOs of square planar H_4

The problem of constructing the molecular orbitals of square planar H_4 is formally analogous to that for the H_2 molecule. It is a case of the interaction of two degenerate orbitals. The $1s$ orbitals of each H atom are simply replaced by the σ_{H_2} (or $\sigma^*_{H_2}$) orbitals of each H_2 unit.

The interaction of the two orbitals of the fragments, $\sigma_{H_2}(1)$ and $\sigma_{H_2}(2)$ leads to the formation of two MOs. ϕ_1 is the bonding, and ϕ_2 the antibonding combination of the orbitals of the fragments (4-5). In the same way, the interaction of the orbitals $\sigma^*_{H_2}(1)$ and $\sigma^*_{H_2}(2)$ leads to the MOs ϕ_3 and ϕ_4. This analysis therefore allows the

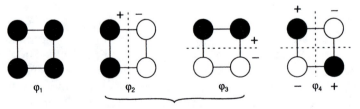

4-5

unambiguous determination of the relative signs of the coefficients of the four MOs of H_4. Since the interactions which take place involve degenerate orbitals of the fragments ($\sigma_{H_2}(1)$ and $\sigma_{H_2}(2)$, for example, are of the same energy) the mixtures contain the fragment orbitals with equal weights. Thus correct to within a normalization constant,

$$\phi_1 = \sigma_{H_2}(1) + \sigma_{H_2}(2)$$

$$\phi_2 = \sigma_{H_2}(1) - \sigma_{H_2}(2)$$

$$\phi_3 = \sigma_{H_2}^*(1) + \sigma_{H_2}^*(2) \tag{1}$$

$$\phi_4 = \sigma_{H_2}^*(1) - \sigma_{H_2}^*(2)$$

The result is that within each MO, the coefficients of each of the hydrogen $1s$ orbitals are of equal magnitude. We can obtain via this route, without a numerical calculation, a precise description of the form of the molecular orbitals of square planar H_4 (4-5). The energetic ordering of the levels can be simply deduced from the interaction diagram of Figure 4.1. Orbital ϕ_1, totally bonding between all pairs of atoms is the most stable; ϕ_4 antibonding between all pairs of neighboring hydrogen atoms is the least stable. At an intermediate energy lie ϕ_2 and ϕ_3. These two orbitals are degenerate, each characterized by two bonding interactions and two antibonding interactions between neighboring pairs of atoms. Their energy is intermediate between those for the fragment orbitals σ_{H_2} and $\sigma_{H_2}^*$, and therefore close to that of an isolated $1s$ orbital. One can refer to such orbitals as being overall *nonbonding*, their bonding and antibonding characters exactly cancelling. We note too that the energies of the MOs track the number of nodal surfaces (4-5). ϕ_1 with no such surfaces lies deepest in energy; ϕ_2 and ϕ_3 lie higher in energy and have one such nodal plane; ϕ_4 lies highest in energy and has two nodal planes. Thus a simple way to remember the energy ordering of the molecular orbitals is to count the number of nodal planes.

Finally it is interesting to comment on the degeneracy found for ϕ_2 and ϕ_3. In molecules which contain a three-fold or higher rotation axis, degenerate orbital may occur (see Appendix). In square-planar H_4 it is the rotation axis of order four which leads to this result. This is a facet of a very general result. As the symmetry of a molecule increases, i.e., as the number of symmetry elements increases, the possibility of degenerate energy levels increases. For tetrahedral molecules the symmetry is so high that triply-degenerate levels occur, as we will see in CH_4. It is easy to see the origin of the degeneracy for square-planar H_4 but it will be more difficult in some of the molecules which follow.

(e) Calculation of the coefficients

Square planar H_4 is a system where one can calculate the coefficients of the AOs in each MO by simply using the normalization condition $\langle \phi_i \mid \phi_i \rangle = 1$ ($i = 1$-4). In effect we have been able to show that in each MO the coefficients are of equal magnitude. In order to obtain the relative signs of the coefficients one needs just to

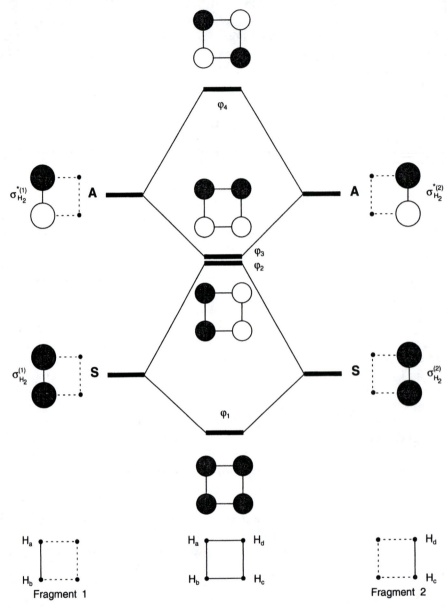

Figure 4.1. Construction of the MOs of square planar H_4.

write the MOs in the following form

$$\phi_1 = N_1(1s_a + 1s_b + 1s_c + 1s_d)$$
$$\phi_2 = N_2(1s_a + 1s_b - 1s_c - 1s_d)$$
$$\phi_3 = N_3(1s_a - 1s_b - 1s_c + 1s_d)$$
$$\phi_4 = N_4(1s_a - 1s_b + 1s_c - 1s_d)$$

For each MO there is thus just a single coefficient (N_i) to be found. If S and S' represent the overlap integrals between $1s$ orbitals situated respectively on the neighboring atoms and on the atoms further away (related by the diagonal of the square), then

$$\langle 1s_a \mid 1s_b \rangle = \langle 1s_a \mid 1s_d \rangle = \langle 1s_b \mid 1s_c \rangle = \langle 1s_c \mid 1s_d \rangle = S$$

$$\langle 1s_a \mid 1s_c \rangle = \langle 1s_b \mid 1s_d \rangle = S'$$

Certainly $S > S'$. Since the $1s$ orbitals themselves are normalized one has

$$\langle 1s_i \mid 1s_i \rangle = 1 \quad (i = a, b, c, d)$$

Using the normalization condition for the ϕ_i we obtain

$$\langle \phi_1 \mid \phi_1 \rangle = 1 = N_1^2(4 + 8S + 4S')$$

From which we get

$$N_1 = \frac{1}{2(1 + 2S + S')^{1/2}}$$

In the same way we find

$$N_2 = N_3 = \frac{1}{2(1 - S')^{1/2}} \qquad N_4 = \frac{1}{2(1 - 2S + S')^{1/2}}$$

The expressions for the coefficients simplify considerably for the case where the overlap integrals S and S' are small compared to 1. This would be the case for a large distance between the atoms. For this condition all the N_i become equal to $\frac{1}{2}$ and all the atomic orbital coefficients become equal to $\pm\frac{1}{2}$. In **4-5** this result is pictorially represented by drawing circles of equal diameter to represent these coefficients. It is however more realistic to include the overlap integrals with values more consistent with the real distances expected in a molecule of this type. By way of an example for a square with a side of 100 pm one finds $S = 0.469$ and $S' = 0.263$, values hardly negligible compared to 1. For these values of the overlap integrals we obtain for the coefficients; $N_1 = 0.337$, $N_2 = N_3 = 0.582$ and $N_4 = 0.877$.

An interesting trend is that these molecular orbital coefficients become larger as the number of nodal surfaces increases. This tendency, already found in H_2 for the obital σ_{H_2} and $\sigma_{H_2}^*$ (Chapter 3) occurs too in the orbitals of more complex molecules. Inclusion of overlap in the calculation tends to reduce the size of the coefficients in bonding orbitals and increase them in antibonding ones.

4.1.2. Rectangular H_4

From a purely geometrical point of view, we will suppose that the transformation from the square planar to rectangular geometry occurs by increasing the distances

H_a—H_d and H_b—H_c. The strategy we will use for the construction of the MOs of rectangular H_4 is exactly analogous to that used for the square plane. The molecule can be considered as being assembled by the union of the two H_2 fragments (H_a—H_b and H_c—H_d, **4-6**), where the plane P passes through the middle of the fragments and

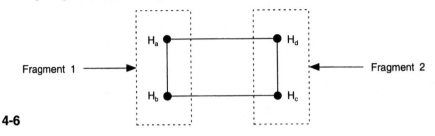

4-6

perpendicular to the molecular plane, always being a plane of symmetry for the two fragments. The interaction diagram for the fragment orbitals, Figure 4.2, is thus very similar to that shown in Figure 4.1, the only difference arising from the fact that the

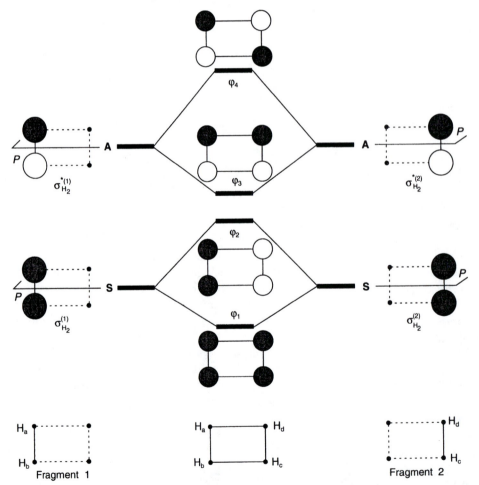

Figure 4.2. Construction of the MOs of rectangular H_4.

overlap integrals between the orbitals of the fragments are smaller than in the previous case. This comes about since the distance between the two fragments is larger than that within the fragments. (H_a—H_d, H_b—H_c are longer than H_a—H_b, H_c—H_d.) All other things being equal, the bonding combinations ϕ_1 and ϕ_3 are less stabilized and the antibonding combinations ϕ_2 and ϕ_4 less destabilized than in the square planar geometry. The major consequence of the distortion from the square planar to the rectangular geometry is therefore the *lifting of the degeneracy* of the orbitals ϕ_2 and ϕ_3. In ϕ_2 the bonding interactions are between the close atoms and the antibonding interactions between the atoms further away. This situation is energetically more favorable than that found for ϕ_2 in square H_4. The opposite is true for ϕ_3; the bonding interactions develop between the more distant atoms and the antibonding ones between the close atoms. This orbital is therefore destabilized more than in the square planar geometry.

4.1.3. Linear H₃

The fragmentation which takes into account the symmetry of a linear H_3 unit consists of pairing the two terminal hydrogen atoms (H_a and H_c) as one fragment and the central H_b atom as the second (**4-7**). The plane P, perpendicular to the internuclear axis, and passing through the central atom is therefore a plane of symmetry for both fragments.

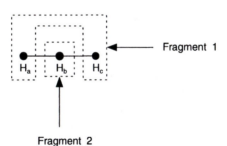

4-7

The fragment $H_a \ldots H_c$ can be considered as a very stretched H_2 molecule. The orbitals associated with this fragment are always in-phase (σ_{H_2}) and out-of-phase ($\sigma_{H_2}^*$) combinations of the atomic orbitals $1s_a$ and $1s_c$. But, because of the large distance separating the atoms H_a and H_c, the overlap integral between $1s_a$ and $1s_c$ is small, such that the 'bonding' and 'antibonding' orbitals are close in energy and approximately equal to the energy of an isolated $1s$ orbital. The fragment H_b is characterized by the single atomic orbital $1s_b$.

The orbitals σ_{H_2} and $1s_b$ are symmetric (**S**) and $\sigma_{H_2}^*$ is antisymmetric (**A**) with respect to the plane of symmetry P. The fragment orbital $\sigma_{H_2}^*$ remains unchanged in form and in energy (ϕ_2, Figure 4.3), since there is no orbital of the same symmetry present on the second fragment. However an interaction does develop between σ_{H_2} and $1s_b$, both of symmetry **S**, which leads to the formation of a bonding orbital ϕ_1, and an antibonding orbital ϕ_3. Since these fragment orbitals start off with approximately the same energies, they mix together in these two orbitals with close to equal weights. Now, the σ_{H_2} orbital is delocalized over two centers (H_a and H_c) so that the

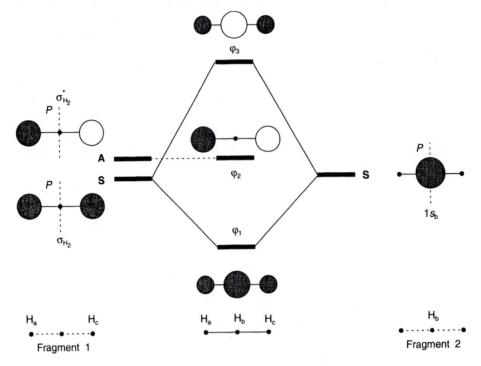

Figure 4.3. Construction of the MOs of linear H_3.

coefficients of H_a and H_c are less than 1 (in the neighborhood of $2^{-1/2}$), and the coefficient of $1s_b$ is equal to 1 since this is a pure atomic orbital. As a result the coefficient of the central atom (H_b) in the MOs ϕ_1 and ϕ_3 is larger in absolute magnitude than those on the terminal atoms (by a factor close to $2^{1/2}$). The energetic ordering of the three molecular orbitals can be simply understood by reference to the MO diagram of Figure 4.3. ϕ_1, a totally bonding orbital is the most stable and ϕ_3, antibonding between both pairs of neighboring atoms is the least stable. ϕ_2 lies at an intermediate energy. In this orbital, although the coefficients of the two terminal hydrogen atoms are out of phase, the two atoms are not bonded to each other and lie far apart. The destabilization of ϕ_2 relative to the energy of an isolated $1s$ orbital is then very small, being identically zero if one neglects the overlap integrals between nonbonded atoms. We describe ϕ_2 as a *nonbonding orbital*. Finally we note that the energy of the MOs increases with the number of nodal surfaces; ϕ_1 lies lowest with no nodal surfaces, ϕ_3 lies highest in energy with two and ϕ_2 lies at an intermediate energy with one.

4.1.4. Linear H_4

The fragmentation of the linear H_a—H_b—H_c—H_d system which conserves an element of symmetry common to the two fragments consists of pairing the terminal atoms $H_a \ldots H_d$ with each other and the inner atoms H_b—H_c with each other **(4-8)**. The plane P, perpendicular to the internuclear axis and passing through the middle of

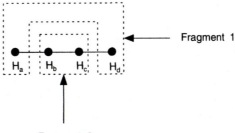

4-8   Fragment 2

H_b—H_c (and indeed $H_a \ldots H_d$) is clearly the symmetry plane for the two fragments. The first, $H_a \ldots H_d$ consists of two atoms of hydrogen quite separated in distance. The two orbitals associated with a fragment of this type have energies which are close together and close to that of an isolated $1s$ orbital. (These are orbitals which are virtually nonbonding.) One, $\sigma_{H_2}(1)$ is symmetric (**S**) with respect to the plane P, and the other, $\sigma_{H_2}^*(1)$ is antisymmetric (**A**) as shown in Figure 4.4. The second fragment

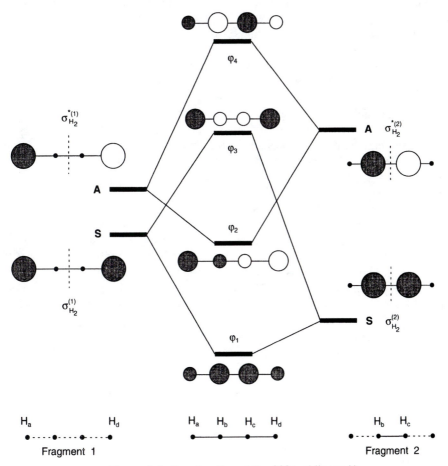

Figure 4.4. Construction of the MOs of linear H_4.

H_b—H_c also comes with two orbitals, $\sigma_{H_2}(2)$ and $\sigma^*_{H_2}(2)$, but at very different energies since the atoms H_b and H_c are close together. The interaction diagram which controls the construction of the MOs of linear H_4 thus contains two interactions between two pairs of non-degenerate orbitals. One interaction is between symmetric orbitals and the other between antisymmetric orbitals. The first interaction between $\sigma_{H_2}(1)$ and $\sigma_{H_2}(2)$ leads to the formation of a bonding combination (ϕ_1) and an antibonding combination (ϕ_3). In the same way $\sigma^*_{H_2}(1)$ and $\sigma^*_{H_2}(2)$ lead to two MOs, ϕ_2 and ϕ_4. The starting orbitals are non-degenerate. As a result ϕ_1 is largely located on $\sigma_{H_2}(2)$ and ϕ_3 on $\sigma_{H_2}(1)$. The rule we follow here is that the character of an MO resulting from interaction of ψ_A and ψ_B will contain more ψ_A than ψ_B character if it lies energetically closer to ψ_A than ψ_B, and more ψ_B character if it lies energetically closer to ψ_B. As a result the larger coefficients in ϕ_1 are on the inner atoms H_b, H_c, but the larger coefficients in ϕ_3 are found on the terminal atoms H_a and H_d. The same analysis for the interaction of the antisymmetric orbitals $\sigma^*_{H_2}(1)$ and $\sigma^*_{H_2}(2)$ shows that in the bonding combination ϕ_2 the coefficients are larger on H_a and H_d than on H_b and H_c, the opposite situation being observed in the antibonding combination ϕ_4 (Figure 4.4).

The energetic ordering of the four molecular orbitals doesn't pose any particular problems. Orbital ϕ_1, bonding between all atoms is the lowest in energy, and ϕ_4 antibonding between all atoms is the highest one. Orbital ϕ_2 is bonding between H_a and H_b and between H_c and H_d, but antibonding between the central atoms. However, this last interaction is weak since the out-of-phase coefficients are small. This MO is then, overall, bonding, that is to say its energy lies deeper than that of an isolated $1s$ orbital. In ϕ_3 the situation is the opposite since there are now two antibonding interactions and one weak bonding one between the central atoms. Overall this MO is antibonding with an energy lying higher than that of an isolated $1s$ orbital. As in the earlier examples the molecular orbitals can be classified in terms of the number of nodal surfaces: ϕ_1 (0), ϕ_2 (1), ϕ_3 (2) and ϕ_4 (3).

4.1.5. Triangular H_3

The determination of the MOs of a system containing three hydrogen atoms arranged at the vertices of an equilateral triangle proceeds according to a strategy analogous to that used for linear H_3. The fragmentation divides the system into a H_a—H_c entity and an atom of hydrogen, H_b (**4-9**), the two fragments being symmetric with respect to the plane P perpendicular to H_a—H_c and passing through H_b. The only difference compared to the case of linear H_3 is that the two atoms H_a and H_c are now close together; that is to say that the orbitals of this fragment (σ_{H_2} and $\sigma^*_{H_2}$) have quite different energies (Figure 4.5).

4-9

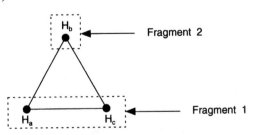

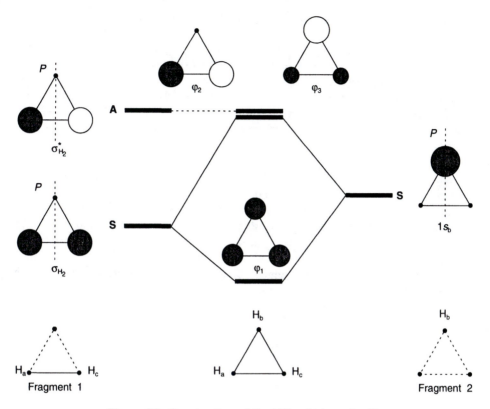

Figure 4.5. Construction of the MOs of triangular H_3.

(a) MOs of triangular H_3

The symmetry properties of the fragment orbitals are identical to those found for linear H_3, and so there is only one interaction to consider, that between the symmetric (S) orbital σ_{H_2} and $1s_b$. But here the interaction is between two orbitals whose energies are very different. The MO ϕ_1 is a bonding combination of σ_{H_2} and $1s_b$ of the type

$$\phi_1 = \gamma \cdot \sigma_{H_2} + \mu \cdot 1s_b$$

Following the rule for the interaction of non-degenerate orbitals $\gamma > \mu$ since σ_{H_2} lies deeper in energy than $1s_b$. The coefficients of H_a and H_c in σ_{H_2} are of the order of $2^{-1/2}$ (an orbital delocalized over two centers) and are therefore smaller than that of H_b in $1s_b$ (which of course is equal to 1 since it is an isolated atomic orbital). These two effects actually compensate for each other and the result is that the three atomic orbital coefficients of $1s_a$, $1s_b$ and $1s_c$ in ϕ_1 are equal, leading to an MO which is totally symmetric. The situation is different in the antibonding MO, ϕ_3, which evolves from the higher energy fragment orbital $1s_b$. We may write an expression analogous to the above of

$$\phi_3 = \gamma^* \cdot \sigma_{H_2} - \mu^* \cdot 1s_b$$

where now $\mu^* > \gamma^*$. The coefficients of H_a and H_c are in the neighborhood of $\gamma^* 2^{-1/2}$, that of H_b equal to μ^*. Since $\mu^* > \gamma^*$ the coefficient on H_b is larger in absolute value than those on H_a or H_c. So rather than the two effects compensating for each other as they did in ϕ_1, here they work in the same direction. Calculations show that the coefficient of H_b is exactly a factor of -2 times that on H_a or H_c. The orbital ϕ_2 is identical in form and energy to the fragment orbital $\sigma^*_{H_2}$ since it is the only antisymmetric orbital of the system.

Concerning the energetic ordering of the MOs (Figure 4.5), ϕ_1 bonding everywhere is clearly the most stable. ϕ_2 is antibonding between H_a and H_c but ϕ_3 is antibonding between H_b and H_a (H_c) and bonding between H_a and H_c. However this last interaction is weak since the coefficients of H_a and H_c are small so that overall the orbital is antibonding. An important claim that we make is that the orbitals ϕ_2 and ϕ_3 have identically the same energy. One can show, either by calculation, or by the use of group theory that this type of degeneracy is always present in molecules containing a three-fold or higher rotation axis. By this we mean that a rotation of the molecule by $360°/3 = 120°$ leads to the same molecule—except, of course, that the labels H_a, H_b, H_c are interchanged (4-10). In the case of triangular H_3, this

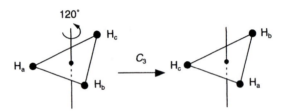

4-10

rotation axis lies perpendicular to the plane of the molecule and passes through the center of gravity of the molecule. In contrast to the case of square-planar H_4, it is difficult to see the origin of the degeneracy on constructing the orbital diagram as in Figure 4.5. Finally, we remark that the energies of the MOs increase with the number of nodal surfaces, 0 for ϕ_1, 1 for ϕ_2 and ϕ_3 (4-11).

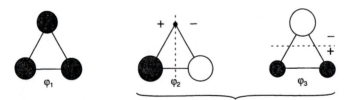

4-11

(b) Calculation of the coefficients

We can calculate the orbital coefficients in ϕ_2 by simply using the normalization condition. Since the symmetry properties of ϕ_2 show that the coefficient of $1s_b$ is zero and those of $1s_a$ and $1s_c$ of opposite sign we have

$$\phi_2 = N_2(1s_a - 1s_c)$$

$$\langle \phi_2 \mid \phi_2 \rangle = 1 = N_2^2(2 - 2S)$$

where S is the overlap integral between two $1s$ orbitals. From this we obtain

$$N_2 = [2(1 - S)]^{-1/2}$$

Regarding the orbitals ϕ_1 and ϕ_3 we have decided that the three coefficients are equal in ϕ_1 but in ϕ_3 the coefficient $1s_b$ is twice the magnitude of $1s_a$ and $1s_c$. So we can write

$$\phi_1 = N_1(1s_a + 1s_b + 1s_c)$$
$$\phi_3 = N_3(1s_a - 2 \cdot 1s_b + 1s_c)$$

From the normalization condition

$$\langle \phi_1 \mid \phi_1 \rangle = 1 = N_1^2(3 + 6S)$$
$$\langle \phi_3 \mid \phi_3 \rangle = 1 = N_3^2(6 - 6S)$$

which leads to

$$N_1 = [3(1 + 2S)]^{-1/2} \qquad N_3 = [6(1 - S)]^{-1/2}$$

If we neglect the overlap integrals S in the normalization process (if the atoms are far apart) one obtains the following expressions for the MOs.

$$\phi_1 = 3^{-1/2}(1s_a + 1s_b + 1s_c) \qquad 3^{-1/2} = 0.577$$
$$\phi_2 = 2^{-1/2}(1s_a - 1s_c) \qquad 2^{-1/2} = 0.707$$
$$\phi_3 = 6^{-1/2}(1s_a - 2 \cdot 1s_b + 1s_c) \qquad 6^{-1/2} = 0.408$$

In a more realistic model where S is not negligible (for example $S = 0.469$ for an interatomic separation of 100 pm) we get

$$\phi_1 = 0.415(1s_a + 1s_b + 1s_c)$$
$$\phi_2 = 0.970(1s_a - 1s_c)$$
$$\phi_3 = 0.560(1s_a - 2 \cdot 1s_b + 1s_c)$$

It is important to note once again that the introduction of overlap in the calculation of the normalization coefficients tends to reduce the coefficients in bonding MOs and increase them in antibonding MOs.

4.1.6. Tetrahedral H$_4$

The H$_4$ species where each hydrogen atom occupies a vertex of a regular tetrahedron can be decomposed into two H$_2$ fragments (**4-12**); H$_a$—H$_b$ located in the xy plane and a parallel to the y axis, and H$_c$—H$_d$ located in the xz plane and lying parallel

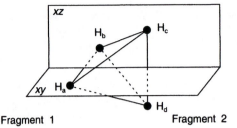

4-12   Fragment 1   Fragment 2

to the z axis. In this problem the symmetry of the orbitals σ_{H_2} and $\sigma^*_{H_2}$ of the two fragments will be indexed with respect to their behavior on reflection in the two planes xz and xy (Figure 4.6). The bonding orbitals $\sigma_{H_2}(1)$ and $\sigma_{H_2}(2)$ are symmetric (S) with respect to these two planes, and are labeled SS. The orbital $\sigma^*_{H_2}(1)$ is antisymmetric (A) with respect to xz and symmetric (S) with respect to xy since both atoms H_a and H_b lie in this plane. The symmetry of this orbital is thus labeled AS. In the same way the $\sigma^*_{H_2}(2)$ orbital, symmetric with respect to the xz and anti-symmetric with respect to xy is of SA symmetry. The MOs of tetrahedral H_4 are simply obtained by letting the orbitals of the same symmetry interact (Figure 4.6). A single interaction links the orbitals $\sigma_{H_2}(1)$ and $\sigma_{H_2}(2)$ of symmetry SS leading to the formation of bonding, ϕ_1, and antibonding, ϕ_2, combinations. In each of these two orbitals the coefficients on the four hydrogens are equal in absolute magnitude since the MOs derived from two degenerate fragment orbitals have equal contributions from each. The fragment orbitals $\sigma^*_{H_2}(1)$ (AS) and $\sigma^*_{H_2}(2)$ (SA) are the only ones of their symmetry type and remain untouched on formation of the orbitals of tetrahedral H_4. Concerning the energetic ordering of the MOs it is clear that ϕ_1, bonding between all pairs of atoms, lies deepest in energy and that ϕ_3 (AS) and ϕ_4 (SA) are antibonding, lie higher in energy and are degenerate. The energetic location of ϕ_2 is a little problematic for us. We note that this orbital contains four antibonding interactions but only two bonding ones. Overall therefore it must be antibonding in character. We claim, but do not prove here that it is exactly equal in energy with ϕ_3 and ϕ_4, the other two antibonding orbitals. The nodal properties of the MOs are detailed in **4-13**; ϕ_1 does not contain a nodal surface but ϕ_2, ϕ_3 and ϕ_4, higher in energy each possess a nodal surface.

It is interesting to ask whether a single plane would be sufficient to derive these

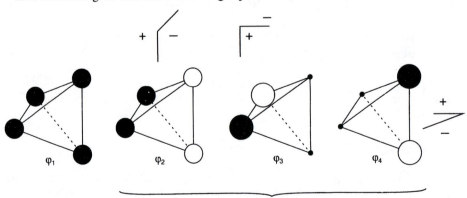

4-13

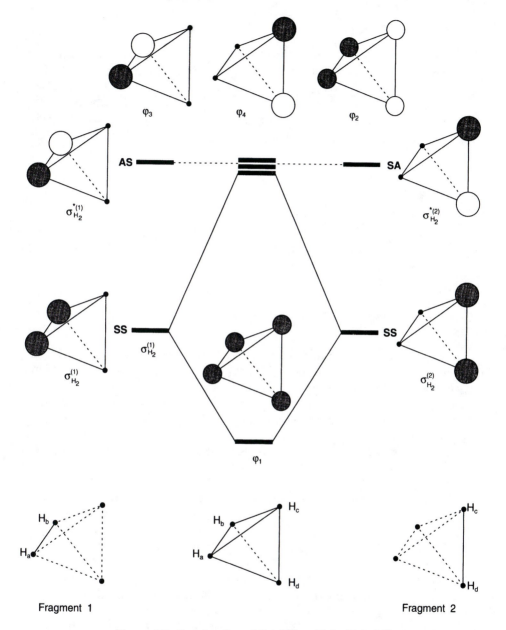

Figure 4.6. Construction of the MOs of tetrahedral H_4

results, just as in our earlier examples. Here we used the two symmetry planes xz and xy. If only xz is used it is easy to see that the orbitals $\sigma_{H_2}(1)$ and $\sigma^*_{H_2}(2)$ would be of the same symmetry (**S**) since both H_c and H_d lie in the xz plane. Not taking into account the symmetry properties of these two orbitals with respect to the other plane (xy) leads to a loss of information which penalizes us considerably in our analysis.

4.1.7. Hexagonal H_6

The hexagonal H_6 molecule can be divided into the two fragments, shown in **4-14**, whose orbitals are already known. One fragment consists of the four atoms H_b, H_c, H_e and H_f arranged in a rectangle and the other is made up of the two atoms $H_a \ldots H_d$, essentially a stretched H_2 molecule. The rectangular H_4 orbitals, labeled ψ_i ($i = 1$–4) and those of H_2 are shown in Figure 4.7 along with their symmetry properties with respect to the two planes, P_1 and P_2 of **4-15**. From the energetic point

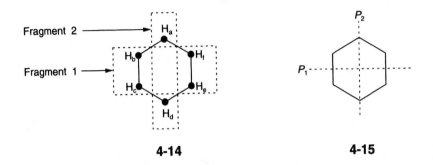

4-14   **4-15**

of view σ_{H_2} and $\sigma_{H_2}^*$ are quasi-degenerate since the two hydrogen atoms are far apart. Their energies are close to that of an isolated hydrogen $1s$ orbital, the latter lying halfway between the orbitals ψ_2 and ψ_3 of the H_4 fragment. Examination of their symmetry properties with respect to the planes P_1 and P_2 shows that the orbitals ψ_2 (**SA**) and ψ_4 (**AA**) are the only ones of their symmetry. They remain unchanged during the assembly of the diagram for H_6 and become ϕ_2 and ϕ_5. Contrarily there are interactions between two pairs of orbitals of the same symmetry, ψ_1 and σ_{H_2} (**SS**), and ψ_3 and $\sigma_{H_2}^*$ (**AS**) leading to the bonding MOs ϕ_1 and ϕ_3 and to the antibonding MOs ϕ_4 and ϕ_6.

The form of the MOs one gets by allowing the fragment orbitals to interact are given in **4-16**. In the lowest orbital, ϕ_1, the contribution of ψ_1 is larger than that of σ_{H_2}. However, the coefficients in ψ_1 (delocalized over four centers) are smaller than those in σ_{H_2} (delocalized over two centers), leading to the result that the coefficients in the deepest lying MO are equal. This result is analogous to the ones found in triangular H_3 and square planar H_4. In the antibonding partner ϕ_4 this is not true. Here σ_{H_2} makes the larger contribution and the coefficients are weighted more heavily on the two atoms H_a and H_d. The same reasoning allows us to understand that in the orbital ϕ_3, H_a and H_d have the largest coefficients ($\sigma_{H_2}^*$ makes the larger contribution) but in the completely antibonding orbital ϕ_6 all of the coefficients have the same magnitude (but alternate in sign).

As far as the energies of these orbitals are concerned, it is clear that ϕ_1, bonding between all neighboring pairs of atoms, lies deepest in energy and that ϕ_6, analogously antibonding, lies highest in energy. We make the claim here that ϕ_2 and ϕ_3 and also ϕ_4 and ϕ_5 are degenerate. The origin of these degeneracies lies in the presence of a higher than two-fold rotational axis just as for triangular H_3 and square planar H_4. Here the rotation axis is of order six. Notice that, as before, the energies of the orbitals

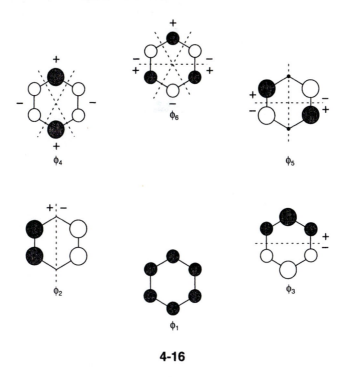

4-16

increase with the number of nodal planes (**4-16**). Thus ϕ_1 has no nodal planes, ϕ_2 and ϕ_3 have one, the degenerate pair ϕ_4 and ϕ_5 have two and ϕ_6 has three. We can see that ϕ_2 and ϕ_3 are bonding overall, since within each orbital the bonding interactions are larger than the antibonding ones. This is easy to see in ϕ_2 where the two bonding interactions are between adjacent atoms but the antibonding interactions between next-nearest neighbors and thus further away. In ϕ_3 there are two weak antibonding interactions between adjacent atoms (these are weak because of the small atomic coefficients) but four strong bonding interactions between atoms containing large and small coefficients. The MOs ϕ_4 and ϕ_5 are overall antibonding for similar reasons, the bonding interactions in ϕ_2 and ϕ_3 being replaced by antibonding ones in ϕ_4 and ϕ_5 and vice versa.

To conclude, we have shown in our study on the model H_n systems that the fragment orbital model combined with symmetry ideas allows construction of molecular orbital diagrams in a qualitative way which allows an understanding of the relative energies of the levels concerned. Although the systems we studied have from three to six hydrogen atoms and thus from three to six atomic and molecular orbitals, the problem of MO construction has been considerably simplified by setting up the interaction scheme in terms of just two fragment orbitals. Sometimes these are degenerate and sometimes they are not. In the following chapter we will show how we can construct molecular orbital diagrams of systems which are apparently much more complex (linear AH_2, trigonal AH_3 and tetrahedral AH_4 molecules) without increasing very much the difficulty of analysis.

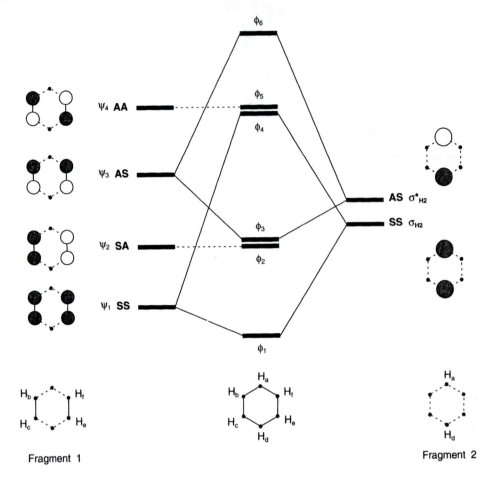

Figure 4.7. Construction of the MOs of hexagonal H_6.

4.2. Influence of electronegativity on the form and energy of the molecular orbitals

In all of the examples treated so far, the atoms have all been of the same type (H) and the AOs used in orbital construction have correspondingly been identical ($1s_H$). It is interesting to analyze the consequences in the description and energy of the MOs of varying the electronegativity of one of the atoms. We will study an example simplified to the extreme, namely triangular H_2X where we will assume that X is more, or less electronegative than H. Moreover the atom X will be such that it will only possess a single s-type orbital (labeled s_X). We will use a fragmentation scheme identical to that used for the triangular H_3 species, one H_2 fragment carrying the orbitals σ_{H_2} and $\sigma^*_{H_2}$, and a fragment X carrying a single s_X orbital. The symmetry of the orbitals will be defined with respect to the plane P passing through X and the middle of the H—H bond. The orbitals σ_{H_2} and $\sigma^*_{H_2}$ are of quite different energies since the two hydrogen atoms are close to each other.

When X is more electronegative than H, s_X lies lower in energy than the energy of an isolated $1s_H$ orbital. We will fix it close to the energy of σ_{H_2} (4-17). Since $\sigma^*_{H_2}$

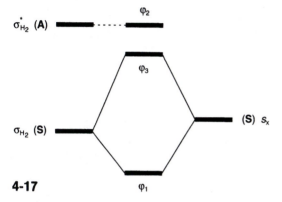

4-17

is the only orbital antisymmetric (A) with respect to P it remains unchanged during assembly of the MO diagram and becomes the MO ϕ_2 of H_2X. Contrarily σ_{H_2} and s_X (S) interact to form the MOs ϕ_1 (bonding) and ϕ_3 (antibonding). The form and energy of these orbitals are strongly influenced by the electronegativity of the X atom. The lower the energy of the s_X orbital, the lower the energies of ϕ_1 and ϕ_3, both of which have a non-zero atomic orbital contribution from s_X. Importantly, relative to the reference structure of triangular H_3 the degeneracy of the MOs ϕ_2 and ϕ_3 is lifted and the symmetric partner ϕ_3 drops to lower energy. Concerning the evolution of the form of the MOs, as the energy of s_X drops its contribution to ϕ_1 increases. At the same time the contribution of the hydrogen atom orbitals decreases. Contrariwise, in the highest energy MO (ϕ_3) the coefficient on X decreases and those of the hydrogen atom orbitals increases. (See the right-hand side of Figure 4.8.)

For the case where X is less electronegative than H, the energy of the s_X orbital lies higher than that of an isolated $1s_H$ orbital. Whereas the antibonding MO ϕ_2 remains at the same energy as in triangular H_3 (4-18) ϕ_1 and ϕ_3 lie higher too. The

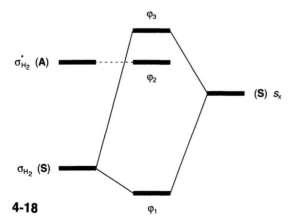

4-18

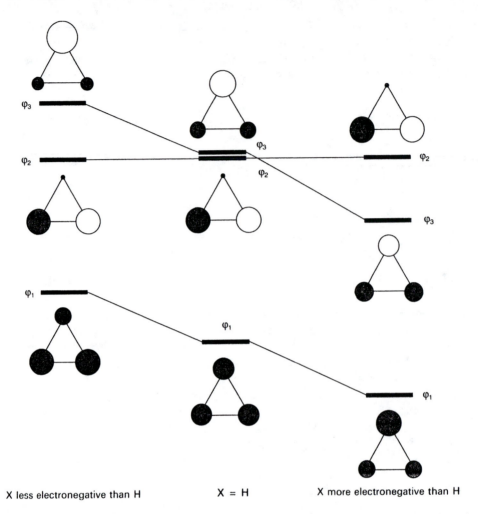

Figure 4.8. Variation in the energy and form of the MOs of triangular H_2X as a function of the relative electronegativities of X and H.

degeneracy of ϕ_2 and ϕ_3 is once again lifted, this time the antisymmetric partner lying deeper in energy. The changes in the orbital descriptions are just opposite to those just described. In the deepest lying MO the coefficient of s_X decreases but those of the hydrogen atoms increase. The tendency is reversed in the antibonding MO ϕ_3. (See the left-hand side of Figure 4.8.)

In summary the electronegativity perturbation induced by X does not influence the energy of ϕ_2 because this orbital in the H_3 reference structure has a zero atomic orbital coefficient at this site. This is not the case for ϕ_1 and ϕ_3 where this site is represented. When X is more electronegative than H then the s_X contribution increases in the deeper lying MO and decreases in the higher lying MO. The opposite is true when the electronegativity of X is less than that of hydrogen.

EXERCISES

Calculation of coefficients

4.1 Calculate the atomic orbital coefficients for the MOs of tetrahedral H_4 (**4-13**), (i) neglecting overlap between the $1s_H$ orbitals in the normalization process and (ii) taking into account this overlap when the atoms are 100 pm apart. Use the numerical result that the overlap integral between two $1s_H$ orbitals on atoms at this distance is 0.469.

4.2 Calculate the atomic orbital coefficients for the MOs of rectangular H_4 (Figure 4.2), (i) neglecting overlap between the $1s_H$ orbitals in the normalization process and (ii) taking overlap into account. Use the following numerical result:

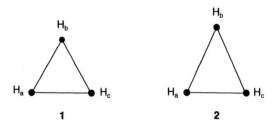

$$S = \langle 1s_a \mid 1s_b \rangle = \langle 1s_c \mid 1s_d \rangle = 0.469$$

$$S' = \langle 1s_a \mid 1s_d \rangle = \langle 1s_b \mid 1s_c \rangle = 0.231$$

$$S'' = \langle 1s_a \mid 1s_c \rangle = \langle 1s_b \mid 1s_d \rangle = 0.143$$

4.3 *MOs of isosceles triangular H_3*
Consider H_3 in an isosceles triangular geometry (**2**). With reference to the equilateral triangular structure (**1**) studied in this chapter (**4-9**) this geometry is generated by elongation of the distances H_b—H_a and H_b—H_c.

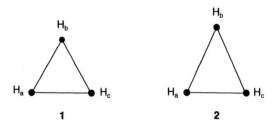

         1   **2**

(i) Select a pair of fragments for construction of the MOs.
(ii) Find a symmetry element which will allow a separation of orbitals into two sets, those symmetric and those antisymmetric with respect to this operation.

(iii) Show how the fragment orbital interaction diagram is of the same type as that obtained for equilateral triangular H_3. By analysis of the overlaps between the interacting orbitals, what is the electronic difference between the two geometries.

(iv) Compare the form and relative energy of the molecular orbitals with those found for the equilateral triangular geometry (Figure 4.5). Show in particular that the degeneracy of the orbitals ϕ_2 and ϕ_3 is lifted (removed).

4.4   MOs of square planar H_4

Construct the MO diagram for a square planar H_4 molecule using the given coordinate system where the hydrogen atoms lie along the x and y axes, starting off with the orbitals of the two fragments $H_a \ldots H_c$ and $H_b \ldots H_d$.

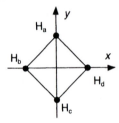

(i) On an energy diagram draw out the energies of the fragment orbitals.

(ii) Define the symmetries of these orbitals with respect to the xz and yz planes.

(iii) Which interactions between the fragment orbitals are allowed by symmetry?

(iv) Give the form and relative energies of the MOs.

(v) Compare the molecular orbitals with those of Figure 4.1. Are they equivalent? (See Appendix for help.)

4.5   MOs of 'butterfly' H_4

Construct the MOs of 'butterfly' H_4. (H_a—H_b is the 'body' of the butterfly and the planes $H_a H_b H_c$ and $H_a H_b H_d$ are the 'wings'.) All of the hydrogen atoms are assumed to be equidistant from the origin (O). The angles $H_c OH_d$ and $H_a OH_b$ are thus 180° and 120° respectively.

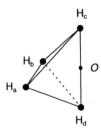

(i) Suggest a fragmentation process for the molecule.

(ii) Select two planes of symmetry and derive the symmetry properties of the fragment orbitals with respect to these planes.

(iii) Construct the fragment orbital interaction diagram.

(iv) Give the form of the MOs you have derived. In particular show the relative sizes of the atomic orbital contributions to each MO.

4.6 *Influence of electronegativity differences in tetrahedral H_2X_2*

Consider the model tetrahedral system H_2X_2 in which the X atom only comes with an *s* type orbital (s_X).

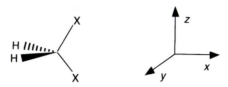

(i) Suggest a fragmentation route for the construction of the molecular orbitals.

(ii) Choose two symmetry planes.

(iii) Assuming that X is more electronegative than H (a) on an energy diagram draw out the energies of the fragment orbitals. Give their form in terms of the constituent AOs and their symmetry with respect to the two planes you chose in (ii). (b) Construct the MOs via interaction of the fragment orbitals. (c) Compare their composition and energy with those of the tetrahedral H_4 molecule (Figure 4.6).

(iv) Repeat part (iii) for the case where X is *less* electronegative than H.

APPENDIX

Degenerate orbitals

In several of our model systems, including square H_4 and triangular H_3, some of the orbitals are degenerate, i.e., they have the same energy. We find degenerate orbitals such as these in geometries where there is a rotation axis of order higher than two. Obviously the triangle and the square have axes of order three and four respectively. Besides their energetic degeneracy these MOs have a number of other characteristics.

(i) As a result of a symmetry operation an MO is usually transformed into itself (symmetric) or into minus itself (antisymmetric). So for H_4 a rotation of $2\pi/4 = \pi/2$ around the z axis (this is the four-fold rotation axis perpendicular to the plane of the molecule) transforms ϕ_1 into ϕ_1 (**4-19**) and ϕ_4 into $-\phi_4$ (**4-20**). The situation

4-19

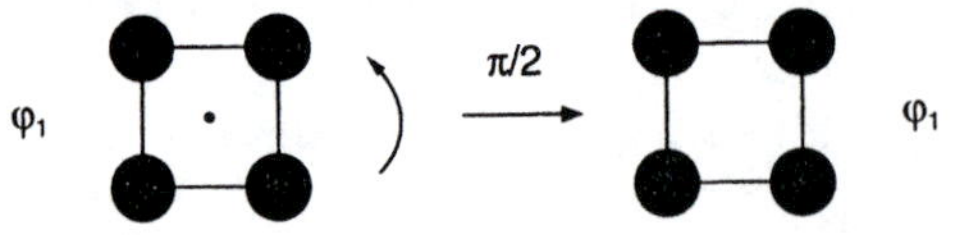

4-20

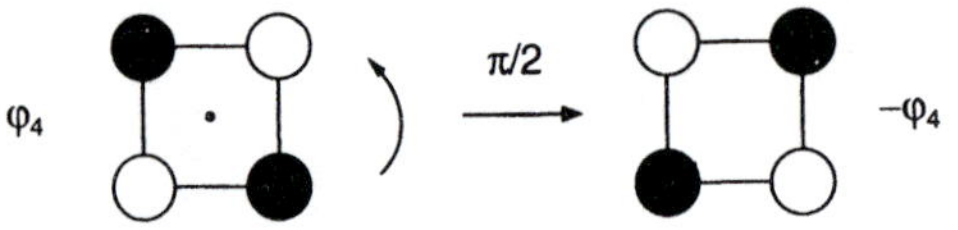

is different for the degenerate orbitals ϕ_2 and ϕ_3. They are transformed into each other, namely $\phi_2 \to -\phi_3$ (**4-21**) and $\phi_3 \to \phi_2$ (**4-22**). The same holds for p_x and p_y orbitals located at the center of the square ($p_x \to p_y$ and $p_y \to -p_x$). In a very

4-21

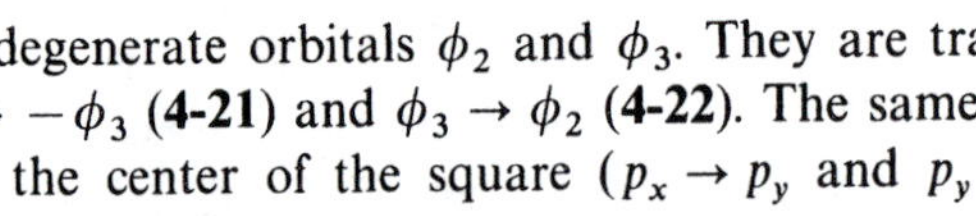

4-22

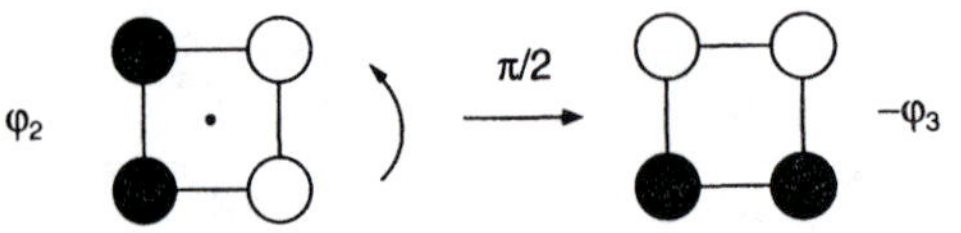

general way certain symmetry operations transform degenerate MOs into linear combinations of the starting functions. Taking, for example, triangular H_3, one can show that a rotation of $2\pi/3$ around the z axis (C_3) takes ϕ_2 to $-\frac{1}{2}\phi_2 + \sqrt{\frac{3}{2}}\phi_3$

(4-23) and ϕ_3 to $-\sqrt{\frac{3}{2}}\phi_2 - \frac{1}{2}\phi_3$ (4-24). Degenerate orbitals must always therefore be treated as a pair and never individually.

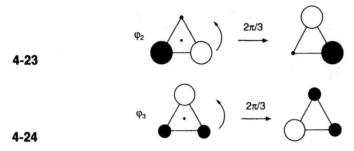

4-23

4-24

(ii) Although the symmetry operation does not transform each degenerate orbital into itself or minus itself, the new orbital has the same energy as the starting one. In effect, if ϕ_i and ϕ_j are two functions of the same energy then all normalized linear combinations of the type $\lambda\phi_i + \mu\phi_j$ are equally good functions with the same energy (see Section 2.1.2c).

(iii) This leads us to conclude that the degenerate orbitals which we obtained for systems such as square planar H_4 (Figure 4.1) and triangular H_3 (Figure 4.5) represent just one solution out of a whole host of possibilities. In general one can replace such pairs of orbitals with a pair of linear combinations of the form

$$\phi_i' = \phi_i \cos \theta + \phi_j \sin \theta$$
$$\phi_j' = -\phi_i \sin \theta + \phi_j \cos \theta$$

We can easily show that the MOs ϕ_i' and ϕ_j' are normalized and orthogonal just like ϕ_i and ϕ_j themselves. In the cases of square planar H_4 (4-21 and 4-22) and triangular H_3 (4-23 and 4-24) described above, rotation about the C_4 or C_3 axis respectively transforms the initial pair of degenerate orbitals into an exactly equivalent pair which may be derived by using $\theta = \pi/2$ and $2\pi/3$ respectively in these formulae.

5 Interactions between two fragment orbitals: linear AH₂, trigonal AH₃ and tetrahedral AH₄

When combined with symmetry ideas the fragment orbital method leads to the determination of the molecular orbital diagrams of many simple molecules. In this chapter we will study some molecules which have orbital diagrams which may be assembled by the interaction of pairs of orbitals, one from each fragment. We will derive the level structures of linear AH_2, trigonal planar AH_3 and tetrahedral AH_4 molecules (5-1) in which A is an element from the second or third row of the periodic

5-1

table. We assume that all the A—H distances are equal and only use the valence orbitals on the atoms concerned. These are thus the $1s$ orbitals on the hydrogen atoms and the ns and np orbitals ($n = 2$ or 3) for the A atoms. The core orbitals are ignored. As we have noted earlier since they lie very deep in energy and their overlap integrals with other orbitals is tiny, their influence on bond formation is negligible.

A vital aspect of our analysis concerns the symmetry properties of the orbitals concerned. We will gradually introduce the symmetry labels for orbitals of various types as the chapter progresses. Although these formally come from group theory, as we will see, no knowledge of the mathematics behind them is needed.

The molecules to be studied (5-1) have the common property of being able to be decomposed into the two fragments, A and H_n (H_2, triangular H_3 and tetrahedral H_4) whose orbitals we have already described in Chapter 4. In each case the MOs will be generated by allowing the interaction of pairs of orbitals which have the same symmetry properties. These, as we saw in Chapter 3 are the only pairs of orbitals which have a non-zero overlap integral. If the principle itself is simple, its application sometimes poses problems when some of the fragment orbitals are degenerate as in triangular H_3 and tetrahedral H_4. It turns out that the use of one or two symmetry planes is not sufficient to completely characterize the orbital symmetry. This is a consequence of the rather simple approach to the molecular orbital problem used in this book, but we will be able to make analogies between the symmetry properties of the central atom orbitals and the molecular orbitals of the H_n fragments to produce a readily understandable picture.

5.1. Linear AH₂ molecules

We divide the molecule into two fragments; a pair of non-bonded atoms $H_a \ldots H_b$ which give rise to the orbitals σ_{H_2} (bonding) and $\sigma_{H_2}^*$ (antibonding), and a central atom A on which we keep the valence orbitals s, p_x, p_y and p_z. The internuclear axis is chosen as z. (Although ns orbitals ($n \geq 2$) have radial nodes as shown in **2-6** for the $2s$ function, we shall ignore these in generating our orbital diagrams. Only the overlap with the outermost part of the orbital is chemically important at normal internuclear distances.)

5.1.1. Symmetry properties of the fragment orbitals

Consider the collection of symmetry elements the two fragments have in common. This is effectively the collection of symmetry operations for the linear AH_2 molecule. There are an infinite number of these. For example all planes which contain the z-axis are planes of symmetry for the two fragments (**5-2**). In the same way a rotation of any

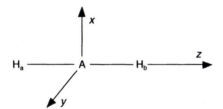

5-2

angle around z leaves the positions of the nuclei unchanged. Other elements of symmetry include the xy plane perpendicular to z and containing the atom A, the inversion center, i, located at A, etc. A general treatment of the symmetry problem would study the behavior of the orbitals as a result of all of these symmetry operations but we will content ourselves here by making a judicious selection of just one symmetry element which will allow us to provide a symmetry classification good enough to be able to decide which pairs of orbitals may interact via non-zero values of their overlap integral.

The p_x orbital on A (**5-3**) is antisymmetric (A) with respect to the yz plane, a nodal plane of this orbital. Contrarily the orbitals σ_{H_2} and $\sigma_{H_2}^*$ are symmetric (S) with

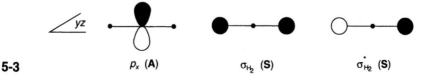

5-3

respect to this plane. Thus there is no interaction between this p orbital and these two hydrogen located orbitals since the overlap integral is zero by symmetry. In the same way their behavior with respect to reflection in the xz plane shows that the p_y orbital (**A**) may interact (**5-4**) with neither σ_{H_2} nor $\sigma_{H_2}^*$ (**S**). This result was discussed

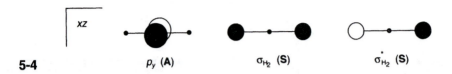

5-4 p_y (**A**) σ_{H_2} (**S**) $\overset{*}{\sigma}_{H_2}$ (**S**)

in Section 3.4.4a; the overlap between a p orbital and an s orbital lying in its nodal plane is zero.

We need now to consider the possible interaction between the s and p_z orbitals on the central atom and the σ_{H_2} and $\sigma_{H_2}^*$ orbitals on $H_a \ldots H_b$. For these we will make use of the xy plane. It is clear to see (**5-5**) that both s and σ_{H_2} are symmetric

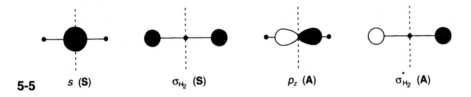

5-5 s (**S**) σ_{H_2} (**S**) p_z (**A**) $\overset{*}{\sigma}_{H_2}$ (**A**)

with respect to this plane, but p_z and $\sigma_{H_2}^*$ are both antisymmetric (**A**). It is simple to show that the overlap integrals associated with these interactions are non-zero. In each case (**5-6** and **5-7**) the overlap integral of the central atom orbital with each of the hydrogen atoms is positive, so that the total overlap integral is different from zero.

5-6 $S \neq 0$ **5-7** $S \neq 0$

The construction of the molecular orbital diagram for the linear AH_2 molecule thus consists of two pairs of interactions between two fragment orbitals (**5-8**). The positions of the AO energy levels depend of course on the nature of A but those shown will be sufficient for our needs. This interaction scheme though applies to any linear AH_2 molecule.

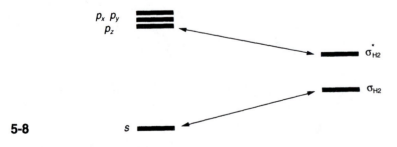

5-8

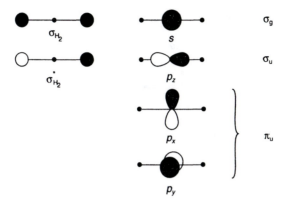

5-9

5-9 shows a common labeling scheme for the set of orbitals under consideration, using the conventions of group theory. The orbitals σ_{H_2}, $\sigma_{H_2}^*$, s and p_z are all symmetric with respect to rotation about the z-axis, that is to say they remain unchanged as a result of any rotation about this axis. They are each labeled by σ. The σ_{H_2} and s orbitals are symmetric with respect to inversion and are thus labeled σ_g; the $\sigma_{H_2}^*$ and p_z orbitals are antisymmetric with respect to inversion and are thus labeled σ_u. The pair of orbitals p_x, p_y are not symmetric with respect to rotation about z and are antisymmetric with respect to inversion. They are labeled as π_u.

5.1.2. MOs for linear AH₂ molecules

The construction of the interaction diagram relies both on the symmetry properties of the fragment orbitals and their relative energies. In the linear AH_2 molecule the hydrogen atoms are far apart. Accordingly since the overlap integral between the two $1s$ orbitals is small the energies of the σ_{H_2} and $\sigma_{H_2}^*$ orbitals are similar and close to that of the energy of an isolated hydrogen $1s$ orbital. The σ_{H_2} orbital (bonding) lies a little lower in energy than $\sigma_{H_2}^*$ (antibonding). The energies of the s and p orbitals depend upon the nature of A. The more electronegative A, the deeper these levels lie. The values used in Figure 5.1 are those appropriate for beryllium ($\varepsilon_{2s} = -9.4$ eV; $\varepsilon_{2p} = -6.0$ eV).

The molecular orbital diagram is assembled simply by pairing up those orbitals on the two framents with the same symmetry. Thus the s and σ_{H_2} orbitals (σ_g) interact to give bonding ($1\sigma_g$) and antibonding ($2\sigma_g$) orbitals. In the same way interaction between p_z and $\sigma_{H_2}^*(\sigma_u)$ leads to a bonding ($1\sigma_u$) and antibonding ($2\sigma_u$) pair. The p_x and p_y orbitals are not changed in energy since they do not find a symmetry matched with the $H_a \ldots H_b$ fragment. They become the degenerate, π_u MOs of the molecule. We still call them molecular orbitals even though they are localized on one atomic center.

The molecular orbitals thus fall into three groups (**5-10**).

    (i) Two MOs bonding between the central atom and the hydrogen atoms, built from the in-phase combination of the fragment orbital pairs s and σ_{H_2}, and p_z and $\sigma_{H_2}^*$. Of these orbitals the lowest, $1\sigma_g$, is that derived from the fragment orbital which lies lowest in energy.

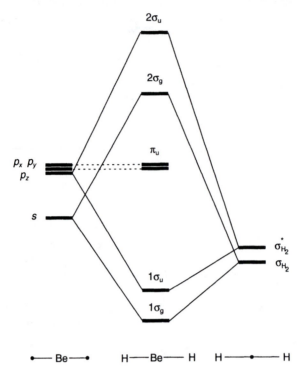

Figure 5.1. Construction of the MOs of a linear AH_2 molecule. (The relative AO energies are appropriate for A = Be.)

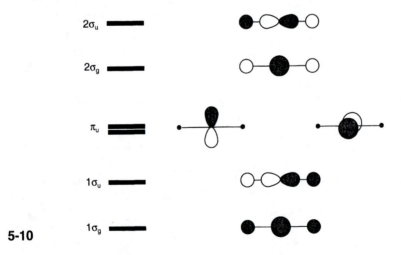

5-10

(ii) Two MOs, antibonding between the central atom and the hydrogen atoms ($2\sigma_g$ and $2\sigma_u$) built from the out-of-phase combinations of these same fragment orbitals.

(iii) Between these two groups, two degenerate MOs completely localized on the central atom, and therefore with no contribution from the hydrogen atom orbitals. Such orbitals are called nonbonding orbitals.

Although the details of the molecular orbital diagram depend upon the nature of the central atom, this general description in terms of two bonding, two nonbonding and two antibonding orbitals is a general one, applicable to all linear AH_2 molecules.

5.1.3. Application to BeH$_2$

BeH_2 is a linear triatomic molecule which has four valence electrons. In its electronic ground state the lowest two orbitals, $1\sigma_g$ and $1\sigma_u$ are therefore doubly occupied to give the configuration $1\sigma_g^2\ 1\sigma_u^2$ (**5-11**). These two orbitals are bonding between the

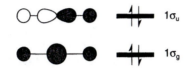

5-11

central atom and the hydrogens. With two bonding pairs of electrons we should expect two Be—H bonds as indeed indicated by the Lewis structure H—Be—H. Notice however, that it is not possible to identify one doubly occupied bonding orbital with one Be—H bond, and the other with the second Be—H bond. Each bonding MO is equally associated with both Be—H bonds. The one $2p$ ($2p_z$) and one $2s$ orbital on beryllium are then equally associated with each Be—H linkage in the bonding orbitals in which they are involved. We say that these orbitals and the electrons in them are *delocalized* over the whole molecule in contrast to the localized viewpoint of the Lewis structure.

The pattern of ionization energies for BeH_2 leads to some further insight into this delocalized view of the bonding problem. From Figure 5.1 it is clear that the ionization energy depends upon the origin of the ejected electron. The ionization energy from the $1\sigma_g$ orbital is larger than that from $1\sigma_u$. From the Lewis viewpoint one might have expected just a single ionization energy. Thus, although the two Be—H bonds are equivalent in every way the molecular orbitals which describe them are not. Later in the Appendix to Chapter 8 we will show an interesting connection between the two viewpoints.

5.2. Trigonal planar molecules

The natural fragmentation of the AH_3 molecule, where the angles between the A—H bonds are $120°$, is into a central A atom and a collection of three H atoms at the corners of an equilateral triangle. We described the energy levels of such an H_3 unit in the previous chapter. The levels of the A atom to be used are just the valence s and p orbitals.

5.2.1. Symmetry properties of the fragment orbitals

The two fragments, A and H_3 have many symmetry elements in common, among them (**5-12**) the molecular plane (xy), three two-fold rotation axes (x being one of them) collinear with the A—H bonds, three symmetry planes perpendicular to the

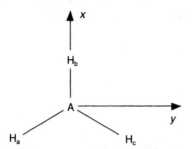

5-12

mirror plane and containing an A—H bond (xz for example) and the z-axis which is a three-fold rotation axis. From all of these symmetry elements we will only keep the molecular plane (xy) and the xz plane in order to characterize the symmetry of the molecular orbitals. In doing this we 'reduce' the symmetry of the system (since the number of symmetry elements has decreased) but as we will see this new set will be sufficient for our needs.

All of our fragment orbitals will be either symmetric (**S**) or antisymmetric (**A**) with respect to reflection in these planes of symmetry. So there is just one **AS** orbital (p_z) which is antisymmetric with respect to xy but symmetric with respect to xz (**5-13**)

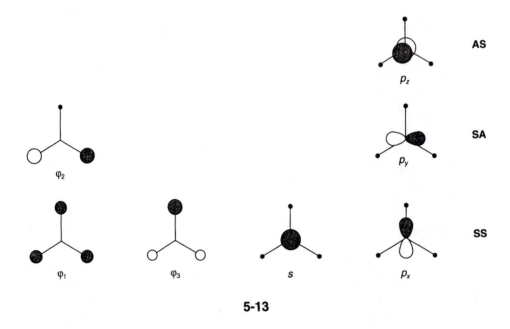

5-13

two **SA** orbitals (ϕ_2 and p_y) and four **SS** orbitals (ϕ_1, ϕ_3, s and p_x). This preliminary analysis allows us to separate the fragment orbitals into three groups (**5-13**). Orbitals in one group may not interact with orbitals from another since their symmetry properties with respect to one or other of the planes xy or xz are different. One can conclude therefore that the p_z orbital, the only orbital of **AS** symmetry, cannot take part in any interaction. We now have to determine whether the overlaps between orbitals from within the same group are different from zero. In effect, if it turns out that a pair of orbitals have different behavior with respect to a symmetry element ignored in the simple treatment, then the overlap is of course zero. The overlap between the orbitals ϕ_2 and p_y (AS) is non-zero since the contributions from p_y–$1s_H$ overlap are both of the same sign (**5-14**). But now consider the case of four orbitals of **SS** symmetry. For the pairs (s, ϕ_1) and (p_x, ϕ_3) the overlaps involved are different from zero since all the contributions between the central orbital and each of the $1s_H$ orbitals are of the same sign (**5-15** and **5-16**). However, consider the overlap integrals between s and ϕ_3 and between p_x and ϕ_1. The first of these (**5-17**) is made up of a positive contribution from s and $1s_b$ and two negative contributions from s and $1s_a$ and $1s_c$. Since the coefficient of $1s_b$ is twice as large, in an absolute sense, than the coefficients of $1s_a$ and $1s_c$ (see Section 4.1.5) the total overlap integral is identically zero. A similar situation holds (**5-18**) for the overlap between p_x and ϕ_1. The three

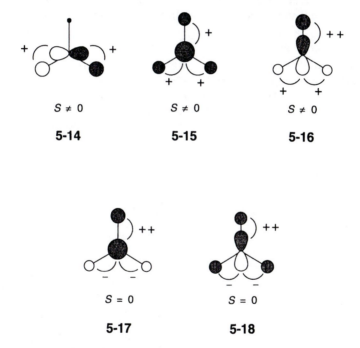

$S \neq 0$	$S \neq 0$	$S \neq 0$
5-14	**5-15**	**5-16**

$S = 0$	$S = 0$
5-17	**5-18**

coefficients on the hydrogen atoms are equal in ϕ_1 but the overlap integral with p_x varies as the cosine of the angle the A—H bond makes with the x-axis (see Section 3.4.3). Thus there is one positive overlap (where this angle is zero) and two negative overlaps (where this angle is $\pm120°$). Thus the total overlap integral is proportional to $\cos(0°) + \cos(120°) + \cos(-120°) = 1 - \frac{1}{2} - \frac{1}{2} = 0$.

We should point out at this stage that the zero overlap integral between orbitals of the 'same symmetry' (**SS**) is a consequence of the reduction in symmetry we used to make this problem tractable. The two pairs of orbitals, (s, ϕ_1) and (p_x, ϕ_3) do in fact have different symmetry if all of the symmetry elements are used.

In order to construct a molecular orbital diagram for the AH_3 molecule we use the same technique employed for the linear AH_2 system, namely only fragment orbitals with non-zero overlap may interact. This reduces the orbital problem to one of interactions between the three pairs of orbitals, s and ϕ_1, p_y and ϕ_2 and p_x and ϕ_3 as shown in **5-19**. The p_z orbital is not involved in any interaction. The actual

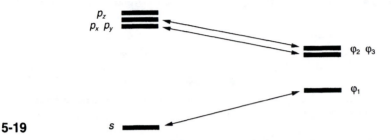

5-19

form of the diagram will vary from one AH_3 molecule to another since the energies of the central atom s and p orbitals depend upon the identity of A, but the interactions shown are the same irrespective of the identity of the system.

As described before for the AH_2 molecule we usually attach group theoretical labels to describe the orbitals of the fragment. The s orbital is labeled a_1'. Here, a describes a non-degenerate level, just as σ did in the linear molecule, and the single prime a function symmetric with respect to reflection in the plane perpendicular to the three-fold axis (z). The p_z orbital is labeled a_2'', antisymmetric with respect to reflection. The pairs of degenerate levels carry an e label, just like the label π in linear molecules. Both p_x, p_y and ϕ_2, ϕ_3 are labeled e' (**5-20**) although we will want to distinguish e_x' (p_x and ϕ_3) and e_y' (p_y and ϕ_2).

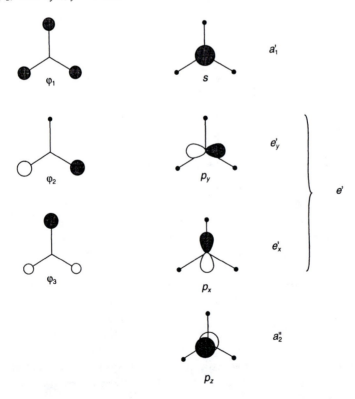

5-20

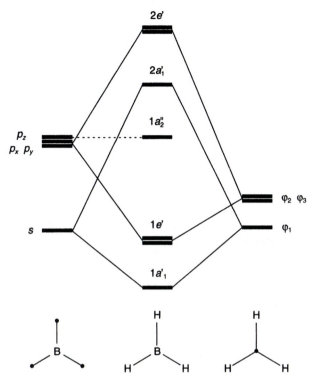

Figure 5.2. Construction of the MOs of a trigonal planar AH_3 molecule. (The relative AO energies are appropriate for A = B).

5.2.2. MOs of trigonal planar AH_3

The fragment orbital interaction diagram of Figure 5.2 corresponds to that for BH_3 where $\varepsilon_{2s} = -14.7$ eV and $\varepsilon_{2p} = -5.7$ eV. The energies of the H_3 fragment orbitals straddle the energy of an isolated $1s_H$ orbital, since ϕ_1 (bonding) lies a little below and ϕ_2 and ϕ_3 (antibonding) a little above. Their splitting is small since the interaction between the hydrogen $1s$ orbitals is small as a result of the large H—H separation. First of all we readily see that the p_z orbital ($1a_2''$) is unchanged in energy. The orbitals ϕ_1 and $2s$, both of a_1' symmetry interact to give a bonding orbital ($1a_1'$) and an antibonding orbital ($2a_1'$). Similarly the orbital pairs, ϕ_3 and p_x (e_x'), and ϕ_2 and p_y (e_y') interact to give a bonding pair ($1e_x'$ and $1e_y'$) and an antibonding pair ($2e_x'$ and $2e_y'$). The orbitals $1e_x'$ and $1e_y'$ are degenerate, as are $2e_x'$ and $2e_y'$; as shown in exercise 5.1, since the overlap integrals associated with the x and y partners of a degenerate pair are equal the resultant molecular orbitals are degenerate. The origin of this degeneracy comes just as in triangular H_3 from the presence of a three-fold rotation axis in the molecule.

The MOs of trigonal planar molecules thus divide into three groups (**5-21**).

   (i) Three MOs bonding between the central atom and the hydrogen atoms. These are $1a_1'$, $1e_x'$ and $1e_y'$, in-phase combinations of the fragment orbitals (ϕ_1, s),

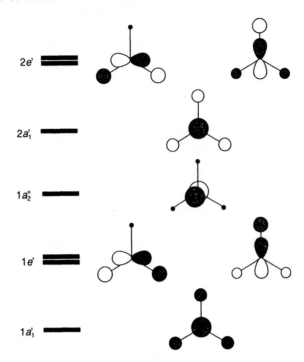

5-21

(ϕ_2, p_y) and (ϕ_3, p_x) respectively. The deepest lying orbital, $1a_1'$ is the one that comes via interaction of the deepest lying fragment orbitals.

(ii) Three MOs antibonding between the central atom and the hydrogen atoms. These are $2a_1'$, $2e_x'$ and $2e_y'$ and are the out-of-phase combinations of the fragment orbitals.

(iii) Between these two groups there is a non-bonding orbital ($1a_2''$), completely localized on the central atom without any hydrogen atom contribution.

5.2.3. Application to the electronic structure of BH₃

BH_3 is a short-lived molecule, rapidly dimerizing to give B_2H_6, although many of the reactions of the latter may be understood via an equilibrium between the two but lying very much in favor of the dimer. It is a trigonal planar molecule with six valence electrons. In its electronic ground state the three lowest energy levels are doubly occupied to give (**5-22**) $1a_1'^2 \, 1e_x'^2 \, 1e_y'^2$. These three orbitals are bonding between the central atom and the hydrogen atoms thus providing a connection to the Lewis structure with three B—H bonds each made up of two electrons. However, as before for AH_2, although it is not possible to identify the two electrons in one particular MO with a particular B—H bond, the *collection* of three doubly occupied bonding orbitals gives rise to three chemical bonds. Obviously in $1a_1'$ each of the hydrogen atoms are bonded equally to the central atom since the corresponding overlap integrals are equal. The situation is more complex in the $1e'$ pair. $1e_y'$ is only bonding between the boron atom and two of the hydrogen atoms (H_a and H_c) since the coefficient on H_b is zero. Contrariwise the $1e_x'$ orbital is largely bonding between boron and H_b. In this orbital the coefficient on H_b is twice as large as those on H_a

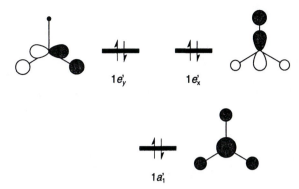

5-22

and H_c. In addition the $2p_x$ orbital points directly at H_b and so its overlap will be larger than with H_a or H_c. In fact if one considers the orbitals $1e'_x$ and $1e'_y$ as a pair one can show that they lead to equal bonding character between the central atom and each of the hydrogen atoms. Thus consideration of the trio of doubly occupied bonding orbitals leads to the conclusion that the three B—H bonds are equivalent in every way. However the three molecular orbitals which lead to this picture are not energetically equivalent (**5-22**). It is easier to eject an electron from the $1e'$ level than it is from the $1a'_1$ level. There are then two different ionization energies for the molecule which depend upon the origin of the ionized electron. We must once again clearly recognize the equivalence of three B—H bonds which arise via the occupation of three clearly non-equivalent orbitals.

A final point merits mention. The lowest unoccupied orbital in BH_3 is a non-bonding p orbital ($1a''_2$). This orbital, vacant and low in energy, is susceptible to donation by a pair of electrons. If this electron pair is associated with an H^- ion such that $BH_3 + H^- \rightarrow BH_4^-$, then we can readily see the origin of the Lewis acid properties of such a species.

5.3. Tetrahedral AH₄ molecules

The natural decomposition for a tetrahedral AH_4 molecule is into a central A atom and a tetrahedron of hydrogen atoms. The levels of the latter, a tetrahedral H_4 unit were studied in the previous chapter. The fragment orbitals for the atom A are just its valence s and p orbitals.

5.3.1. Symmetry properties of the fragment orbitals

The fragments A and H_4 have many symmetry elements in common, among them (**5-23**) six planes of symmetry containing two A—H bonds (xz and xy are two

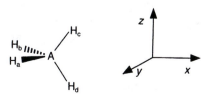

5-23

examples), three C_2 axes which bisect opposite pairs of H—A—H angles (x for example) and four C_3 axes collinear with the A—H bonds. As before we will just retain two planes (xz and xy) in order to distinguish between the orbitals concerned.

There are two **SA** orbitals (ϕ_4 and p_z) symmetric with respect to xz and antisymmetric with respect to xy, two **AS** orbitals (ϕ_3 and p_y) and four **SS** orbitals (ϕ_1, ϕ_2, s and p_x) as shown in **5-24**. The fragment orbitals thus separate into three

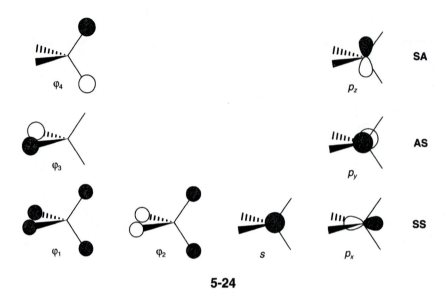

5-24

groups. Recall that orbitals belonging to different groups may not interact, but we still have to look carefully at the overlaps between orbitals within each group. We can see that overlap between the p_z and ϕ_4 orbitals is non-zero since the individual overlaps between p_z and $1s_H$ orbitals are of the same sign (**5-25**). The same is true for p_y and ϕ_3 (**5-26**). For the orbitals of **SS** symmetry we need to consider the two pairs (s, ϕ_1) and (p_x, ϕ_2). In each case since all of the individual overlaps between $1s_H$ orbitals and the central atom orbital are of the same sign (**5-27** and **5-28**) the total overlap integral is non-zero and the orbitals within each pair may interact. This is not the case for the overlap between (s and ϕ_2) and (p_x and ϕ_1). Just as we showed for the related AH_3 case, the overlap integrals between these pairs are identically zero (**5-29** and **5-30**). In both cases the two positive overlap integrals are exactly cancelled by the two negative overlap integrals. Also, as in AH_3, these zero overlap integrals between orbitals of the 'same symmetry' come about because of the reduction of the tetrahedral symmetry to just the two planes xz and yz. Use of the full symmetry removes this problem. In conclusion, just as in all of the preceding examples, only the orbitals of the same symmetry, with non-zero overlap may interact.

The construction of the molecular orbital diagram for tetrahedral AH_4 thus reduces to a question of the four pairs of interactions of **5-31**, the variation from one molecule to another being set by the central atom s and p orbital energies dependent upon the identity of A. We will use in what follows the group theoretical labels for these orbitals (**5-32**). Both s and ϕ_1 are of a_1 symmetry and the trios (p_x, p_y, p_z) and (ϕ_2, ϕ_3, ϕ_4)

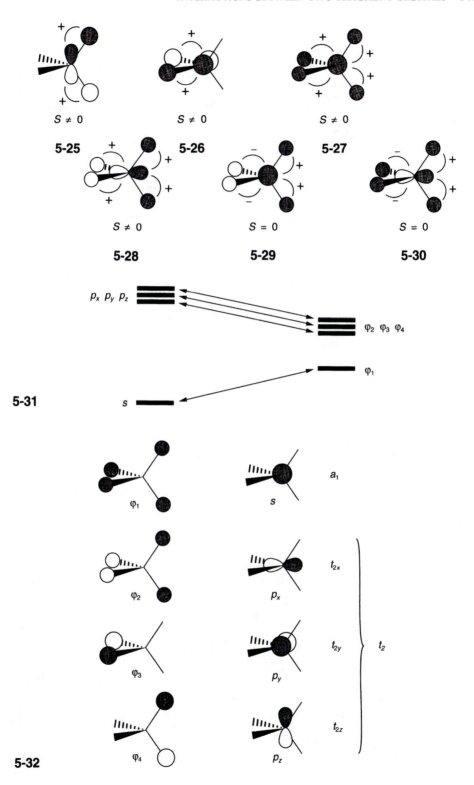

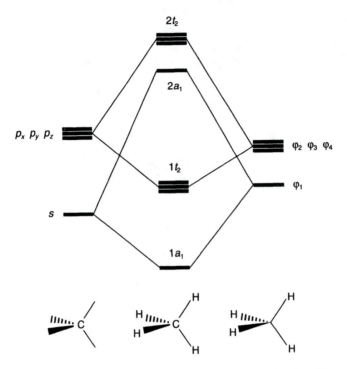

Figure 5.3. Construction of the MOs of a tetrahedral AH_4 molecule. (The relative AO energies are appropriate for A = C).

are of t_2 symmetry. (The label t is used for triply degenerate levels.) We may distinguish t_{2x} (p_x and ϕ_2), t_{2y} (p_y and ϕ_3) and t_{2z} (p_z and ϕ_4). Notice that here (and in the AH_3 molecule too) there is no center of symmetry unlike the situation in linear AH_2. Accordingly the subscript g or u which described the behavior with respect to inversion is inappropriate here.

5.3.2. MOs of tetrahedral AH_4 molecules

The fragment orbital interaction diagram of Figure 5.3 corresponds to the case of CH_4 where $\varepsilon_{2s} = -19.4$ eV and $\varepsilon_{2p} = -10.7$ eV. The H_4 fragment levels lie just below (ϕ_1 is H—H bonding) and just above (ϕ_2, ϕ_3 and ϕ_4 are H—H antibonding) the energy of an isolated $1s_H$ orbital (-13.6 eV). The orbitals ϕ_1 and s (a_1) interact to give bonding ($1a_1$) and antibonding ($2a_1$) partners. Interaction between ϕ_2 and p_x, between ϕ_3 and p_y and between ϕ_4 and p_z leads to the formation of three bonding MOs, $1t_2$ ($1t_{2x}$, $1t_{2y}$, $1t_{2z}$) and three antibonding MOs $2t_2$ ($2t_{2x}$, $2t_{2y}$, $2t_{2z}$). The set of $1t_2$ levels is degenerate, as is the set $2t_2$. As shown in exercise 5.2 the pairwise overlap integrals between orbitals of each degenerate set, ϕ_2, ϕ_3 and ϕ_4 on H_4 with respectively p_x, p_y and p_z on A, are equal. This triple degeneracy comes about because of the high symmetry of the tetrahedral molecule.

The molecular orbitals of tetrahedral AH_4 molecules divide into two sets (**5-33**).

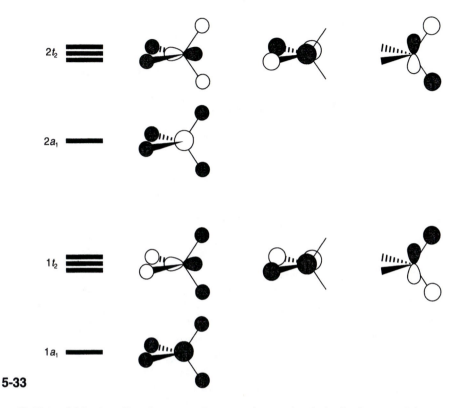

5-33

(i) Four MOs bonding between the central atom and the hydrogens. These are $1a_1$, $1t_{2x}$, $1t_{2y}$, $1t_{2z}$, in-phase combinations of the fragment orbitals (ϕ_1, s), (ϕ_2, p_x), (ϕ_3, p_y) and (ϕ_4, p_z) respectively. The deepest lying orbital, $1a_1$, arises via interaction with the deepest lying fragment orbitals. (The three $1t_2$ orbitals are degenerate.)

(ii) Four MOs antibonding between the central atom and the hydrogens. These are $2a_1$, $2t_{2x}$, $2t_{2y}$ and $2t_{2z}$ out-of-phase combinations of the same fragment orbitals. (The three $2t_2$ orbitals are degenerate.)

5.3.3. Application to the electronic structure of CH_4

In the methane molecule, with a total of eight valence electrons, the lowest four molecular orbitals are doubly occupied in the electronic ground state to give the electronic configuration $1a_1^2\ 1t_{2x}^2\ 1t_{2y}^2\ 1t_{2z}^2$ or $1a_1^2\ 1t_2^6$ as in **5-34**. These four occupied bonding orbitals correspond to the four C—H bonds of the Lewis structure. The central atom uses one s and three p orbitals to form these bonds. Just as in our earlier AH_2 and AH_3 examples it is not possible to make a one-to-one correspondence between a single delocalized molecular orbital and a particular C—H bond. In the $1a_1$ orbital the bonding character is the same between the central atom and each of the hydrogens since the hydrogen coefficients are all equal. The same is true for the $1t_{2x}$ orbital. Here all of the coefficients are equal in absolute magnitude and each of the A—H bonds make the same angle (one half of the 'tetrahedral' angle, 109.5°/2)

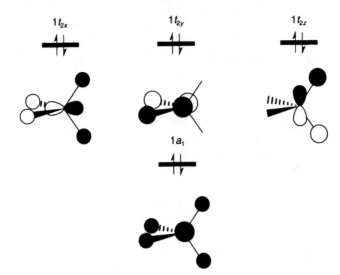

5-34

with the axis (x) of the p_x orbital. On the other hand the $1t_{2y}$ orbital is bonding only between carbon and H_a and H_b, and the $1t_{2z}$ orbital is bonding in the same way between carbon and H_c and H_d. It is the collection of four occupied MOs taken together which lead to four equivalent C—H bonds.

Finally, all of the occupied MOs are not of the same energy, $1a_1$ lying deeper than $1t_2$. This prediction from molecular orbital theory is confirmed experimentally via the photoelectron spectrum. There are two ionization energies which differ by about 10 eV. To conclude, just as in the earlier examples it is necessary to distinguish between the equivalence of the four C—H bonds and the non-equivalence of the four occupied molecular orbitals (split into the two sets $1a_1$ and $1t_2$).

EXERCISES

5-1 *Overlap integrals between fragment orbitals in AH_3*
We will consider the orbitals ϕ_2 and ϕ_3 on the triangular H_3 unit and the p_x and p_y orbitals on the A atom which lead to the levels of the trigonal planar AH_3 molecule. The values of the coefficients in the orbitals ϕ_i are those that were calculated by including the overlaps between the $1s_H$ orbitals in Section 4.1.5b. S is the overlap integral between two $1s_H$ orbitals.

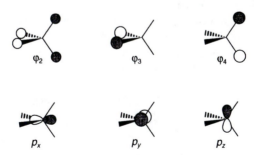

Show that the overlap integrals $S_2 = \langle \phi_2 \mid p_y \rangle$ and $S_3 = \langle \phi_3 \mid p_x \rangle$ are equal. (Call S_0 the overlap integral between a p orbital and a $1s_H$ orbital lying along the p orbital axis at a distance $d = $ A—H.)

5.2 *Overlap integrals between fragment orbitals in AH_4*
Show in the same way that the overlap integrals between the pairs of fragment orbitals $S_2 = \langle \phi_2 \mid p_x \rangle$, $S_3 = \langle \phi_3 \mid p_y \rangle$ and $S_4 = \langle \phi_4 \mid p_z \rangle$ in tetrahedral AH_4 molecules are equal. The values of the coefficients in the orbitals ϕ_i are those that were calculated in the exercise 4.1. Note that the angle between the bonds in a tetrahedron (α) is 109.5° and verify that $\cos(\alpha/2) = 1/\sqrt{3}$.

5.3 *Star-shaped H_4*
Construct the molecular orbitals of star-shaped H_4 starting from the two fragments, triangular H_3 and a single central H atom.

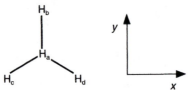

(i) Give the symmetry properties of the fragment orbitals with respect to the yz plane. Deduce immediately the form of one of the MOs of star-shaped H_4.

(ii) Analyze the overlap integrals between the symmetric orbitals of the two fragments and derive a second MO.

(iii) Construct a complete molecular orbital diagram given the fact that the highest MO is non-degenerate. Give the form of all of the MOs.

5.4 *Trigonal bipyramidal H_5*

Construct the energy level diagram of trigonal bipyramidal H_5. The atoms H_d and H_e lie along the z-axis and H_a, H_b and H_c lie at the vertices of a trigonal plane. Make all of the distances of these hydrogen atoms to the origin equal and construct the MOs of this system from the two fragments, triangular H_3 ($H_a H_b H_c$) and linear H_2 ($H_d \ldots H_e$) units.

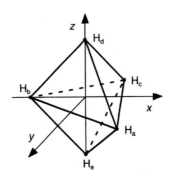

(i) Describe the relative energies of the fragment orbitals, taking into account the distances between the hydrogen atoms.

(ii) Use the xy plane of symmetry to find one of the MOs of H_5.

(iii) Use the xz plane to determine a second MO.

(iv) Analyze the overlap integrals between the orbital pairs symmetric with respect to reflection in both of these planes. Hence determine a third MO of H_5.

(v) Construct the complete orbital interaction diagram, given the fact that the highest energy orbital is non-degenerate. Give the form of each MO.

5.5 *Analogy between the orbitals of square planar H_4 and those of a central A atom (see Appendix)*

Decompose a square planar AH_4 molecular into the two fragments, A and square planar H_4.

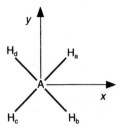

(i) Establish the orbital analogy between the MOs of square planar H_4 (ϕ_1–ϕ_4 of Section 4.1.1) and the AOs (s, p_x, p_y, p_z) of the central A atom.

(ii) Do the same but for the case where the four hydrogen atoms lie along the axes x and y. In this case use the H_4 fragment orbitals determined in exercise 4.4.

APPENDIX

Analogous orbitals

In highly symmetric AH_n molecules it is possible to establish an analogy between the orbitals on A and those of H_n. This analogy allows us to collect together those orbitals which have the same symmetry.

In linear AH_2 molecules, the A atom is situated in the middle of the line connecting the two hydrogen atoms. Consider first of all the s orbital centered on A. The value of the wavefunction depends only on the distance (d) of a point from the nucleus (a spherically symmetrical orbital). Let us call a (5-35) the value of this function found at the two extremities of the line connecting the two hydrogen atoms. Now consider the bonding MO, σ_{H_2}, of the $H_a \ldots H_b$ fragment. This is an in-phase combination of the orbital $1s_a$ and $1s_b$ with equal coefficients on the two atoms (a', 5-36). One can

$+a$ $+a$

5-35

s

$+a'$ $+a'$

5-36

σ_{H_2}

establish an analogy between the equal values of the s function at the extremities of the $H_a \ldots H_b$ line and the equal coefficients of the $1s_H$ orbitals in σ_{H_2}. Thus the σ_{H_2} orbital is a *pseudo-s orbital* (5-37). In the same way the function p_z (antisymmetric

5-37   = pseudo-s

with respect to the xy plane) has values of opposite sign (but of equal magnitude) at the extremities of the $H_a \ldots H_b$ line ($\pm b$ in 5-38). This observation may be coupled with that for the antibonding MO, $\sigma^*_{H_2}$, where the coefficients of the $1s_H$ orbitals are of opposite sign ($\pm b'$ in 5-39). In the same manner as described above for σ_{H_2} and

$-b$ $+b$

5-38

p_z

$-b'$ $+b'$

5-39

$\sigma^*_{H_2}$

s, $\sigma^*_{H_2}$, is a *pseudo-p_z orbital* (5-40). All that is left is consideration of the orbitals p_x and p_y on A. These orbitals have the yz and xz planes respectively as nodal planes,

5-40   = pseudo-p_z

and thus have a zero amplitude at all points along the z-axis (**5-41** and **5-42**). Thus the p_x and p_y orbitals do not have orbital analogs on the $H_a \ldots H_b$ fragment.

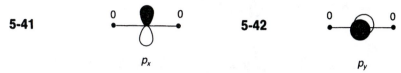

5-41 p_x **5-42** p_y

These results may be compared with those found earlier in this chapter. The only interactions which are possible are those between analogous orbitals, namely s and σ_{H_2} (pseudo-s), and p_z and $\sigma^*_{H_2}$ (pseudo-p_z). The p_x and p_y orbitals, which have no analog on $H_a \ldots H_b$ are not involved in any interaction.

A similar analysis can be developed for the trigonal AH_3 and tetrahedral AH_4 molecules. In the same manner as before we can calculate the amplitudes of the s, p_x, p_y and p_z orbitals on A at the points where the hydrogen atoms are located. The following correspondence between the orbitals of two fragments can then be established once again as in **5-43**.

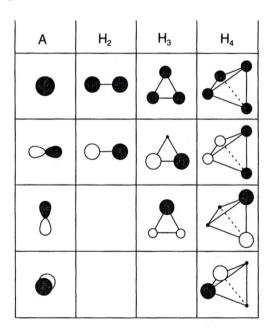

5-43

6 Interactions between three fragment orbitals: AH, bent AH$_2$ and pyramidal AH$_3$

For the majority of molecules, among them many of the simplest ones, the task of constructing a molecular orbital diagram is more complex than the examples we have used up to now, in that there are often more than just two orbitals of each symmetry species to consider. We then need to take into account all of the interactions simultaneously. In general the solution to this problem involves recourse to a numerical calculation which uses, amongst other things, the different overlap integrals between all the orbitals with each other. It is however, relatively simple to obtain in a qualitative way the form of the MOs generated via the interaction of three fragment orbitals of the same symmetry, two on one fragment and one on the other. AH, bent AH$_2$ and pyramidal AH$_3$ molecules fall into this category. A is an element which only contains valence s and p orbitals, the so-called main group elements. In Chapter 7 we will look at the case of four orbitals of the same symmetry.

6.1. Rules for the interaction of three orbitals

6.1.1. Outline of the problem

Let us consider the case where an orbital (χ_1) on fragment 1 can interact with two orbitals (χ_2 and χ_3) on fragment 2 (**6-1**). This implies that the overlap integrals between χ_1 and χ_2 and between χ_1 and χ_3 are non-zero. These are labeled as the interactions 1 and 2 in **6-1**. The orbitals χ_2 and χ_3 are orthogonal to each other ($S = 0$) since they derive from orbitals on the same fragment. The result is the generation of three MOs labeled ϕ_1, ϕ_2 and ϕ_3 in order of increasing energy (**6-2**).

6.1.2. Rules for the construction of the MOs

The resulting molecular orbitals are linear combinations of the fragment orbitals and in general may be written as

$$\phi_i = \lambda_i \chi_1 + \mu_i \chi_2 + \nu_i \chi_3 \quad (i = 1, 2, 3)$$

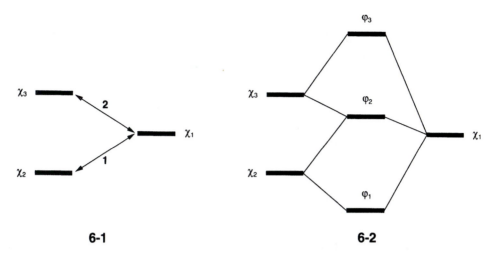

6-1   **6-2**

Each MO is characterized as before by its energy, and also by the sizes and relative signs of the coefficients, λ_i, μ_i and ν_i. It is difficult to give simple rules for the sizes of the coefficients in each molecular orbital, for they depend, amongst other things, upon the relative positions of *three* fragment orbitals. Contrarily it is possible to determine the relative signs of these coefficients and to give some indication as to the energetic placement of the levels relative to those of the starting orbitals*.

(a) The signs of the coefficients

Depending on the signs of these coefficients the interactions within an MO between χ_1 on fragment 1 and χ_2 and χ_3 on fragment 2 will be either bonding or antibonding. The rules for molecular orbital construction are quite simple.

   (i) In the deepest lying MO, ϕ_1, the interactions between χ_1 and χ_2 and between χ_1 and χ_3 are both bonding.

  (ii) In the highest lying MO, ϕ_3, the interactions between χ_1 and χ_2 and between χ_1 and χ_3 are both antibonding.

 (iii) In the orbital of intermediate energy, ϕ_2, there is a bonding interaction between χ_1 and χ_3 and an antibonding interaction between χ_1 and χ_2.

These mixing rules are shown schematically in **6-3**. A + sign is used for a bonding, and a − sign for an antibonding interaction. When the energy of χ_1 is intermediate between those of χ_2 and χ_3 (**6-2**), it is easy to see how the orbitals are mixed to give ϕ_2. In the absence of χ_3, ϕ_2 will be an antibonding combination of χ_1 and χ_2. If, though χ_2 is absent, ϕ_2 will be a bonding combination of χ_1 and χ_3. In the presence of both χ_2 and χ_3 the two effects add together.

---

* These rules are derived using perturbation theory as described in *Orbital Interactions in Chemistry*, T. A. Albright, J. K. Burdett, M.-H. Whangbo, Wiley, New York, 1985, p. 32.

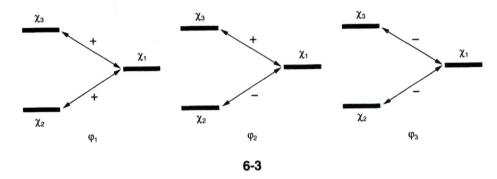

6-3

(b) The energies of the orbitals

The orbital energy depends both on the nature of the bonding and antibonding character of the molecular orbitals and on the relative energies of the atomic orbitals from which they are derived.

   (i) The most stable orbital, ϕ_1, lies deeper in energy than the deepest lying of the starting orbitals, χ_2 in **6-2**. This stabilization comes about since both of the interactions are bonding.

   (ii) The highest energy orbital, ϕ_3, lies at a higher energy than the highest of the starting orbitals, χ_3 in **6-2**. This destabilization is the result of the presence of two antibonding interactions.

  (iii) The energy of the intermediate orbital of the starting set of orbitals is not well-defined. It usually lies close in energy to the intermediate starting orbital, χ_1 of **6-2**. It will lie slightly higher or deeper than χ_1 itself depending on the relative strengths of the bonding and antibonding contributions to its character.

6.2. Electronic structure of AH molecules

6.2.1. Outline of the problem

The MOs of the AH molecule can be constructed via interaction between the valence s and p orbitals on atom A with the $1s_H$ orbital on hydrogen. If z is the internuclear axis, the p_x and p_y orbitals on A have a zero overlap with the $1s_H$ orbital since the hydrogen atom is located in their nodal plane (**6-4**). These orbitals therefore do not

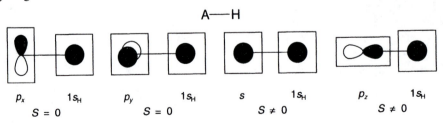

6-4

take part in any interaction. The overlap between the $1s_H$ orbital and the s and p_z orbitals on A is, however, non-zero (**6-4**). This leads to a three-orbital interaction problem between the orbitals of σ symmetry (symmetric with respect to rotation about the z-axis) as shown in **6-5**. The two interactions, 1 and 2 lead to three σ MOs,

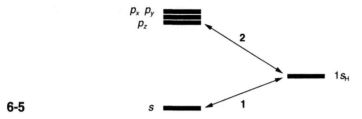

6-5

linear combinations of the orbitals s, p_z and $1s_H$. The relative energies of these starting orbitals depend upon the nature of the element A. If A is very electropositive (e.g. Li) the s and p valence orbitals lie high in energy and lie above that of $1s_H$ (**6-6**). If, on the other hand A is very electronegative (e.g. F) the s and p orbitals lie lower than $1s_H$ (**6-8**). Finally for elements of intermediate electronegativity (e.g. B, C and N) the $1s_H$ orbital lies between the s and p levels (**6-7**).

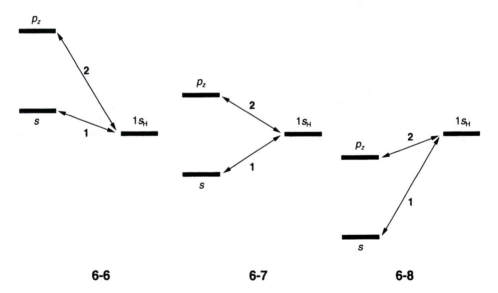

6-6   **6-7**   **6-8**

6.2.2. Form of the MOs

The 1σ MO is the linear combination of $1s_H$, s and p_z such that the overlap integrals between $1s_H$ and s and between $1s_H$ and p_z lead to bonding interactions (**6-9a**). We get the form of the MO by superposing the contributions from the three fragment orbitals. On atom A the addition of the two functions s and p leads to an increase in the amplitude on one side of the atom (where they have amplitudes of the same sign) and a decrease in amplitude on the other side (where they have amplitudes of opposite sign). The conventional representation of this s/p orbital mixture on atom

A depends upon the relative sizes of the s and p_z coefficients. If they are of the same order of magnitude, or the coefficient of p is larger, then this leads to a strongly asymmetric p-like function with one small and one large lobe (**6-9b**). Such admixture of s and p orbitals of this type gives rise to a *hybrid* orbital whose large lobe points toward the hydrogen atom. If the coefficient of the s orbital is much larger than that of p_z the shape of the s orbital is little affected and we can often neglect the p_z contribution in our pictorial description (**6-9c**). (In order to simplify the pictorial

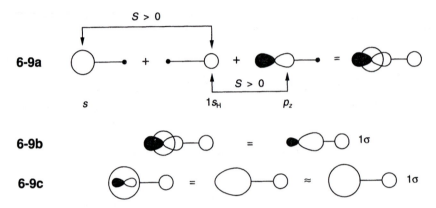

representation we have shown the nodal surface passing through the atom A. In fact it is located within the larger lobe.) It is important to note that the s and p orbital contributions to the hybrid will vary from system to system depending upon the weights of the s and p orbital coefficients.

In the 2σ orbital the fragment orbitals are combined in such a way that an antibonding interaction takes place between $1s_H$ and s and a bonding interaction between $1s_H$ and $2p_z$ (**6-10**). The s and p_z coefficients are, in general, of the same order

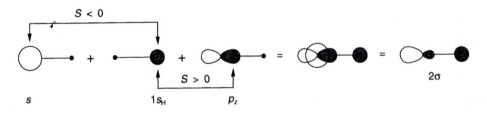

6-10

of magnitude so that the result is a hybrid orbital on A whose large lobe points away from the hydrogen atom. Effectively the sum of two overlap integrals of different signs between A and H orbitals leads overall to a nonbonding situation between the two atoms.

Finally in the highest energy orbital, 3σ, the overlap between the two fragments are both antibonding in character (**6-11**). Once again the orbital mixture on A leads to an s/p hybrid orbital with its large lobe in this case pointing towards the hydrogen, thus accentuating the antibonding character in this orbital. So the σ orbitals of the AH molecule comprise a bonding orbital, 1σ, an orbital, 2σ, which is essentially

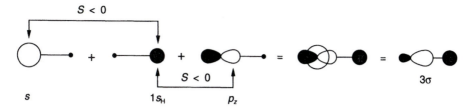

6-11

nonbonding, and an antibonding orbital, 3σ. We will illustrate the results of these orbital interactions by studying the MOs of three molecules (A = Li, B and F) which differ both in the number of electrons and in the relative electronegativity of A and H.

6.2.3. Electronic structure of LiH

Lithium is an electropositive element whose valence orbitals lie high in energy ($\varepsilon_{2s} = -5.4$ eV, $\varepsilon_{2p} = -3.5$ eV) and therefore above the hydrogen $1s$ orbital as shown in **6-6**. On the other hand the strength of the interactions, 1 and 2 between $1s_H$ and $2s$ and $2p_z$ are comparable. Recall that the interaction energy varies as $S^2/\Delta\varepsilon$. Although the energy difference favors interaction 1 ($\Delta\varepsilon = 8.2$ eV) over interaction 2 ($\Delta\varepsilon = 10.1$ eV) the overlap integrals favor interaction 2 ($S = 0.48$) over interaction 1 ($S = 0.36$) at an Li—H distance of 160 pm. Overall $S^2/\Delta\varepsilon$ is thus 0.016 for interaction 1 and 0.023 for interaction 2.

(a) Interaction diagram and form of the MOs

The interaction diagram for the valence orbitals of Li and the hydrogen $1s$ orbital is shown in Figure 6.1. Just as in all molecules of this type the heavy atom p_x and p_y orbitals do not interact with $1s_H$ and give rise to a degenerate pair of π orbitals in LiH. They are nonbonding orbitals localized on the lithium atom. As far as the σ orbitals are concerned, the deepest lying, 1σ, is largely derived from the deepest lying fragment orbital, namely $1s_H$. The small contributions from lithium $2s$ and $2p_z$ are practically equal as we have just shown. The 2σ and 3σ orbitals are largely derived from the lithium orbitals, i.e., the coefficients of $2s$ and $2p_z$ are most important.

Figure 6.1 shows the energetic ordering of the five molecular orbitals of LiH. 1σ lies deeper than $1s_H$ and is the most stable. 2σ, characterized by a bonding interaction between $1s_H$ and $2p_z$, and an antibonding interaction between $1s_H$ and $2s$ has an energy intermediate between the two AOs $2s$ and $2p_z$. It is therefore situated energetically below the nonbonding π type MOs. Finally the highest lying MO is 3σ with an energy higher than that of the lithium $2p$ orbitals.

(b) Electronic structure of LiH

The LiH molecule only has two valence electrons and so in its lowest electronic state the 1σ orbital only is doubly occupied ($1\sigma^2$, **6-12**). This orbital, bonding between lithium and hydrogen corresponds to the single bond in the Lewis structure, Li—H, of this molecule. The polarity of the Li—H bond follows immediately from the form

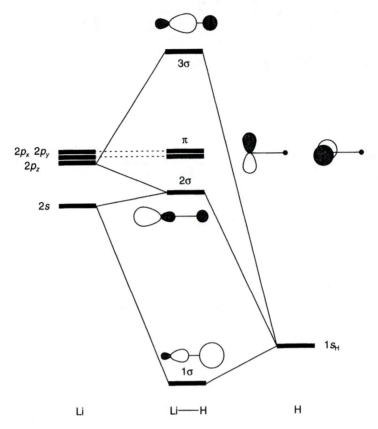

Figure 6.1. Molecular Orbital Diagram for Li—H. (For the neutral molecule, with two valence electrons, 1σ is doubly occupied. The diagram is also appropriate, of course for excited states and ions too.)

6-12

of the occupied MO. Since each isolated atom has a single valence electron, and the electron density in 1σ is not equally distributed between the two atoms, there has to be a charge redistribution on formation of the bond. Since the 1σ orbital is located more on hydrogen than on lithium (**6-12**) in the molecule there is an 'excess' of electrons on hydrogen and a 'deficiency' of electrons on lithium, leading to a polarization of the electrons in the molecule as $Li^{\delta+}H^{\delta-}$. This bond therefore has quite a bit of ionic character which leads to its large dipole moment (5.88 D).

6.2.4. Electronic structure of BH

The energies of the valence orbitals of boron are at $\varepsilon_{2s} = -14.7\,\text{eV}$ and $\varepsilon_{2p} = -5.7\,\text{eV}$. Thus the $1s_H$ orbital lies between the two as illustrated in **6-7**, but closer to $2s$ than to $2p$.

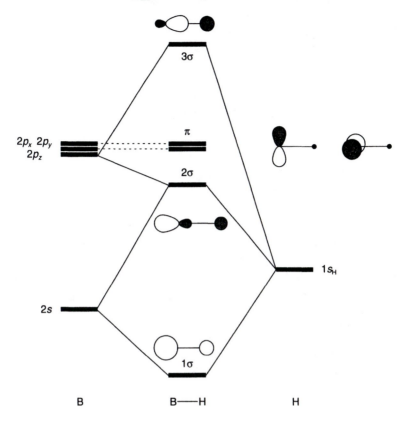

Figure 6.2. Molecular Orbital Diagram for B—H. (With a total of four valence electrons, the lowest two orbitals (1σ, 2σ) are occupied.)

(a) Interaction diagram and form of the MOs

The orbital interaction diagram for the hydrogen $1s$ orbital and the valence orbitals of boron is given in Figure 6.2. As before the $2p_x$ and $2p_y$ MOs constitute the degenerate π orbitals of the BH molecule. Concerning the σ orbitals, the deepest lying MO, 1σ, bonding between the two atoms is largely localized on the deepest lying fragment orbitals ($2s$ and $1s_H$). The $2p_z$ coefficient is small and we often ignore it in the representation of this orbital shown in Figure 6.2. Its contribution becomes smaller as the energy of the $2s$ orbital on the heavy atom becomes deeper in energy. In the 2σ and 3σ MOs the s and p orbital coefficients on boron are of the same order of magnitude such that 2σ is nonbonding and 3σ is strongly antibonding.

From the energetic point of view, the 1σ MO lies deeper than the deepest energy fragment orbital ($2s$). In 2σ there is a bonding interaction between $1s_H$ and $2p_z$, and an antibonding one between $1s_H$ and $2s$. The energy of this level depends on the relative strength of the two interactions. The $1s_H$ level is closer to $2s$ than to $2p_z$ (1.1 eV compared to 7.9 eV) and the overlap integrals (calculated for a B—H distance of 120 pm) are similar (0.48 and 0.53 respectively). This means that the antibonding interaction of $1s_H$ with $2s$ is stronger than the bonding interaction with $2p_z$ (remember the interaction energy is proportional to $S^2/\Delta\varepsilon$) and the resultant energy of the 2σ is above that of $1s_H$ and below that of π (Figure 6.2).

(b) Electronic structure of BH

The BH molecule possesses four valence electrons, so that in the electronic ground state the 1σ and 2σ orbitals are doubly occupied ($1\sigma^2 \, 2\sigma^2$, **6-13**). We can establish

6-13

an approximate correspondence between this description and that implied by the Lewis Structure |B—H which indicates a single bond between the two atoms and a lone pair on the boron. The pair of electrons in 1σ, strongly bonding between the two atoms, essentially lead to the single bond between boron and hydrogen. On the other hand the 2σ orbital is practically nonbonding between the two atoms since it has a large lobe on boron pointing away from hydrogen (**6-13**). The two electrons in it correspond therefore to the boron lone pair of the Lewis structure. The nonbonding character of this orbital is verified by studies on the ion BH^+ where one of these 2σ electrons is missing. Since the bond length change from BH (124 pm) to BH^+ (121 pm) is tiny, the orbital from which it came must be close to being exactly nonbonding.

6.2.5. Electronic structure of FH

The energies of the valence orbitals of fluorine are $\varepsilon_{2s} = -40.1 \text{ eV}$, $\varepsilon_{2p} = -18.6 \text{ eV}$. We illustrated this case in **6-8** where both of the orbitals of the very electronegative fluorine atom lie deeper than $1s_H$ (-13.6 eV). Because the $2s$ orbital lies a long way away from the $1s_H$ orbital ($\Delta\varepsilon = 26.5$ eV) their mutual interaction (1) is very weak and the dominant interaction that of 2 between $1s_H$ and $2p_z$ ($\Delta\varepsilon = 5.0$ eV). Thus a simplified version of the interaction diagram for this molecule is just a two, rather than three, orbital one. This approximation is sufficient to correctly place the energies of the three σ orbitals on the diagram, and thus to describe in broad terms the electronic structure of the molecule. However, in Section 6.2.5c we will need to include interaction 1 to get a more precise description of the molecule. This is due to the fact that, although the relative energy separations do favor interaction 2 between $1s_H$ and $2p_z$, the converse is true in terms of the overlap integrals. For interactions 1 and 2 these are $S = 0.46$ and 0.38 respectively at an F—H distance of 92 pm.

(a) Simplified interaction diagram and form of the MOs

A simplified interaction diagram for the hydrogen $1s$ and valence orbitals of fluorine is given in Figure 6.3. As in both of our previous examples the $2p_x$ and $2p_y$ orbitals are not perturbed in the process and form a pair of degenerate π nonbonding orbitals. On our simplified model where interaction 1 is negligible the $2s$ orbital also

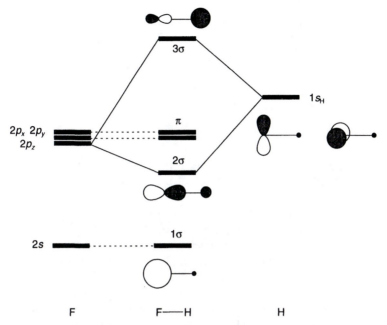

Figure 6.3. Simplified Molecular Orbital Diagram for F—H. (With a total of eight valence electrons, all but the highest orbital are doubly occupied.)

remains unchanged and becomes the 1σ MO of FH. In character it is clearly a nonbonding molecular orbital completely localized on fluorine. The remaining interaction, 2, between $1s_H$ and $2p_z$ leads to a bonding (2σ) and antibonding (3σ) pair. Given the relative energies of the $2p_z$ and $1s_H$ orbitals, the bonding 2σ orbital is more heavily developed on fluorine and the antibonding 3σ orbital more heavily developed on hydrogen. There are no problems in understanding the relative energetic ordering of the MOs. 1σ lies deepest by reason of its atomic $2s$ character, the bonding orbital 2σ lies deeper than the fluorine $2p_x$ and $2p_y$ orbitals followed at highest energy by the most antibonding orbital 3σ. Notice however, that this description is a little different from the general rules of Section 6.1.2a imply. Recall that there we stated that the three orbital problem reduced to bonding, nonbonding, and antibonding orbitals in order of increasing energy. As is readily appreciated the difference is understandable given the simplifications we have introduced into the diagram.

(b) Electronic structure of FH

The molecule FH, hydrogen fluoride, (usually written as HF) has eight valence electrons and in its electronic ground state the four deepest lying MOs are occupied ($1\sigma^2\ 2\sigma^2\ \pi^4$, **6-14**) and the highest energy orbital (3σ) is empty. Among the four occupied MOs only 2σ, bonding between H and F, can contribute to the chemical bond between the two atoms. The three other MOs are nonbonding orbitals. This makes a direct link with the single bond of the Lewis structure $|\overline{\text{F}}$—H, with its three fluorine lone pairs ($1\sigma^2$, π^4) and its single FH bond ($2\sigma^2$). Importantly, notice

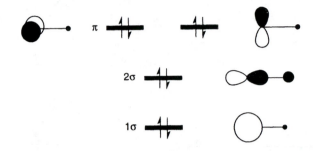

6-14

that the three lone pairs are not equivalent in the orbital approach, neither in form or energy. The situation is not much different if the interaction between $1s_H$ and $2s$ is included.

The polarity of the molecule may be decided by comparing the number of valence electrons in the isolated atoms (one for hydrogen, seven for fluorine) with their distribution in the molecule using the scheme of **6-14**. Six fluorine electrons occupy the nonbonding orbitals 1σ and π localized on fluorine. The last two electrons which occupy 2σ are not equally distributed between the two atoms, but are strongly localized on fluorine since the $2p_z$ coefficient is larger than that of $1s_H$. There are therefore 'more' than seven electrons on fluorine and 'less' than one electron on hydrogen. The polarity of the bond is thus $F^{\delta-}$—$H^{\delta+}$, a result which comes directly from the relative energetic location of the AOs which form the bond (2σ) between the two atoms. The partial ionic character in this bond contributes both to the stability of the molecule and of course is the origin of the large dipole moment (1.82 D). Finally we note that the dipole moment is of the opposite sign to that found in LiH ($Li^{\delta+}$—$H^{\delta-}$) simply because the energies of the AOs concerned are the other way around ($\varepsilon_{1s}(H) < \varepsilon_{2s}(Li)$ in Figure 6.1) compared to the FH case. In LiH the bonding orbital is largely hydrogen located, in FH it is largely fluorine located.

(c) Exact interaction diagram

Ignoring the interaction (1) between $1s_H$ and $2s$ certainly led to a simplification of the orbital picture. Now we see how the interaction diagram is perturbed by switching on interaction 1 (Figure 6.4). The 1σ orbital is pushed down in energy since it is bonding between all three fragment orbitals. It remains heavily localized on fluorine $2s$, the deepest lying fragment orbital and we neglect the very small $2p_z$ coefficient in our pictorial representation. The bonding interaction between $1s_H$ and $2p_z$ in the 2σ orbital was included in Figure 6.3, but now we include the antibonding interaction between $2s$ and $1s_H$ which will tend to destabilize this orbital. This is quite small and 2σ is only raised in energy by a small amount, still remaining deeper in energy than the π nonbonding orbitals. Some $2s$ character is mixed in here. Finally 3σ is pushed up in energy compared to Figure 6.3 as a result of this extra antibonding interaction. There are some interesting differences between the two models of Figures 6.3 and 6.4. The 1σ orbital has acquired some bonding character but at the same time 2σ has lost some. One cannot now strictly talk about a single σ bonding MO and a single σ nonbonding MO, thus complicating the ready correspondence with the Lewis picture. Overall there is a single bond but the bonding character is shared over two

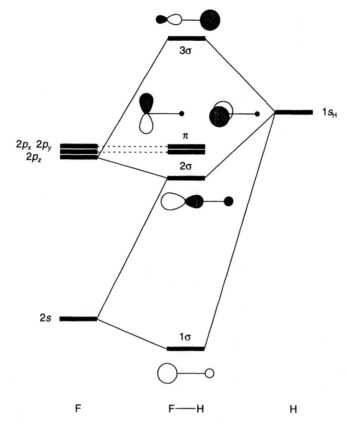

Figure 6.4. Orbital Interaction Diagram for F—H obtained by including all orbital interactions.

orbitals. Inclusion of the $2s$ orbital does not change the arguments concerning the polarity of the molecule. 1σ now contains a little bit of hydrogen $1s_H$ character, but 2σ has increased its fluorine contribution via admixture of a little $2s$ character.

6.2.6. Conclusions for AH molecules

(a) Principal characteristics of the MOs

The use of some simple rules allows ready construction of the MOs of AH molecules of different types and we can collect a certain number of common features from the examples discussed above.

    (i) The energetic ordering is always (from the deepest) 1σ, 2σ, π, 3σ.

    (ii) The two degenerate π orbitals are nonbonding, being the atomic orbitals p_x, p_y, completely localized on the heavy A atom.

    (iii) If all interactions are included, the 1σ and 3σ orbitals are strongly bonding and antibonding combinations of an A-centered s/p hybrid orbital with the

hydrogen $1s$ orbital. In these orbitals the large lobe of the hybrid points towards the hydrogen atom.
(iv) In the middle σ orbital (2σ) the large lobe of the s/p hybrid points away from the hydrogen atom.

Some other characteristics depend on the electronic properties of the element A.

(i) The relative values of the $1s_H$, s and p_z coefficients in the three σ MOs depend upon the position of the $1s_H$ orbital relative to those of the s and p orbitals of atom A. For example, the $1s_H$ coefficient in 1σ is small in FH but large in LiH.
(ii) The degree of s/p mixing on A depends also on the electronegativity of this atom. It is largest, i.e., the s and p coefficients are closest in magnitude, when the s and p levels are closest in energy. This occurs when A is an electropositive atom.
(iii) For a given element A we can show by calculation that usually the most important s/p mixing occurs in the 2σ and 3σ orbitals. In 1σ the coefficient of p_z is always much less than that of s (with the exception of A = Li where the s and p orbital energies are close.)

(b) Polarization of A by hydrogen

An important characteristic of the MOs formed by AH molecules is the simultaneous participation of the s and p_z orbitals on A leading to the formation of s/p hybrid orbitals. At first sight this is surprising since the two orbitals, centered on the same atom, are orthogonal. Such hybrids are only possible if there exists on another atom (H in this case) a third orbital ($1s_H$) which has a non-zero overlap integral with both s and p_z. We say that the interactions 1 ($1s_H/s$) and 2 ($1s_H/p_z$) lead to a polarization (s/p_z mixture) of the A atom orbitals (**6-15**). This polarization is large if the s and p_z orbitals are close in energy (A = Li for example) but weak if they are of very different energies (A = F).

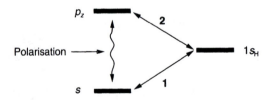

6-15

6.3. Bent AH₂ molecules

AH₂ molecules (**6-16**) can be decomposed into two fragments, one consisting of the two hydrogen atoms, $H_a \ldots H_b$, and the other the central A atom. The $H_a \ldots H_b$ orbitals are just the in-phase (σ_{H_2}) and out-of-phase ($\sigma_{H_2}^*$) combinations of $1s_a$ and $1s_b$. Since the hydrogen atoms are usually well-separated (160 pm in H_2O) the overlap

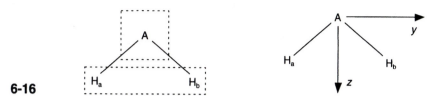

6-16

between the $1s_H$ orbitals is weak and the energetic separation between σ_{H_2} (bonding) and $\sigma_{H_2}^*$ (antibonding) rather small. The orbitals of the second fragment are just the valence orbitals of atom A.

6.3.1. Symmetry of the fragment orbitals

Bent AH_2 molecules have three symmetry elements which will also serve as symmetry elements of the fragments defined above (see Section 3.4.4b). If z is defined as the axis which bisects the HAH angle and yz as the molecular plane (**6-16**) there are three elements of symmetry which leave the nuclei unchanged; (i) a two-fold rotation axis around z (C_2^z), (ii) a mirror plane xz and (iii) a mirror plane yz (the molecular plane).

We may simply analyze the symmetry properties of the orbitals with respect to all of these elements (**6-17**). The orbitals σ_{H_2}, s and p_z are symmetric with respect to all

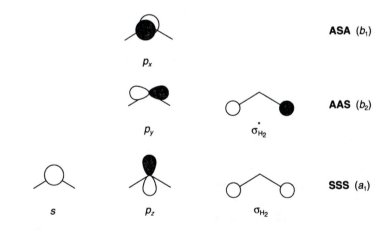

6-17

three elements (**SSS**). $\sigma_{H_2}^*$ and p_y are antisymmetric with respect to the two-fold rotation and the xz plane, but symmetric with respect to the molecular plane yz (**AAS**). Lastly the p_x orbital is antisymmetric with respect to the two-fold rotation, symmetric with respect to the xz plane and antisymmetric with respect to the plane yz (**ASA**). The major difference between the orbital pictures of linear (Section 5.1.2) and bent AH_2, is that in the bent form p_z has the same symmetry as s and σ_{H_2}. In effect, since the molecule is bent, the hydrogen atoms leave the nodal plane of p_z and the overlap with σ_{H_2} increases from zero (**6-18**).

The group theoretical labels for these three symmetry types, **SSS**, **AAS** and **ASA** are a_1, b_2 and b_1 respectively (**6-17**). As before, a knowledge of the orbital symmetries allows us to readily identify the fragment orbital interactions necessary to assemble

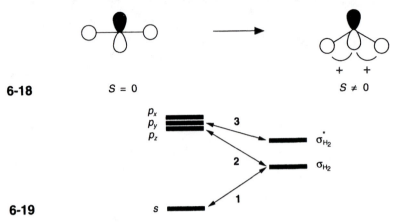

6-18

$S = 0$ $S \neq 0$

6-19

the molecular orbital diagram. They are (**6-19**):

(i) The p_x orbital (b_1) does not find a symmetry match with a fragment orbital of $H_a \ldots H_b$, and remains strictly nonbonding.

(ii) Interaction 3 between the two orbitals of b_2 symmetry, $\sigma^*_{H_2}$ and p_y.

(iii) Interactions 1 and 2 between the orbitals of a_1 symmetry, namely σ_{H_2} and the s and p_z orbitals on atom A. This leads to a three-orbital problem analogous to the one we met in the AH case.

We must note that these conclusions are based only on the symmetry properties of the fragment orbitals, and are independent of the energetic location of the orbitals.

6.3.2. Interaction diagram and form of the MOs: H₂O as an example

The relative energies of the fragment orbitals depend of course on the identity of the central atom orbital and its electronegativity. Using water (H_2O) as an example their relative energies are given in Figure 6.5. For oxygen $\varepsilon_{2s} = -32.4$ eV, $\varepsilon_{2p} = -15.9$ eV, and since the hydrogen atoms are quite far apart, the energy difference between σ_{H_2} and $\sigma^*_{H_2}$ is small. The antibonding combination, $\sigma^*_{H_2}$, thus lies a little higher than the energy of a free $1s_H$ orbital (-13.6 eV) and σ_{H_2}, the bonding combination, a little lower and in the neighborhood of the oxygen $2p$ orbital.

The principal characteristics of the interaction diagram (Figure 6.5) obtained from the interaction scheme of **6-19** are the following:

(i) The $2p_x$ orbital, which is not implicated in any orbital interaction remains a nonbonding orbital ($1b_1$) localized completely on oxygen.

(ii) The orbitals $\sigma^*_{H_2}$ and $2p_y$ combine to give a bonding MO ($1b_2$) and an antibonding MO ($2b_2$), in phase and out of phase respectively (**6-20**).

(iii) The three orbitals, $2s$, $2p_z$ and σ_{H_2} combine to give the MOs $1a_1$, $2a_1$ and $3a_1$. Examination of the relative energies of the three fragment orbitals suggests that the interaction between the σ_{H_2} and $2p_z$ orbitals is considerably stronger than that between the σ_{H_2} and $2s$. However the overlap integrals suggest the opposite. That between σ_{H_2} and $2s$ is about twice as large as that between σ_{H_2} and $2p_z$ (0.56 compared to 0.30 for an HAH angle of 104.5°). So we must take into account both interactions to derive the 'exact' picture.

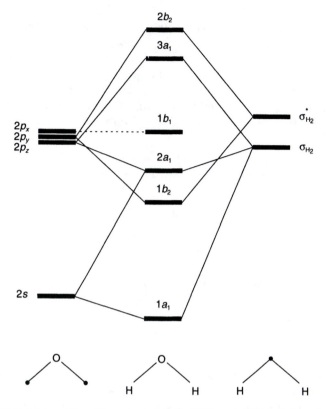

Figure 6.5. Orbital Interaction Diagram for H_2O. (With a total of eight valence electrons the four lowest energy orbitals only are occupied.)

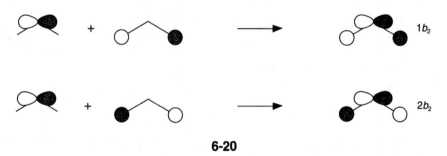

The $1a_1$ orbital is a linear combination of all of the fragment orbitals such that σ_{H_2} is bonding with both $2s$ and $2p_z$ (**6-21**). A calculation shows that the $2p_z$ coefficient is very much smaller than that of $2s$, and we ignore it in the pictorial representation of the MO. The energy of this $1a_1$ orbital has to lie deeper than the most stable fragment orbital ($2s$). The antibonding orbital, $3a_1$, is an antibonding combination of σ_{H_2} with $2s$ and $2p_z$ (6-22). The s/p mixing on oxygen ensures that the large lobe points towards the hydrogen atoms thus accentuating this antibonding character between O and H. The $3a_1$ orbital lies higher than the highest energy fragment orbital ($2p_z$). Finally, the intermediate energy orbital, $2a_1$, is characterized by a bonding interaction between σ_{H_2} and $2p_z$ and by an antibonding interaction between σ_{H_2} and $2s$ (6-23). The orbital is more strongly localized on oxygen than hydrogen and the

6-22 $S < 0$ $S < 0$

6-23 $S < 0$ $S > 0$

orbital mixing is such that the large lobe points away from the hydrogen atoms. From the energetic point of view it is more stable than $2p_z$. This orbital is well described as a nonbonding orbital. Thus, as before, we have generated a bonding orbital ($1a_1$), a nonbonding orbital ($2a_1$) and an antibonding orbital ($3a_1$) from the set of three starting orbitals.

The energetic ordering is given in Figure 6.5. In order of increasing energy there are at lowest energy two orbitals ($1a_1$ and $1b_2$), strongly bonding between O and H. Then follow $2a_1$, essentially nonbonding and $1b_1$, strictly nonbonding. Finally at highest energy are the two orbitals ($2b_2$ and $3a_1$) strongly antibonding between O and H. This general scheme, two bonding, two nonbonding and two antibonding orbitals is a general scheme for bent AH_2 molecules.

6.3.3. Electronic structure of H_2O

Water with a total of eight valence electrons has the electron configuration $1a_1^2\ 1b_2^2\ 2a_1^2\ 1b_1^2$ (6-24) in its electronic ground state. Of these four occupied MOs, two are bonding and two are nonbonding. The two antibonding MOs are empty. Occupation of two orbitals, bonding between O and H leads immediately to two O—H bonds. Just as in the case of the BeH_2 molecule of Chapter 5 it is not possible to link occupation of one particular MO with a given O—H bond since the orbitals are delocalized over the whole molecule. Analogously it is *the two occupied orbitals taken together which represent the two O—H linkages*. In each orbital the overlap between the central atom orbital ($2s$ in $1a_1$ and $2p_z$ in $1b_2$) and each hydrogen atom $1s_H$ orbital is identical, so that the two O—H bonds are completely equivalent. However,

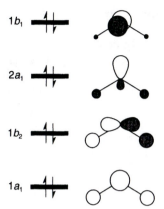

6-24

once again, just as in BeH_2 the two bonding MOs differ both in form and in energy. The same is true of the two nonbonding orbitals. The $2a_1$ orbital is largely directed away from the hydrogen atoms so that it well approximates an oxygen-localized lone pair orbital. Of course the $1b_1$ orbital is strictly a lone pair orbital on oxygen. Also it is possible to establish a rough correspondence between the electronic description of the molecule via molecular orbital theory and the simple ideas of Lewis which show two bond pairs and two lone pairs. A similar comment though applies to the lone pair orbitals as to the bonding orbitals. They are of distinctly different form and therefore energy. This result is confirmed experimentally. Photoelectron spectroscopy shows two lone pairs, associated with ionization from $1b_1$ and $2a_1$, and separated in energy by 2.1 eV. Thus molecular orbital theory gives a simple interpretation of this experimental result mentioned initially in Section 1.5 as one of the problems with the classical description of molecules.

6.4. Pyramidal AH₃ molecules

Pyramidal AH_3 molecules may be decomposed into two fragments (**6-25**), an equilateral triangle of hydrogen atoms and the atom A. We have already described the orbitals of the triangular H_3 unit in Chapter 4 but as a reminder they are shown at the right-hand side of **6-30**. They comprise a bonding orbital ϕ_1 and two degenerate

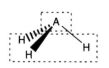
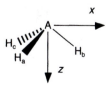

6-25

antibonding orbitals ϕ_2 and ϕ_3. The orbitals of the second fragment are just the valence orbitals (s and p) of the atom A.

6.4.1. Symmetry of the fragment orbitals

Distortion of the planar AH_3 molecule to a pyramidal geometry involves the loss of some symmetry elements, amongst them being the xy plane, the molecular plane of the planar AH_3 molecule. The main consequence of this result is that the overlap between p_z and ϕ_1, which was exactly zero in the planar structure, is now non-zero **(6-26)**. Since the s orbital is spherically symmetrical the overlaps between s and $1s_H$

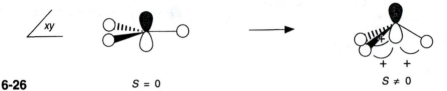

6-26 $S = 0$ $S \neq 0$

are independent of the angular geometry and are always different from zero **(6-27)**. As a result of the geometry change the three orbitals, s, p_z and ϕ_1, now all have the same symmetry properties. They are symmetric with respect to the mirror planes which contain an AH linkage and bisect an opposite HAH angle **(6-25)**. The xz plane is an example. In addition they remain unchanged as a result of a rotation of $\pm 2\pi/3$ around the three-fold rotation axis C_3^z. We label such orbitals as a_1 **(6-30)**. The behavior of the pairs of orbitals (p_x, p_y) on A and (ϕ_2, ϕ_3) on H_3 is similar to that described in the planar case. They are degenerate simply because of the presence of the three-fold axis. In this pyramidal molecule they carry the label e and it is simple to show that the pair of overlap integrals associated with e_y (p_y and ϕ_2 in **6-28**) and e_x (p_x and ϕ_3 in **6-29**) are equal and quite different from zero. Additionally the overlaps between e_x and e_y are identically zero since the e_x orbitals are symmetric with respect to the xz plane and the e_y orbitals antisymmetric.

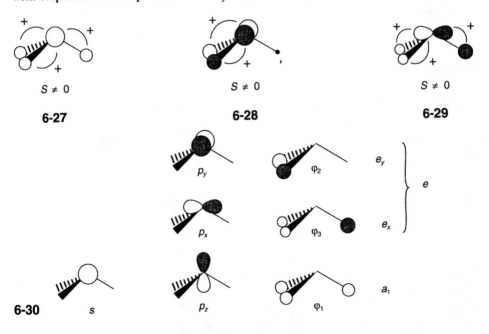

$S \neq 0$ $S \neq 0$ $S \neq 0$

6-27   **6-28**   **6-29**

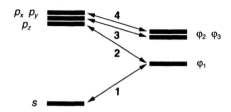

6-31

The interactions to consider for the construction of the MOs of pyramidal AH_3 are thus those given in **6-31** and comprise (a) two interactions between two orbitals; p_y, ϕ_2 (interaction 3) and p_x, ϕ_3 (interaction 4) and (b) the three orbital scheme involving the interactions 1 and 2 between ϕ_1, s and p_z. These conclusions once again are based on orbital symmetry ideas alone and since they are independent of the relative positions of the interacting orbitals, apply to all pyramidal AH_3 molecules.

6.4.2. Interaction diagram and form of the MOs: the example of NH_3

In the case of NH_3 (Figure 6.6) the orbital energies of the central nitrogen atom are $\varepsilon_{2s} = -25.6$ eV and $\varepsilon_{2p} = -12.9$ eV. For the H_3 fragment the bonding orbital (ϕ_1) lies a little deeper than the energy of an isolated $1s_H$ orbital (-13.6 eV) and the energies of ϕ_2 and ϕ_3 a little higher. This comes about since the H ... H distance in NH_3 (160 pm) is quite long. The fragment orbitals ϕ_2 and ϕ_3 lie just above the $2p$ orbitals of nitrogen. The principal characteristics of the interaction diagram are:

(i) Two pairs of degenerate orbitals, e_x and e_y which interact to give two pairs of MOs, one bonding (1e in **6-32**) and one antibonding (2e in **6-33**).

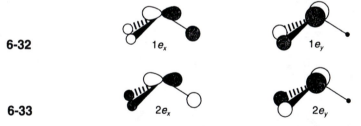

6-32 $1e_x$ $1e_y$

6-33 $2e_x$ $2e_y$

(ii) Three orbitals of a_1 symmetry derived from ϕ_1, $2s$ and $2p_z$, which interact to give the three MOs $1a_1$, $2a_1$ and $3a_1$. Although the interaction 2 (**6-31**) is favored over 1 on energy difference grounds, as in H_2O, interaction 1 is favored on overlap grounds over interaction 2. (The overlap integral is twice as large.) Both interactions have to be taken into account for an accurate orbital picture of the molecule. The $1a_1$ orbital is a linear combination of the fragment orbitals such that ϕ_1 is mixed in a bonding fashion (in phase) with $2s$ and $2p_z$ (**6-34**). The coefficient of $2p_z$, being small, is neglected in our orbital picture. This $1a_1$ orbital lies deeper than the deepest lying fragment orbital ($2s$). Analogously the $3a_1$ orbital is the out of phase, antibonding combination of ϕ_1 with $2s$ and $2p_z$ (**6-35**). The s/p mixing on nitrogen is such that the large lobe points towards the hydrogen atoms, thus maximizing the antibonding character between N and H. The energy of $3a_1$ lies higher than the highest fragment orbital energy ($2p_z$). Finally the orbital of intermediate energy ($2a_1$) is

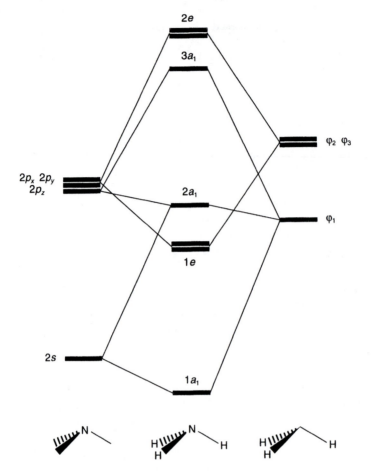

Figure 6.6. Orbital Interaction Diagram for pyramidal NH_3. (With a total of eight valence electrons the four lowest energy orbitals only are occupied.)

6-34

characterized by a bonding interaction between ϕ_1 and $2p_z$ and by an antibonding interaction between ϕ_1 and $2s$ (**6-36**). This hybrid orbital is largely localized on nitrogen and the large lobe is directly away from the hydrogen atoms. So this is an orbital essentially nonbonding between N and H, whose energy lies deeper than that of $2p_z$, just as in the water molecule. As before, for the other three orbital problems, we have constructed a bonding, nonbonding and antibonding trio of orbitals.

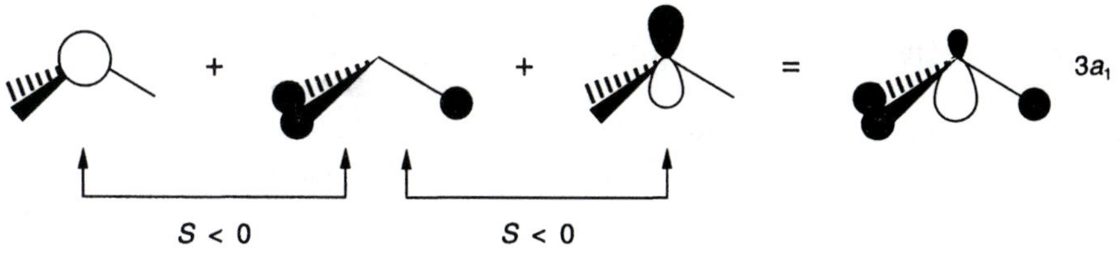

6-35

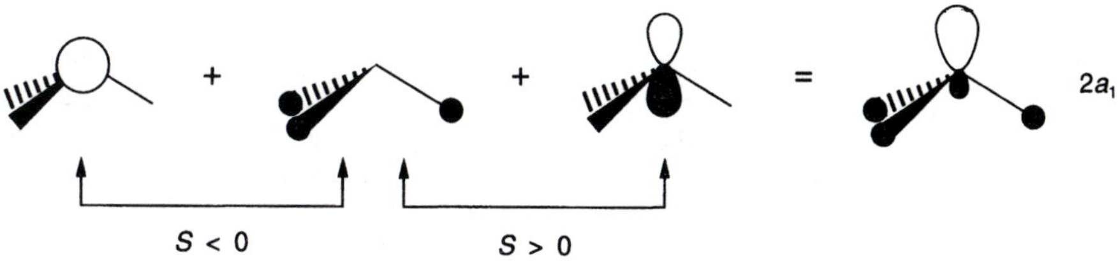

6-36

The energetic ordering of all of the orbitals is shown in Figure 6.6. In order of increasing energy they are the three orbitals which are bonding between N and H, namely $1a_1$ and $1e$ ($1e_x$ and $1e_y$), followed by a single nonbonding orbital, $2a_1$, and the three antibonding orbitals $3a_1$ and $2e$ ($2e_x$ and $2e_y$). This general scheme applies to all pyramidal AH_3 molecules. There could in fact be some argument over the energetic ordering of $3a_1$ and $2e$ (the same comment applies in fact to $3a_1$ and $2b_2$ for AH_2), but which one does lie higher will turn out not to be particularly important here.

6.4.3. Electronic structure of NH_3

The ammonia molecule has a total of eight valence electrons which fill the lowest four molecular orbitals ($1a_1^2\ 1e^4\ 2a_1^2$ of **6-37**) in the configuration appropriate for the electronic ground state. The $2e$ and $3a_1$ orbitals are empty. The three lowest energy occupied MOs are bonding between N and H and lead to the description of the three N—H bonds of the Lewis structure. Once again it is not possible to establish a

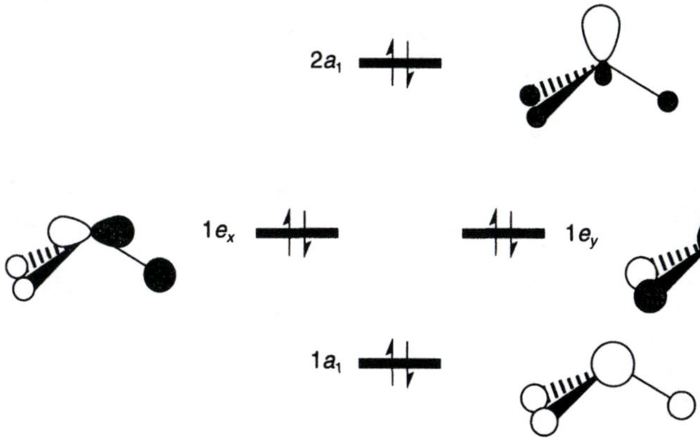

6-37

one-to-one correspondence between a given N—H bond and occupation of a given MO, since the latter are delocalized over all three N—H bonds. It is the collection of doubly occupied bonding orbitals as a group which characterizes the three N—H bonds. Since in the $1a_1$ orbital the overlap between $2s$ and each of the hydrogen $1s$ orbitals is the same, and since, just as in the case of trigonal planar AH_3 molecules of Section 5.2.4, the pair of $1e$ orbitals leads to identical bonding character between nitrogen and each hydrogen orbital, the three N—H bonds are equivalent. In an exactly analogous fashion $1e_y$ is bonding between N and (H_a, H_c) but $1e_x$ is largely bonding between N and H_b. Of the three occupied bonding orbitals, two have a different energy from the third. This shows up in the photoelectron spectrum; ionization from $1e$ occurs at a different energy to ionization from $1a_1$. Finally the $2a_1$ orbital is essentially a lone pair orbital localized on nitrogen and pointing away from the hydrogen atoms. This is thus the Lewis lone pair and, being high in energy, is responsible for the well-known Lewis base properties of ammonia.

EXERCISES

6.1 *The CH radical*
   (i) Why is this species called a radical?
   (ii) Give its Lewis Structure.
   (iii) The AO coefficients in the MOs of CH are given in the Table below, the MO's being given in order of increasing energy.

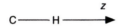

MO	$1s_H$	$2s$	$2p_x$	$2p_y$	$2p_z$
1σ	0.29	0.72	0	0	0.28
2σ	-0.29	0.62	0	0	-0.71
π	0	0	1	0	0
π	0	0	0	1	0
3σ	1.56	-0.96	0	0	-1.08

   (a) Draw out the MOs. Verify that they have the same form as a related species discussed in this chapter.
   (b) Give the electronic configuration of the ground state.
   (c) Establish the correspondence with the Lewis structure.
   (d) Why do we describe CH as a π-radical?
   (iv) On ionization the bond length of the molecule hardly changes (CH; 112 pm, CH^+; 113 pm). Explain this result.

6.2 *BeH radical and its cation*
   (i) Give the Lewis structures for each of these species.
   (ii) Give the electronic configurations appropriate for their ground states. Why is BeH described as a σ-radical?
   (iii) Ionization of BeH leads to a very small change in the bond length (BeH; 134 pm, BeH^+; 131 pm). Explain this result.
   (iv) In contrast to (iii) the excitation $1\sigma \rightarrow 2\sigma$ for the cation leads to a substantial increase in bond length (131 pm to 161 pm). Suggest an explanation.

6.3   *The hydroxyl ion, OH⁻*
   (i) Give its Lewis structure.
   (ii) Give the ground state electronic configuration. Establish a correspondence
   with the Lewis structure.

6.4   *The molecular orbitals of water*
   The coefficients of the AOs in the MOs of water are given in the table below
   and the axis system and atomic positions are given in the figure.

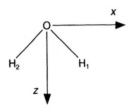

MO	$2s$	$2p_x$	$2p_y$	$2p_z$	s_{H_1}	s_{H_2}
$1a_1$	0.84	0	0	0.14	0.14	0.14
$1b_2$	0	−0.66	0	0	−0.38	0.38
$2a_1$	−0.45	0	0	0.82	0.22	0.22
$1b_1$	0	0	1.0	0	0	0
$3a_1$	−0.88	0	0	−0.72	0.84	0.84
$2b_2$	0	−1.0	0	0	0.89	−0.89

Draw out the MOs, and compare them with the qualitative description of
Section 6.3.2. Verify in particular the form of the a_1 MO given in **6-21, 6-22** and
6-23.

6.5   *The molecular orbitals of ammonia*
   Answer the same questions as those of exercise 6.4 for the MOs of NH_3, whose
   orbital coefficients are given below.

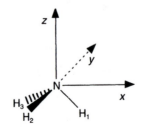

MO	$2s$	$2p_x$	$2p_y$	$2p_z$	S_{H_1}	S_{H_2}	S_{H_3}
$1a_1$	0.73	0	0	-0.15	0.15	0.15	0.15
$1e_x$	0	0.64	0	0	0.44	-0.22	-0.22
$1e_y$	0	0	0.64	0	0	-0.37	0.37
$2a_1$	-0.37	0	0	-0.92	0.08	0.08	0.08
$3a_1$	1.36	0	0	-0.55	-0.77	-0.77	-0.77
$2e_x$	0	1.11	0	0	-1.11	0.55	0.55
$2e_y$	0	0	1.11	0	0	0.95	-0.95

6.6 *Redetermination of the molecular orbitals of linear H_3*
The MOs of linear H_3 were determined in Chapter 4 by using a fragmentation which paired the two terminal atoms as one fragment and used the central atom as the other. Use the following pair of fragments to construct the MOs of the molecule.

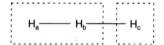

   (i) What are the fragment orbitals?
  (ii) What interactions can occur between the fragment orbitals? Does this fragmentation provide a better way to assemble the molecular orbital diagram of H_3?
 (iii) In spite of everything try to generate the MOs of linear H_3.

6.7 *The MOs of cyclic H_5*
Construct the MOs of cyclic H_5 using the division shown into the two fragments H_4 (H_b—H_c—H_d—H_e) and H_a. Since the atoms H_b and H_e are far apart and the overlaps between two s orbitals dependent only upon internuclear separation and independent of their relative orientation, the orbitals of the H_4 fragment will be very much like those of linear H_4 of Chapter 4.

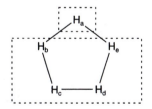

   (i) Draw out the fragment orbitals on an energy diagram and identify their symmetry with respect to a symmetry element of your choice.
  (ii) Which interactions are possible by symmetry?
 (iii) Construct the interaction diagram. (Hint: there are two pairs of degenerate orbitals).
 (iv) Give the form of the MOs.

7 Interactions between four fragment orbitals: the diatomic molecules A₂ and AB

The set of homonuclear diatomic molecules are among the simplest species of chemical interest, yet they provide a valuable testbed of electronic structure models. As we noted at the very beginning of this book in Chapter 1 the Lewis approach runs into several unsurmountable difficulties in this area. One of the most worrisome examples is that of oxygen, O_2, whose Lewis structure (**7-1**) indicates the presence of a double bond between the atoms and the existence of four lone pairs, two per atom. This structure is the most favorable one energetically that one can draw, since the octet rule is obeyed and neither atom has a formal charge. Importantly the Lewis structure indicates that in the electronic ground state, all of the electrons are paired up. However, such a description is not in accord with the paramagnetism of the molecule, the measured magnetic moment unequivocally shows the presence of two unpaired electrons with their spins parallel. One could describe this state of affairs by using the Lewis structure of **7-2** where two electrons are unpaired.

$$\overline{O}\!\!=\!\!=\!\!\overline{O}$$

7-1

$$|\dot{O}\!-\!-\!\dot{O}|$$

7-2

Here, however, the octet rule is not obeyed and there is now only a single bond between the oxygen atoms. This is unacceptable too. The experimentally determined value of the O—O distance in the O_2 molecule is 121 pm, much shorter than that of a typical O—O single bond such as the 148 pm in H_2O_2. It thus proves impossible to describe the properties of this molecule using the Lewis approach. One of the first successes of molecular orbital theory was to give a correct electronic description of this molecule, one which allowed a double bond for a paramagnetic molecule.

7.1. Homonuclear diatomics, A_2

7.1.1. Outline of the problem

A is an element from the second or third row of the periodic table and so the fragment orbitals are simply the s and p valence orbitals of the two atoms A(1) and A(2) arranged as in **7-3** with z chosen as the internuclear axis.

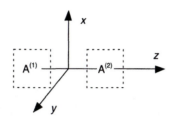

7-3

Let us consider first the $p_x(1)$ orbital on the first atom. It is clear that the overlap integrals of $p_x(1)$ with the orbitals $s(2)$, $p_y(2)$ and $p_z(2)$ on the second atom are identically zero by symmetry (**7-4**). The $p_x(1)$ orbital is antisymmetric with respect to the yz plane but the other three orbitals are symmetric. The only orbital with a non-zero overlap with $p_x(1)$ is its parallel partner on A(2), namely $p_x(2)$ (**7-5**). (One could envisage the contributions to the overlap from the two halves of the p orbitals adding together to give a non-zero overlap overall.) This type of overlap is called lateral (or sideways) overlap. The same arguments apply to the pair of p_y orbitals (**7-6**) which also overlap in a lateral fashion, and in an identical way to the p_x pair.

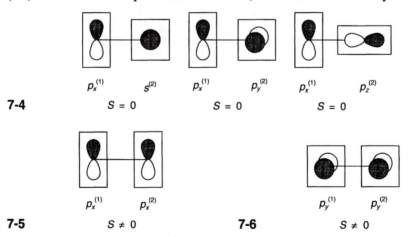

7-4

$p_x^{(1)}$ $s^{(2)}$ $p_x^{(1)}$ $p_y^{(2)}$ $p_x^{(1)}$ $p_z^{(2)}$

$S = 0$ $S = 0$ $S = 0$

7-5 $p_x^{(1)}$ $p_x^{(2)}$ $S \neq 0$ **7-6** $p_y^{(1)}$ $p_y^{(2)}$ $S \neq 0$

This pair of orbitals, p_x, p_y on each atom, have π symmetry. It remains to consider the interactions between the s and p_z orbitals carried by each atom. These orbitals all have σ symmetry (they are symmetric with respect to rotation about the internuclear axis) and so the overlap integrals between these orbitals located on atom 1 and those located on atom 2 will be non-zero (**7-7**). As a result the construction of the molecular orbital diagram for the A_2 diatomic involves the valence orbitals of the fragments in two ways (**7-8**).

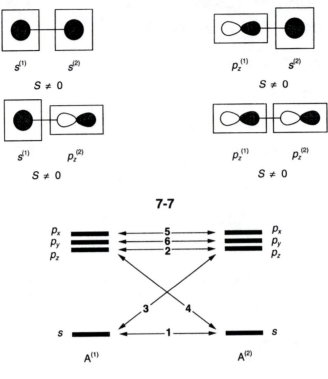

7-7

7-8

(i) Two interactions between two orbitals. The π interactions, 5, between $p_x(1)$ and $p_x(2)$ and 6, between $p_y(1)$ and $p_y(2)$.

(ii) An interaction scheme for the σ orbitals involving interactions 1–4, between $s(1)$, $s(2)$, $p_z(1)$ and $p_z(2)$. In this scheme we can distinguish between the interactions 1 and 2 which involve degenerate interactions (s–s and p–p) and those (3, 4) which do not (s–p).

7.1.2. Construction of the π-type MOs

The interactions 5, 6 between the π orbitals p_x and p_y do not present any particular problems. In each case (**7-9** and **7-10**) in phase and out of phase combinations

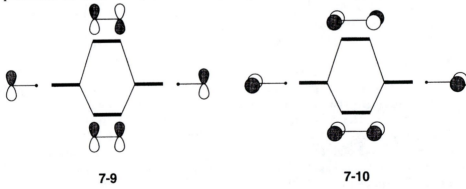

7-9   **7-10**

of the starting orbitals lead to bonding and antibonding orbitals respectively. In each case the coefficients on both atoms are equal in absolute magnitude since the interacting orbitals are degenerate. The bonding orbital is labeled π_u and the antibonding orbital π_g, the subscripts g, u describing the symmetric or antisymmetric behavior with respect to inversion. The inversion center is located midway between the two atoms. Notice that the pair of orbitals π_u (π_u^x, π_u^y) are degenerate as is to be expected if the starting orbitals have the same energy and same overlap integrals (**7-5**, **7-6**). The same is true of π_g (π_g^x, π_g^y).

7.1.3. Construction of the σ-type MOs

The difficulty of constructing the σ MOs of the A_2 molecule is associated with the fact that there are four of them, s and p_z on each atom. We have already noted that two of the interactions are between degenerate orbitals (1, s–s and 2, p–p) which suggests an obvious simplification.

(a) Simplified approach

In the simplest approach we ignore the cross-interactions 3 and 4 and only include those between degenerate orbitals (**7-11**). This approximation can be justified by

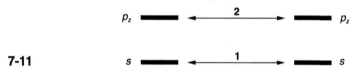

7-11

noting that other things being equal, interaction between a degenerate orbital pair where $\Delta\varepsilon = 0$, is stronger than between non-degenerate orbitals. So we retain on this model just the s–s and p–p interactions. It is clear that such an approach will be most valid when the s–p interactions are weak and we will return later to this point in Section 7.1.4a to study the conditions under which this is appropriate.

Within the framework of this model the construction of the σ MOs reduces to two interactions between two pairs of orbitals leading to bonding and antibonding orbital pairs as in **7-12**. In each MO the weights of the AOs (s or p) are equal since these are degenerate orbital interactions. The bonding orbitals are labeled σ_g and the antibonding ones σ_u according to their symmetry with respect to inversion. In this way the s orbitals lead to $1\sigma_g$ and $1\sigma_u$ and the p_z orbitals to $2\sigma_g$ and $2\sigma_u$. Note that the bonding orbital $2\sigma_g$ corresponds to the mixing of the two p_z orbitals of opposite sign, namely $p_z(1) - p_z(2)$ correct to within a normalization constant. Although the p orbitals are mixed with opposite signs it is clear from the picture in **7-12** that this is the arrangement when the lobes of the orbitals mix in phase, i.e., with a positive overlap between them. Similarly the p_z orbitals are mixed out of phase in $2\sigma_u$ which implies here that the p_z orbitals are added together, $p_z(1) + p_z(2)$.

In terms of the relative energies of the MOs we point out the difficulty of knowing whether $1\sigma_u$ lies below or is above $2\sigma_g$. Obviously $1\sigma_u$ lies above the energy of $2s$ and $2\sigma_g$ below the energy of p. We just have to point out that for all the systems we shall study the order of these two orbitals is as given in **7-12**.

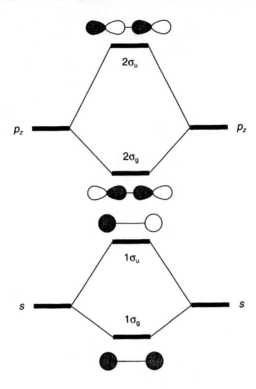

7-12

(b) Exact approach

Taking into account the interactions between the s orbital on one atom and the p_z orbital on the other will perturb the form and energy of the σ orbitals obtained using the simplified model above. The π levels being of different symmetry remain unchanged. The four-orbital problem can in fact be tackled rather simply in two steps.

   (i) First of all only include *the degenerate interactions between the fragment orbitals*. We did this in the simplified model above by including only s–s and p–p interactions. We will call these the principal interactions and label the resulting orbitals *provisional orbitals* $n\sigma^\circ$ ($n = 1$–4).

   (ii) To obtain the *actual MOs*, which we will label as $n\sigma$, we interact the σ° MOs constructed from the s orbitals ($1\sigma_g^\circ$ and $1\sigma_u^\circ$) with the σ° MOs constructed from the p_z orbitals ($2\sigma_g^\circ$ and $2\sigma_u^\circ$). This takes into account the secondary interactions between $s(1)$ and $p_z(2)$ etc.

At first sight the problem does not seem to have simplified very much since we still have four orbitals to consider in (ii). However the provisional MOs are either symmetric or antisymmetric with respect to inversion and so in this second step we only need to include interactions between $1\sigma_g^\circ$ and $2\sigma_g^\circ$ and between $1\sigma_u^\circ$ and $2\sigma_u^\circ$. The problem is thus reduced to the classic one of interactions between two orbitals. It is important to note however that in **7-13** the interaction scheme is a little different from those described earlier in the sense that the interacting orbitals are not located on different fragments. **7-13** represents the effect of 'switching on' the s–p interaction.

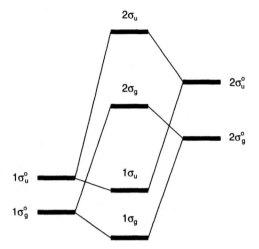

7-13

However the same rules apply as before. Each interaction leads to the formation of a new MO lying deeper in energy than the lower of the two starting orbitals (bonding combination) and a new MO lying higher in energy than the higher of the two starting orbitals (antibonding combination).

To determine the form of the new orbitals (σ_g or σ_u) it is necessary to identify the way the σ_g^o or σ_u^o mix together to give rise to bonding and antibonding combinations. This is not immediately obvious since the orbitals in question are delocalized over two atoms. Take, for example, the $1\sigma_g^o$ and $2\sigma_g^o$ shown in **7-14**. (For clarity the orbitals are shown separated but it's easy to visualize them as being on the same atoms.) Their overlap can be decomposed into four terms, two describing overlap between s and p_z orbitals on the same atom (wavy lines), which are identically zero since the orbitals are orthogonal. The other two terms (solid lines) involve the overlap between an s orbital on one atom and a p orbital on the other. These are certainly different from zero and are set by the s–p interactions. In the mixing shown in **7-14** for $1\sigma_g^o$ and $2\sigma_g^o$ it is clear to see that the overlap is the sum of two positive terms. This, therefore is the bonding combination which leads to $1\sigma_g$. Similarly the overlap in **7-15** is the sum of two negative terms and so represents the antibonding

7-14 $S > 0$ **7-15** $S < 0$

combination leading to $2\sigma_g$. The same analysis applies to the σ_u^o orbitals. They give rise to a bonding combination in **7-16** leading to formation of $1\sigma_u$ and an antibonding combination in **7-17** leading to formation of $2\sigma_u$.

Now we can determine without any ambiguity the form of the four σ orbitals. The deepest lying orbital $1\sigma_g$ results from the appropriate mixture (**7-14**) of $1\sigma_g^o$ and $2\sigma_g^o$

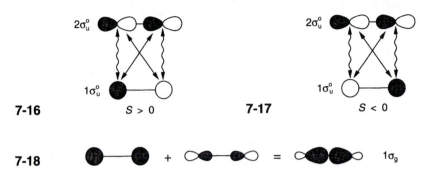

7-16 $S > 0$ **7-17** $S < 0$

7-18
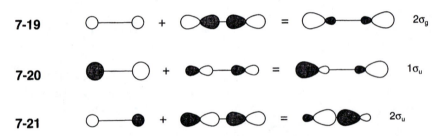

and lies deeper in energy than $1\sigma_g^o$. It is largely made up of $1\sigma_g^o$ since this is the energetically closest starting orbital (**7-18**). The s/p mixture on each is such that the large lobes point toward each other, thus enhancing the interaction between the two. The opposite is true of $2\sigma_g$. This orbital lies above $2\sigma_g^o$ in energy and is derived by the admixture of $1\sigma_g^o$ and $2\sigma_g^o$ shown in **7-19**. The mixing is such that the large lobes of the s/p hybrids point away from each other. Compared to the $2\sigma_g^o$ orbital this orbital has lost bonding character. In the same way we can construct the new orbitals $1\sigma_u$ and $2\sigma_u$, the bonding (**7-20**) and antibonding (**7-21**) combinations of $1\sigma_u^o$ and

$2\sigma_u^o$. In $1\sigma_u$ the antibonding interaction is reduced and the orbital is stabilized. Now only the small lobes of the s/p hybrids point at each other (**7-20**). Contrarily the antibonding interaction of $2\sigma_u$ is reinforced and the orbital destabilized (**7-21**). *So the introduction of the s–p interactions into the orbital problem has important consequences for the form and energy of the σ orbitals. They are summarized in* **7-22** *and show the striking effect of s/p mixing on the bonding or antibonding character of the MOs.*

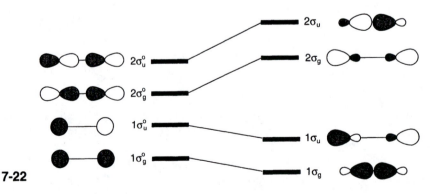

7-22

7.1.4. MO diagrams for A_2 molecules (A = Li, ..., Ne)

(a) Extent of s/p mixing

It should be pointed out that although the MOs calculated numerically for A_2 molecules always take into account this s/p mixing in the way we have described, it is useful to distinguish those cases where such mixing is weak so that the state of affairs is described by the left-hand side of **7-22**, from those cases where it is strong and the picture at the right-hand side of **7-22** more appropriate.

The extent of the mixing between $1\sigma_g^\circ$ and $2\sigma_g^\circ$ and between $1\sigma_u^\circ$ and $2\sigma_u^\circ$ depend upon two factors determined by the nature of the element A = Li, ..., Ne.

(i) The energy separations between the $n\sigma^\circ$ orbitals. Given that these orbitals are derived from the valence s and p orbitals a good measure of the energy separation between $1\sigma_u^\circ$ and $2\sigma_u^\circ$, and between $1\sigma_g^\circ$ and $2\sigma_g^\circ$ might be the atomic s–p separation itself. This separation changes in a relatively smooth fashion, but quite dramatically across the periodic table. $\Delta\varepsilon(s–p) = 1.9$ eV for lithium but increases on moving to the right reaching $\Delta\varepsilon(s–p) = 21.5$ eV for fluorine. Since the interaction between two orbitals decreases as their energy separation increases we expect that s/p mixing decreases in the order $Li_2 > Be_2 > \cdots > F_2 > Ne_2$. This is assuming that the s–s and p–p overlap integrals remain constant such that the corresponding interaction energies do not change. (This in fact is not true as we will see below.)

(ii) The overlap between the $n\sigma^\circ$ orbitals. This is directly related to the overlap between an s orbital on one center and a p orbital on the other. We expect that the s/p mixing will increase as the overlap increases. The change of this factor across the periodic table is difficult to analyze because of two factors.

(a) The electronegativity of A. As the electronegativity increases the atomic orbitals become more contracted. The result is a decrease in their mutual overlap. This effect favors strong s/p mixing for the elements at the left-hand side of the periodic table where the electropositive atoms possess diffuse orbitals.

(b) The internuclear separation (R_e). This varies considerably as shown in Table 7.1 for example for 267 pm for Li_2 to 110 pm for N_2. This term therefore favors strong s/p mixing for most of all the elements in the middle of the row.

Table 7.1: Internuclear separations in the A_2 diatomics

A_2	Li_2	Be_2	B_2	C_2	N_2	O_2	F_2	Ne_2
R_e (pm)	267	—	159	124	110	121	142	—

The number of factors at work here, added to the fact that they do not all vary in the same manner, means that we cannot give a simple rule for all the examples of the A_2 series. Numerical calculation, however, shows that s/p mixing is important for the elements at the left of the periodic table up to $A = N$. Mixing is particularly important for B_2, C_2 and especially so for N_2. Past this point its importance rapidly diminishes. We can see why. Up to nitrogen the decrease in R_e more than compensates for the increase in s–p separation and the contraction of the orbitals. Beyond nitrogen all three factors work against strong s/p mixing. In general terms we should use the right-hand side of **7-22** (the 'exact' result) for $A = Li, \ldots, N$ and the left-hand side of **7-22** (the 'approximate' result) for $A = O, F, Ne$.

(b) Complete MO diagram

The complete orbital interaction diagram is obtained by superposing the π interactions between the p_x, p_y orbitals and the σ interactions between the s and p_z orbitals. The energetic ordering is the same for all the A_2 molecules with one important exception. As Figure 7.1 shows there are two level ordering schemes for the $2\sigma_g$ and π_u orbitals. The π_u orbital is a bonding orbital arising via lateral overlap of two p orbitals **(7-23)** and the $2\sigma_g$ a bonding orbital arising via axial overlap of two p orbitals **(7-24)**. At equilibrium distances, overlap of σ type (S_a) is always larger than that of π type (S_l) and thus we would expect $2\sigma_g$ to lie below π_u if no s/p mixing

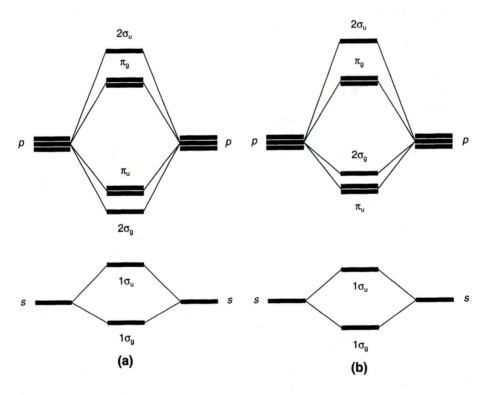

Figure 7.1. Orbital Interaction Diagrams for A_2 molecules: (a) the diagram applicable to $A = O, F (Ne)$ where $2\sigma_g$ lies lower than π_u and (b) the diagram applicable to $A = Li, \ldots, N$ where $2\sigma_g$ lies higher than π_u.

7-23 S_1 **7-24** S_a

occurs. The actual ordering depends on the magnitude of the s/p interactions. If these are weak then the $2\sigma_g$ orbital is not pushed up very much in energy and lies (Figure 7.1a) below the π_u orbital. This would be the case for O_2 and F_2. If the mixing is strong then the $2\sigma_g$ orbital may be pushed above π_u (Figure 7.1b). This would be the case for $Li_2, \ldots, N_2$. There is no similar problem with the ordering of $2\sigma_u$ and π_g. $2\sigma_u$ lies above π_g on overlap grounds (S_1 vs S_a again) before s/p mixing pushes it to higher energy.

7.1.5. Electronic structure of the A_2 molecules (A = Li, . . . , Ne)

(a) Number of 'bonds'

The electronic configuration obtained by placing electrons in the lowest energy MOs available to them, two electrons per orbital are readily obtained for all these molecules. We must remember that the MO diagram only shows the valence orbitals of A and thus for each atom there are also two electrons ($1s^2$) which are not shown.

Among the occupied MOs of these molecules, some are bonding ($1\sigma_g$, $2\sigma_g$, π_u) and others are antibonding ($1\sigma_u$, $2\sigma_u$, π_g). Occupation of bonding orbitals by electrons strengthens, and occupation of antibonding orbitals, weakens the A—A bond. Thus the net number of electrons holding the nuclei together is just

$$n_b - n_a \tag{1}$$

where n_b and n_a are respectively the number of electrons in bonding and antibonding orbitals. By analogy with Lewis' idea of two electrons per bond we may define the *bond order* between two atoms as

$$n = (n_b - n_a)/2 \tag{2}$$

In what follows we shall take this idea a little further and distinguish between electrons in orbitals of σ and π types.

(b) Ground state electron configuration

Li_2 (2 electrons). Only the deepest lying orbital is occupied, $(1\sigma_g)^2$. Since these two electrons occupy a bonding orbital the bond order is equal to one, in accord with the Lewis structure Li—Li. Note that the bond is a σ bond since it arises via occupation of a σ orbital.

Be_2 (4 electrons). Here both bonding ($1\sigma_g$) and antibonding ($1\sigma_u$) orbitals are occupied, $(1\sigma_g)^2 (1\sigma_u)^2$. The bond order is zero just as in the case of He_2 described in Chapter 3. Experimentally, two beryllium atoms do not form a Be_2 molecule, in accord with this result*.

B_2 (6 electrons). The next two electrons occupy the π_u levels which lie below $2\sigma_g$ in Figure 7.1b. The most stable arrangement is obtained by placing one electron in each of the degenerate orbitals to give $(1\sigma_g)^2 (1\sigma_u)^2 (\pi_u^x)^1 (\pi_u^y)^1$ with their spins parallel (7-25). This is just the extension of Hund's rule to molecules. With a total spin which

$$\pi_u^x \;\; \underline{\uparrow} \qquad\qquad \underline{\uparrow} \;\; \pi_u^y$$

$$\underline{\uparrow\downarrow} \;\; 1\sigma_u$$

7-25

$$\underline{\uparrow\downarrow} \;\; 1\sigma_g$$

is non-zero the B_2 molecule is a *paramagnetic species*. Although the B_2 molecule is short-lived this experimentally determinable property is proof of the relative ordering of the $2\sigma_g$ and π_u levels. If π_u lay higher in energy then the highest two electrons would be spin paired in the $2\sigma_g$ orbital and the molecule would be diamagnetic.

Concerning the bond order here, the contribution of the σ electrons is zero overall since there is one pair in $1\sigma_g$ and one pair in $1\sigma_u$. However two electrons occupy a π-bonding orbital whose antibonding partner is empty. Thus there is a π bond between the two atoms, or more exactly, two half-bonds associated with the orthogonal π_u^x and π_u^y components. The Lewis representation for the molecule $|B{-}B|$ indeed shows the single bond between the atoms and the two lone pairs represent the four σ electrons. This representation is however not very good since it indicates that the two bonding electrons are spin paired. B_2 is an unusual molecule (as is C_2) in that it is a species held together by a single π bond. Bonds of this type are usually found in conjunction with σ bonds in molecules.

C_2 (8 electrons). The next two electrons complete the filling of the π_u orbital, $(1\sigma_g)^2 (1\sigma_u)^2 (\pi_u^x)^2 (\pi_u^y)^2$. This molecule is thus diamagnetic, a property which may be verified experimentally. Once again it confirms the energetic ordering of π_u and $2\sigma_g$. If $2\sigma_g$ lay lower in energy then two electrons would have to occupy π_u and the molecule would be paramagnetic.

As before for B_2, the σ bond order is zero since both bonding $(1\sigma_g)$ and antibonding $(1\sigma_u)$ levels are occupied. However the π bond order is two since there are four bonding π_u electrons while the π_g level is empty. The Lewis structure associated with this description is therefore $|C{=}C|$ with two π bonds and two σ lone pairs. The expected Lewis structure for C_2 is, however, $C{\equiv}C$ and indicates a quadruple bond between the two atoms. We have mentioned this particular problem before at the end of Chapter 1. Although it satisfies all of the criteria for being the most favorable Lewis structure, such as obeying the octet rule and the absence of formal charges,

---

* Actually this is not quite true. The Be_2 molecule is very weakly bound indeed with a large internuclear distance. It is probably best described as being held together by van der Waals forces.

it is considerably at variance with the orbital result. In C_2, since two electrons occupy an antibonding orbital ($1\sigma_u$) a quadruple bond is just not possible. In fact since the antibonding character of this pair of electrons approximately cancels the bonding character contributed by the two electrons in $1\sigma_g$ (see Appendix to Chapter 7) a double bond results.

N_2 (10 electrons). Here the $2\sigma_g$ orbital is occupied at last to give the configuration $(1\sigma_g)^2 (1\sigma_u)^2 (\pi_u)^4 (2\sigma_g)^2$. The two electrons in the $2\sigma_g$ bonding orbital increase the σ bond order by one to give a total bond order of three. There is then a good correlation with the Lewis structure $|N\equiv N|$, although the orbital approach tells us that there is one σ bond and two π bonds. Examining the form of the MOs it is clear that the character of the σ orbitals is strongly influenced by s/p mixing (7-26). In fact

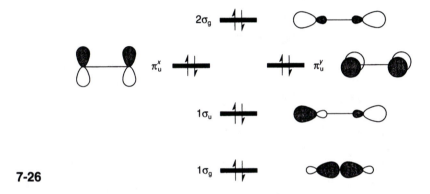

7-26

one could say that the σ bond is best described by the occupation of the $1\sigma_g$ orbital, strongly bonding between the two atoms, since $1\sigma_u$ and $2\sigma_g$ are somewhat more weakly antibonding and weakly bonding respectively between the two nitrogen atoms. This is due to the fact that in both $1\sigma_u$ and $2\sigma_g$ it is the small lobes which point towards each other. A good characterization of the molecule is thus to regard the electron pairs in these two orbitals as two lone pairs of electrons. They are however quite different in energy, $-15.5\,eV$ for $2\sigma_g$ and $-18.8\,eV$ for $1\sigma_u$ from photoelectron spectroscopy. Interestingly, notice that of these two 'lone pair' orbitals the bonding (albeit weakly) combination, $2\sigma_g$ lies higher in energy.

Finally we note that the Lewis structure $|N\equiv N|$ actually describes the molecule quite well. It satisfies all of the rules (the octet rule, absence of formal changes) and above all supports the molecular orbital description of the molecular of two π bonds, one σ bond and two nonbonding σ lone pairs.

O_2 (12 electrons). For A = O the diagram of Figure 7.1a is appropriate with the $2\sigma_g$ level lying lower than π_u. Starting with this element we begin to fill the antibonding orbitals originating from the p orbital interactions and the bond order will decrease accordingly. On moving from N_2 to O_2 the extra two electrons have to be placed in the pair of degenerate π_g orbitals. The lowest energy arrangement is achieved by using Hund's rule, filling each component, π_g^x and π_g^y with a single electron arranged so that their spins are parallel (7-27). The electronic configuration appropriate for the ground state is thus $(1\sigma_g)^2 (1\sigma_u)^2 (2\sigma_g)^2 (\pi_u)^4 (\pi_g^x)^1 (\pi_g^y)^1$. Notice that this result automatically demands a paramagnetic molecule. As we described at

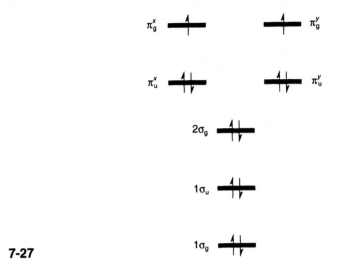

7-27

the beginning of this chapter, the paramagnetism of the oxygen molecule was for some time a puzzle.

Four of the σ electrons lie in bonding orbitals and two in an antibonding orbital leading to a σ bond order of one, with the other electrons best described as lone pairs, one on each atom. Similarly for the π electrons. There are four electrons in π bonding orbitals and two in a π antibonding orbital giving rise to a π bond order of one, the four other electrons being characterized as lone pairs. In this, approximate, way we can retrieve the usual Lewis structure $\overline{O}\!=\!\overline{O}$ for the molecule although it is not completely correct since it arranges all the electrons as pairs and masks the existence of the two unpaired electrons. Thus we see the same problem as found for B_2. Here in O_2 there is not a single π bond but two half π bonds. For both x and y components of π_u and π_g there is a pair of electrons in π_u^x or π_u^y and one electron in π_g^x or π_g^y.

F_2 (14 electrons). Now the antibonding π_g orbitals are all filled giving the electronic configuration $(1\sigma_g)^2 (1\sigma_u)^2 (2\sigma_g)^2 (\pi_u)^4 (\pi_g)^4$. Since both bonding and antibonding π levels are filled, the total π bond order is zero. Thus the eight π electrons do not contribute to bonding between the fluorine atoms at all, and form four of the six lone pairs of the Lewis structure. As far as the σ system is concerned this is a molecule where s/p mixing is weak and the orbitals are described by the left-hand side of **7-22**. Since the bonding character of $1\sigma_g$ is cancelled by the antibonding character of $1\sigma_u$ the four electrons in these two orbitals are described as two σ lone pairs, one on each fluorine atom. Since $2\sigma_g$ is the only occupied orbital without an occupied antibonding partner the bond order is equal to one. Thus we may retrieve the Lewis structure $|\overline{F}\!-\!\overline{F}|$ which shows a single F—F bond. Although we have evaluated the bond order by subtracting the number of antibonding electron pairs from the number of antibonding ones, the situation is usually a little more complex. Since the π_u level is stabilized less than the π_g level is destabilized (see Chapter 3) filling both of these orbitals with four electrons each, leads overall to a repulsive situation. If the π interaction is strong, as it is for second row atoms then the bond between the two fluorine atoms (formally a single σ bond) should be weaker than expected. In fact the F_2 molecule does have a small dissociation energy when compared to Cl_2 and Br_2.

Ne_2 (16 electrons). Now all the molecular orbitals derived from the valence orbitals are doubly occupied, $(1\sigma_g)^2 (1\sigma_u)^2 (2\sigma_g)^2 (\pi_u)^4 (\pi_g)^4 (2\sigma_u)^2$ and since the number of bonding and antibonding electrons are equal, the bond order is zero. Because the stabilization of the bonding orbitals is less than the destabilization of their anti-bonding partners we may readily understand why the inert gases such as neon do not form diatomic molecules.

7.1.6. Bond lengths and bond energies

It is useful to search for correlations between the bond order and the equilibrium internuclear separation (the bond length) and the dissociation energy (bond energy) in these molecules. The theoretical trend is clear, the bond order drops from Li_2 to Be_2, increases to a maximum at N_2, and then drops to zero at Ne_2. Since in general we expect the higher the bond order, the shorter the bond there should be an inverse correlation between the dissociation energy and bond length. These ideas are nicely verified in a qualitative way as shown in Table 7.2. R_e drops from B_2 (159 pm) with $n = 1$ to a minimum at N_2 (110 pm) with $n = 3$ and then increases for O_2 and F_2. Analogously the bond dissociation energy increases from B_2 to N_2 and then decreases for O_2 and F_2. There are some limitations to the correlation which are apparent when comparing parameters for molecules with the same value of n. Thus both the bond length and bond energy of F_2 are quite different from the corresponding values for B_2 and Li_2. All three molecules have $n = 1$. Of course much of this quantitative discrepancy is due just to the fact that the atoms are chemically different. The orbitals of lithium are much more diffuse than those of fluorine.

Analysis of the MO diagrams of Figure 7.1 also allows an understanding of the change in equilibrium distance found when the molecule is ionized ($A_2 \rightarrow A_2^+ + e^-$) or when it captures an extra electron ($A_2 + e^- \rightarrow A_2^-$). The reasoning is based upon the character of the orbital which loses or gains an electron. *If an electron is added to a bonding MO, or removed from an antibonding MO the bond order increases and the bond length decreases. The opposite is also true. If an electron is added to an antibonding MO or is removed from a bonding MO then the bond order decreases and the bond length increases.*

This is nicely demonstrated by the bond lengths in the series derived from O_2. The ion O_2^+ has a bond length of 112 pm which is shorter than O_2 itself (121 pm)

Table 7.2: Bond order from molecular orbital considerations and experimental values of the A—A bond lengths and dissociation energies

A_2	Li_2	Be_2	B_2	C_2	N_2	O_2	F_2	Ne_2
n	1	0	1	2	3	2	1	0
R_e (pm)	267	∞[a]	159	124	110	121	142	∞[a]
E (kJ mol^{-1})	100	—	288	585	940	493	155	—

[a] Be_2 and Ne_2 do not exist as chemically bonded entities

reflecting the loss of an electron from the antibonding π_g orbital of the neutral molecule. The bond length changes are the opposite for the negative ions, found in the solids KO_2 and BaO_2 for example. In O_2^- the bond length has increased to 126 pm and in O_2^{2-} to 149 pm as a result of further populating the antibonding π_g orbital.

7.2. Heteronuclear diatomic molecules, AB

The interaction scheme for the fragment orbitals of A and B which lead to the MO picture for the AB molecule (**7-28**) resembles that obtained for the homonuclear diatomic molecules. There are, as before, two interactions (5, 6) between the pairs of orbitals $(p_x(A), p_x(B))$ and $(p_y(A), p_y(B))$, and a four-orbital problem (1–4) between $(s(A), p_z(A))$ and $(s(B), p_z(B))$. The major difference from the A_2 case is that now none of the interactions is degenerate; the orbitals on the more electronegative atom (B in **7-28**) lie to deeper energy than the corresponding orbitals on atom A.

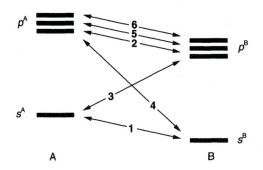

7-28   A   B

7.2.1. Construction of the π MOs

The interactions between the p_x and p_y orbitals lead to the formation of π bonding and antibonding orbitals (**7-29**) labeled $1\pi^x$, $1\pi^y$ and $2\pi^x$, $2\pi^y$ respectively. Since the AB molecule does not have a center of symmetry the g and u labels of the A_2 case are not appropriate here. The bonding MOs are largely localized on the atom whose atomic p orbitals lie deeper in energy (see Chapter 3), in this case the atom B. Analogously the antibonding orbital is largely localized on atom A.

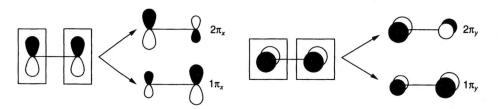

7-29

7.2.2. Construction of the σ MOs

The interation of the s and p_z orbitals on A and B lead to the formation of the set of orbitals $n\sigma$ where $n = 1$–4. Establishing the character of these MOs is considerably more tricky than in the case of the homonuclear diatomic case for two reasons:

(i) Since the starting orbitals are non-degenerate the primary and secondary interactions will be quite different in general. **7-30** shows the interactions between the orbitals for a small electronegativity difference between A and B. Here the situation is expected to be similar to that described for A_2. If the electronegativity difference is large (**7-31**) then the cross-interaction between $s(A)$ and $p_z(B)$ may be very important since the energy difference between the two is small. At the same time the $s(A)$–$s(B)$ and $p_z(A)$–$p_z(B)$ interactions are reduced in importance. We must expect that the form of the MOs will be different in each case.

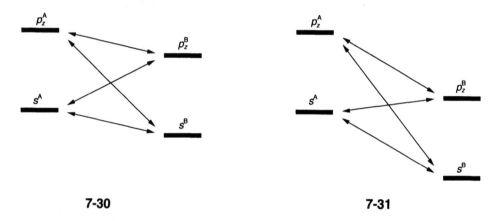

7-30   **7-31**

(ii) Even if the principal interactions (between $s(A)$ and $s(B)$ and between $p_z(A)$ and $p_z(B)$) are similar (**7-30**) to those in the A_2 molecule, leading to the provisional MOs σ^o, these orbitals are not classified into σ_g^o and σ_u^o sets. This means that when considering the effect of the secondary interactions, all four orbitals may mix together rather than as two separate pairs as before.

It is not possible to give a set of rules which will allow the construction of the MOs of all possible AB molecules but one can give some indications.

(i) In the deepest lying MO (1σ) the fragment orbitals mix together in such a way that their overlap integrals are positive. This ensures maximum bonding within this orbital. The larger orbital contribution is associated with the deepest lying starting orbital which lies on the more electronegative atom (**7-32**).

7-32 1σ **7-33** 4σ

(ii) In the highest lying MO (4σ) the opposite is true. The fragment orbitals mix together so that their overlap integrals are negative to ensure maximal antibonding character in this orbital. The larger orbital contribution comes from the less electronegative atom (**7-33**).

(iii) The form of the intermediate orbitals 2σ and 3σ is very sensitive to the effects we have described above associated with the A/B electronegativity difference. In general these orbitals have considerable nonbonding character, but we will illustrate their nature by examination of the specific case of CO. This is a very important molecule since it forms bonds to transition metal atoms and plays a vital role in organometallic chemistry.

7.2.3. MOs and electronic structure of CO

The location of the valence orbitals of oxygen ($\varepsilon_{2s} = -32.4$ eV, $\varepsilon_{2p} = -15.9$ eV) and for carbon ($\varepsilon_{2s} = -19.4$ eV, $\varepsilon_{2p} = -10.7$ eV) imply that oxygen is the more electronegative of the two atoms. We will not describe again the form of π MOs of the molecule (given in **7-29**) but point out that 1π is preferentially located on oxygen and 2π on carbon.

The interaction scheme for the σ MOs is shown in **7-34**. The four interactions we

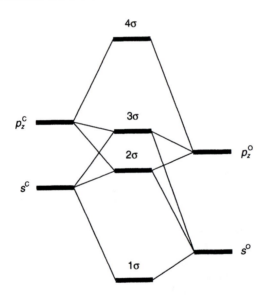

7-34

need to consider are associated with the energy differences $p_z(\text{C})$–$p_z(\text{O})$ (5.2 eV), $s(\text{C})$–$p_z(\text{O})$ (3.5 eV), $s(\text{C})$–$s(\text{O})$ (13.0 eV) and $p_z(\text{C})$–$s(\text{O})$ (21.7 eV). Calculation of the overlap integrals associated with these interactions immediately shows where the problem lies. The two interactions, which on the simplest model we would like to ignore since they are associated with the largest energy differences ($s(\text{C})$–$s(\text{O})$ and $p_z(\text{C})$–$s(\text{O})$) are precisely those which have the largest overlaps. In this case we have no real alternative but to describe the orbital character of 2σ and 3σ as determined by a numerical solution to the problem (**7-35**).

The form of the 1σ and 4σ orbitals are just as described above in the generalizations

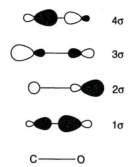

7-35

of Section 7.3.2, being respectively strongly bonding and strongly antibonding between carbon and oxygen. In both 2σ and 3σ the result of s/p mixing leads to the formation of outward pointing hybrid orbitals with small lobes pointing towards the other atom. In the lower energy orbital of the two (2σ) the large lobe is located largely on the more electronegative, oxygen atom. In the higher energy orbital (3σ) the converse is true and it is found on carbon.

An orbital diagram for CO is given in Figure 7.2. Just as in the homonuclear case there is a problem concerning the relative energies of 3σ and 1π. In CO it turns out that the cross-interaction between $s(C)$ and $p_z(O)$ is quite strong ($\Delta\varepsilon = 3.5$ eV) which is sufficient to push the 3σ level above the 1π pair.

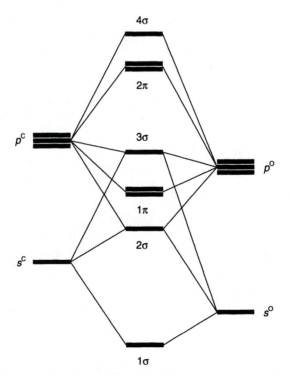

Figure 7.2. Orbital Interaction Diagram for CO. (With ten valence electrons the lowest five orbitals are occupied.)

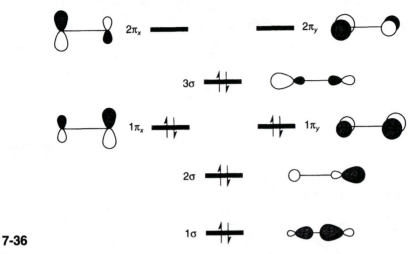

7-36

There are ten valence electrons in CO leading to the ground electronic configuration $(1\sigma)^2 (2\sigma)^2 (1\pi)^4 (3\sigma)^2$ of **7-36**. The two highest occupied σ MOs are weakly bonding. If we try to establish an analogy between this description and that indicated by the Lewis structure $|\overset{\ominus}{C}\!\!\equiv\!\!\overset{\oplus}{O}|$ we can make the correspondence between the two π bonds $((1\pi_x)^2$ and $(1\pi_y)^2)$, one σ bond $((1\sigma)^2)$ and two lone pairs derived from occupation of the almost nonbonding orbitals $(2\sigma)^2$, and $(3\sigma)^2$. It is interesting to note that the two electrons in the highest energy MO, 3σ, are located in an outward pointing hybrid centered on the carbon atom. The presence of this 'high energy' lone pair is responsible for a variety of chemical properties of CO, amongst them coordination as a Lewis base to transition metals to give species such as $Cr(CO)_6$. Another characteristic of CO important in organometallic chemistry is the presence of the low lying empty orbitals 2π. These give CO some Lewis acid character.

One last point concerns the dipole moment of CO. The experimental value is very small (about 0.1 D) in the sense $C^{\delta-}O^{\delta+}$. It is difficult to explain in qualitative terms using our MO diagram, but of course good calculations have no problem in getting it right. The four lowest energy occupied orbitals are preferentially located on oxygen and only the highest is located largely on carbon and we need to sum the contributions of each orbital to find the overall charge distribution. Perhaps the Lewis scheme gives the best qualitative insight. Since there are initially six electrons on oxygen and four on carbon, we could write the molecule as $|\overset{\ominus}{C}\!\!\equiv\!\!\overset{\oplus}{O}|$ indicating donation to carbon. Thus, although oxygen is more electronegative and might be expected to carry the negative charge, to satisfy the octet rule oxygen has to provide both electrons for one of the three bonds.

EXERCISES

7.1 (i) Give the form of the MOs of the Cl_2 molecule. What is its ground electronic configuration?

(ii) Ionization of Cl_2 ($Cl_2 \rightarrow Cl_2^+ + e^-$) leads to a shortening of the interatomic distance (199 pm to 189 pm). Why is this?

(iii) What is the electronic configuration of the first excited electronic state of Cl_2? Explain why the bond length is longer (247 pm) in this excited state than in the electronic ground state.

7.2 Recent calculations on the S_2 molecule and its ions gave the following values for the equilibrium interatomic separation. S_2^{2+} (172 pm), S_2^+ (179 pm), S_2 (188 pm), S_2^- (200 pm) and S_2^{2-} (220 pm). Give an explanation for this trend.

7.3 Recall the MO diagram for Ne_2. Explain why the neutral molecule does not exist but its monocation Ne_2^+ is a stable (but reactive) species with a bond energy of 130 kJ mol^{-1}.

7.4 Explain why the C—O distance varies little on going from CO (113 pm) to CO^+ (112 pm).

7.5 With reference to the NO molecule
  (i) Why is this species a radical?
  (ii) Assume that the MO diagram of this molecule is close in form to that for CO.
    (a) Give the ground electronic configuration for NO.
    (b) How would you expect the N—O distance to change on moving from the neutral molecule to the anion NO^- and cation NO^+?

APPENDIX

The number of bonds (bond order) in diatomic molecules

The decomposition of the number of electrons in bonding and antibonding orbitals allows computation of the bond order between two atoms as in equation (2). In certain cases however the description of an orbital as bonding (or antibonding) or nonbonding is often not clear. It varies considerably depending upon the importance of s/p mixing. For example the $1\sigma_u$ orbital of **7-22** may be truly antibonding if s/p mixing is weak but only weakly antibonding or nonbonding if s/p mixing is strong.

Consider the example of C_2. The usual decomposition leads to the generation of two π bonds between the carbon atoms, (**7-37a**), the bonding character in $1\sigma_g$ being

$$|C{=}\!\!=\!\!C| \qquad |C{\equiv}C \longleftrightarrow C{\equiv}C| \qquad \cdot C{\equiv}C\cdot$$

a   b   c   d

7-37   41%   55%

cancelled by the antibonding character of $1\sigma_u$ (**7-22**). However since s/p mixing is important here, bonding character is enhanced in $1\sigma_g$ and antibonding character reduced in $1\sigma_u$. The result is that there is a residual σ bond in C_2 in addition to the two π bonds. At the limit, where $1\sigma_u$ is nonbonding we would describe the molecule as containing one σ bond, two π bonds, and one lone pair ($1\sigma_u$). Such an analysis is reflected in the importance of the *triply bonded* Lewis structures (**b, c** and **d** of **7-37**) found by calculation.

8 Large molecules

The construction 'by hand' of molecular orbital diagrams for large molecules becomes more difficult as the number of atoms increases, simply because the number of orbitals which need to be included quickly increases. Even for the case of the heteronuclear diatomic molecules with its four σ orbitals these problems are not insignificant as we have seen. Although symmetry helps us out, interaction schemes for larger molecules will generally have large numbers of orbitals of the same symmetry on each fragment. As in the case of the AB molecule it is difficult to qualitatively predict the details of orbital mixing without the aid of a numerical calculation. There are two ways around the problem which we will describe in this chapter.

First, in order to construct an approximate MO diagram for the molecule, we will neglect a large number of interactions between fragment orbitals and only retain those which are most important. In practice we only keep those interactions which are close in energy (or better still degenerate) and, of course, have a non-zero overlap. We will illustrate this approach in this chapter with the molecules acetylene, ethylene and ethane.

A second approach focuses on the construction and orbital properties of a sub-set of the MOs. For example in the second part of this chapter we will study the π orbitals of conjugated molecules. We will be able to do this by making use of the interaction schemes devised for small molecules, and treated exactly earlier in the book. For these conjugated systems we will only study a part of the electronic problem and ignore for the time being the existence of a σ manifold with two electrons per C—C or C—H bond.

8.1. MOs of acetylene, ethylene and ethane

In this section we will construct, in an approximate fashion the MOs of acetylene (**8-1**), ethylene (**8-2**) and ethane (**8-3**) which are the prototypes for three important

8-1   **8-2**   **8-3**

families of organic molecules, namely the alkynes, the alkenes and the alkanes. In each case the molecular symmetry is such that the molecule can be split into two identical fragments, CH, bent CH_2 and pyramidal CH_3 whose orbitals resemble those of the BH, OH_2 and NH_3 molecules we have described in Chapter 6. The problem is similar to that of the assembly of the MO diagram for A_2 molecules (Chapter 7) with the replacement of the A atom by a CH_n unit.

8.1.1. MOs and electronic structure of acetylene

Acetylene may be broken into two CH fragments which carry identical fragment orbitals (**8-4**). There is a pair of degenerate π nonbonding orbitals coming from the p_x and p_y orbitals of carbon and three σ orbitals on each unit. We will specify the C—H bonding character of 1σ, the essentially nonbonding character of 2σ and the antibonding character of 3σ by using the labels σ_{CH}, n_σ and σ_{CH}^*.

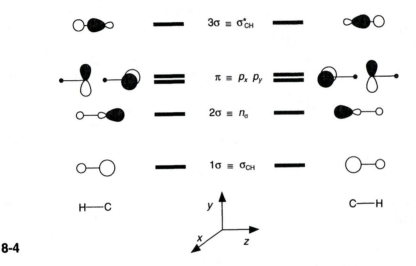

8-4

Since the π MOs (p_x, p_y) on each fragment are antisymmetric with respect to a plane which contains the internuclear axis z (yz or xz), they cannot interact with the σ orbitals of the other fragment, symmetric with respect to this symmetry operation. The relevant overlap integrals are identically zero. This σ/π separation has been described already in Section 3.5.1. Overlap between p_x and p_y and their partners on the other fragment are of course non-zero and the orbital problem the simple one of two interactions between two pairs of orbitals. The situation is more complex for the σ orbitals since there are three on each fragment. We can easily show that none of the relevant overlaps is zero which leads to a total of nine possible interactions. The simplification we will use here to solve this problem in a qualitative way is the same approach we used for the A_2 diatomics in Chapter 7 by ignoring their s–p interactions. In this way we only keep the interactions between degenerate orbitals shown in **8-5**. In addition to the π interactions we have already mentioned, these are the interactions between $\sigma_{CH}(1)$ and $\sigma_{CH}(2)$, $n_\sigma(1)$ and $n_\sigma(2)$ and between $\sigma_{CH}^*(1)$ and $\sigma_{CH}^*(2)$. In this approximation we have therefore ignored the orbital polarization

8-5

$\sigma^{*(1)}_{CH}$ ▬ ◄——————— 5 ———————► ▬ $\sigma^{*(2)}_{CH}$

$p^{(1)}_x \ p^{(1)}_y$ ▬▬ ◄——————— 4 ———————► ▬▬ $p^{(2)}_x \ p^{(2)}_y$

      3

$n^{(1)}_\sigma$ ▬ ◄——————— ———————► ▬ $n^{(2)}_\sigma$

      2

$\sigma^{(1)}_{CH}$ ▬ ◄——————— 1 ———————► ▬ $\sigma^{(2)}_{CH}$

described in Section 6.2.6b. *In toto* this simplification for acetylene has reduced the problem to an orbital situation of five, two-orbital interactions.

Now, assembly of the orbital diagram is easy. We just construct bonding and antibonding orbital pairs for each of these interactions. Since each interaction is a degenerate one the orbital composition is equally weighted by contributions from each CH unit. Starting off with the p_x and p_y orbitals we form bonding and antibonding combinations, π_{CC} and π^*_{CC}, generated by lateral overlap of these fragment orbitals. The π_{CC} orbitals (x or y) are degenerate as are π^*_{CC} (x or y). In the same way σ_{CH} leads to bonding (σ^+_{CH}) and antibonding (σ^-_{CH}) pairs. The σ^-_{CH} orbital, although antibonding between the carbon atoms still lies deep in energy since it is strongly bonding between carbon and hydrogen. The n_σ orbitals lead to the formation of the MOs σ_{CC} and σ^*_{CC}. In Figure 8.1 we have left out the orbitals derived from the σ^*_{CH} on the premise that they are very high in energy (and therefore will not be occupied). We can also see from **8-4** that the overlap between them will be small since the large lobes point towards the hydrogen atoms within each C—H group.

The form of the MOs may be deduced without any difficulty from the fragment orbitals (**8-6**). Since acetylene has a total of 10 valence electrons the electronic

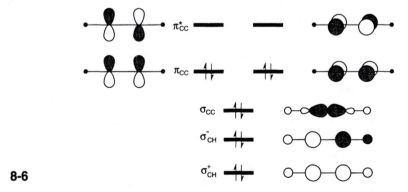

8-6

configuration is just $(\sigma^+_{CH})^2 \ (\sigma^-_{CH})^2 \ (\sigma_{CC})^2 \ (\pi_{CC})^4$, which has immediate ties to the Lewis structure H—C≡C—H. The σ^+_{CH} and σ^-_{CH} orbitals are respectively bonding and antibonding between the two carbon atoms and so overall do not contribute to C—C bonding. However they are both bonding between carbon and hydrogen and together characterize the two C—H bonds. The σ_{CC} orbital is largely bonding between

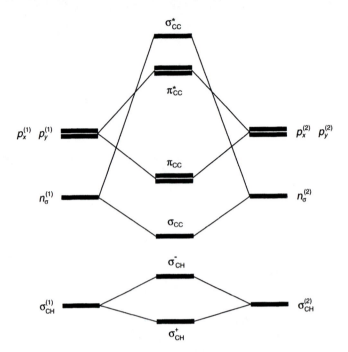

Figure 8.1. The orbital interaction diagram for acetylene, starting from two CH units. (All the orbitals are filled through π_{CC}.)

the two carbon atoms and its double occupancy leads to one of the C—C bonds, of σ type. Notice that it is derived from two fragment orbitals which are well directed towards their partners on the other carbon atom. Finally occupation of the two π_{CC} orbitals, bonding between the two carbon atoms, leads to another two CC bonds, this time of π type. So we immediately retrieve the Lewis structure with its triple C≡C bond. As far as the unoccupied orbitals are concerned the reactivity of the molecule is strongly affected by the presence of the π_{CC}^* levels.

8.1.2. MOs and electronic structure of ethylene

The ethylene molecule may be split into two bent, coplanar CH_2 fragments whose orbitals were given in Figure 6.5. On each fragment there are **(8-7)** two CH bonding orbitals, $1a_1$ and $1b_2$, two nonbonding orbitals, $2a_1$ and $1b_1$, and two antibonding orbitals, $2b_2$ and $3a_1$. Just as for the CH case we will use a notation which describes their bonding, nonbonding or antibonding character with respect to the CH linkages. So $1a_1$, bonding between carbon $2s$ and the hydrogen atoms becomes σ_{CH_2}, and its antibonding partner ($3a_1$) becomes $\sigma_{CH_2}^*$. The $2a_1$ nonbonding orbital ($1b_1$) lying perpendicular to this plane is labeled n_p. Finally the $1b_2$ orbital, bonding between the carbon and hydrogen atoms, and its partner, $2b_2$, are labeled π_{CH_2} and $\pi_{CH_2}^*$ since they have a nodal plane. We note that the labels σ, π are not rigorously correct here since they are the group theoretical labels for linear molecules. However chemical usage of the terms 'σ bonding' and 'π bonding' is so overwhelming that we just

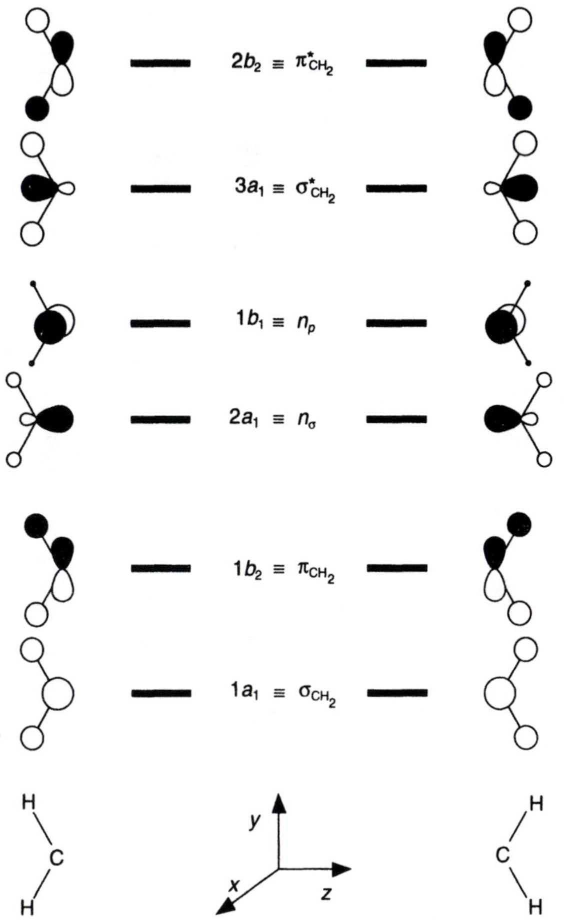

$2b_2 \equiv \pi^*_{CH_2}$

$3a_1 \equiv \sigma^*_{CH_2}$

$1b_1 \equiv n_p$

$2a_1 \equiv n_\sigma$

$1b_2 \equiv \pi_{CH_2}$

$1a_1 \equiv \sigma_{CH_2}$

8-7

have to keep in mind the dual usage of these labels. Note too that there is not an obvious way to generate this new labeling scheme. In fact the labels come about 'in hindsight' after the interaction diagram is complete.

The orbitals which may interact by symmetry are the two nonbonding orbitals n_p (or $1b_1$) which are the only ones antisymmetric with respect to the molecular plane yz, the four b_2 orbitals (π_{CH_2} and $\pi^*_{CH_2}$ on each fragment) antisymmetric with respect to the xz plane and the six a_1 orbitals (σ_{CH_2}, n_σ and $\sigma^*_{CH_2}$ on each fragment) symmetric with respect to both yz and xz planes. Just as in the acetylene molecule we may simplify the problem by only retaining interactions between degenerate orbitals. The problem of MO construction for ethylene thus reduces to six two-orbital interactions. In order to simplify the diagram (Figure 8.2) we have left off the high energy orbitals, $\pi^*_{CH_2}$ and $\sigma^*_{CH_2}$ on each fragment.

Ethylene has twelve valence electrons and thus the ground electronic configuration is $(\sigma^+_{CH_2})^2 \, (\sigma^-_{CH_2})^2 \, (\pi^+_{CH_2})^2 \, (\sigma_{CC})^2 \, (\pi^-_{CH_2})^2 \, (\pi_{CC})^2$ as shown in **8-8**, leading to an approximate correspondence with the Lewis structure $H_2C=CH_2$. Clearly the

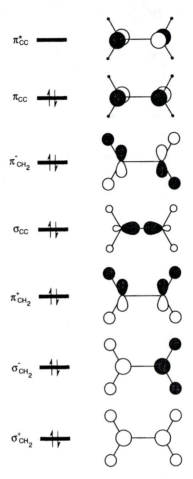

8-8

double occupation of π_{CC} leads to a C—C π bond and double occupation of σ_{CC} leads to a σ bond. The σ_{CC} bond arises via the overlap of two well-directed n_σ orbitals on the two carbon atoms. The other occupied orbitals break down into the C—C bonding and antibonding pair ($\sigma^+_{CH_2}$, $\sigma^-_{CH_2}$) and ($\pi^+_{CH_2}$, $\pi^-_{CH_2}$) which overall lead to no net C—C bonding. However these four orbitals are C—H bonding and are responsible for the four C—H bonds. Thus the ethylene molecule contains a CC double bond, one σ and one π. Finally we note that the lowest energy empty orbital is the π^*_{CC} orbital, antibonding between the two carbon atoms.

An important property of the ethylene molecule, and indeed alkenes in general is the existence of a high barrier to rotation about the C=C which tends to hold the molecule flat. The existence of such a barrier is demonstrated experimentally by the existence of two isomers for the substituted ethylenes RHC=CHR' (*cis* and *trans* of **8-9**) which may be thermally interconverted. We described some of the orbital interactions behind such an observation in Section 3.5.2. Here we can add to that purely π picture by inclusion of the effect of the σ orbitals. The energy of the σ_{CC} and π_{CC} orbitals of **8-8** are set by the size of the relevant σ and π overlap integrals between fragment orbitals. The larger the overlap the larger the stabilization experienced by two electrons in a particular orbital. Rotation by 90° of one CH_2

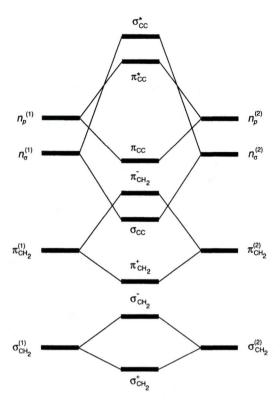

Figure 8.2. The orbital interaction diagram for ethylene, starting off from two CH_2 units. (All the orbitals are filled through π_{CC}.)

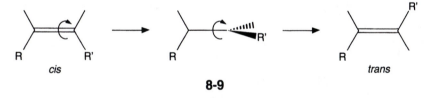

8-9

fragment against the other leads to no change in σ overlap (**8-10a**) since this is an orbital which is axially symmetric with respect to the C—C linkage. Rotation however, destroys the π overlap (**8-10b**) and so the stabilization of the π orbital in the twisted geometry is zero. So the picture we derived earlier is completely correct

8-10a

$S_{\sigma\sigma} \neq 0$ $S_{\sigma\sigma} \neq 0$

8-10b

$S_{pp} \neq 0$ $S_{pp} = 0$

but here we see that it is so because of the energetic insensitivity of the σ_{CC} orbital to rotation about the C—C axis. The activation energy for *cis–trans* isomerization is equal to the π bond energy on this model.

8.1.3. MOs and electronic structure of ethane

The ethane molecule may be split into two pyramidal CH_3 fragments. On each unit **(8-11)** there are three bonding C—H orbitals ($1a_1$, $1e_x$ and $1e_y$), three antibonding ones ($3a_1$, $2e_x$ and $2e_y$) and, at an intermediate energy, an orbital which is essentially nonbonding ($2a_1$). Once again chemists are in the habit of giving particular names to these orbitals which describe their bonding characteristics. The bonding $1a_1$ orbital and its antibonding partner ($3a_1$) are labeled σ_{CH_3} and $\sigma^*_{CH_3}$, and the bonding orbitals ($1e_x$, $1e_y$) and their antibonding analogs ($2e_x$, $2e_y$) are labeled ($\pi^x_{CH_3}$, $\pi^y_{CH_3}$) and ($\pi^{x*}_{CH_3}$, $\pi^{y*}_{CH_3}$). For the last two orbitals the notation reminds us that they are largely composed of p_x (or p_y) character on carbon and antisymmetric with respect to the yz (or xz) plane. Finally, the essentially nonbonding orbital $2a_1$ is labeled n_σ.

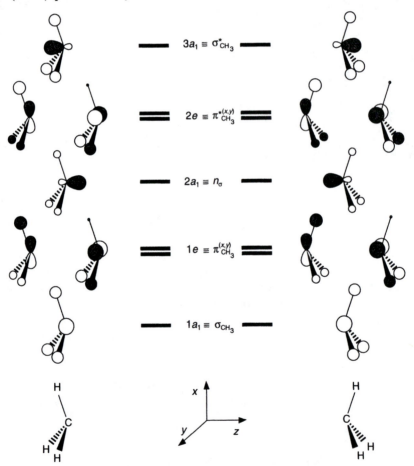

$3a_1 \equiv \sigma^*_{CH_3}$

$2e \equiv \pi^{*(x,y)}_{CH_3}$

$2a_1 \equiv n_\sigma$

$1e \equiv \pi^{(x,y)}_{CH_3}$

$1a_1 \equiv \sigma_{CH_3}$

8-11

As before in order to construct the MOs of ethylene we include only those interactions between orbitals of the same energy and symmetry. First of all there are the three interactions between the a_1 orbitals, σ_{CH_3}, n_σ and $\sigma^*_{CH_3}$, on each fragment. As far as the degenerate orbitals $\pi^x_{CH_3}$ and $\pi^y_{CH_3}$ are concerned they may only interact with the orbital of the same type on the other fragment. This is ensured by the symmetry properties of the two orbitals. One is symmetric with respect to the xz plane ($\pi^x_{CH_3}$) and the other ($\pi^y_{CH_3}$) antisymmetric, so that the overlap between them is exactly zero. The same is true of the antibonding pair $\pi^{x*}_{CH_3}$ and $\pi^{y*}_{CH_3}$. Thus the orbital construction process for ethane in this simple approach is just one of seven two-orbital interactions. In order to simplify the interaction diagram (Figure 8.3) we do not show the MOs derived from the interaction of the highest energy fragment orbitals.

The ethane molecule has a total of fourteen valence electrons leading to the ground electronic configuration (**8-12**), $(\sigma^+_{CH_3})^2 \ (\sigma^-_{CH_3})^2 \ (\pi^{x+}_{CH_3})^2 \ (\pi^{y+}_{CH_3})^2 \ (\sigma_{CC})^2 \ (\pi^{x-}_{CH_3})^2 \ (\pi^{y-}_{CH_3})^2$. Obviously the double occupation of the σ_{CC} orbital leads to a C—C bond, but of the other occupied orbitals, three are C—C bonding (the orbitals labeled with a +) and three are antibonding (labeled with a −). The overall effect is that these six orbitals do not contribute to C—C bonding at all. However they are C—H bonding and are responsible for the formation of the six C—H bonds. Notice that the σ_{CC} orbital comes about from the very favorable overlap between the n_σ orbitals of each fragment, each well directed towards the other carbon atom. The ethane molecule

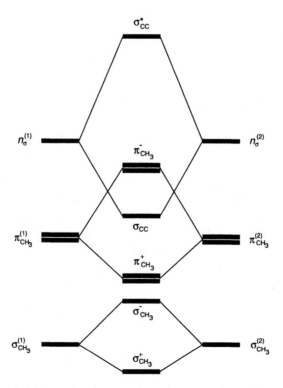

Figure 8.3. The orbital interaction diagram for ethane, starting off from two CH₃ units. (All the orbitals are filled through $\pi^-_{CH_3}$.)

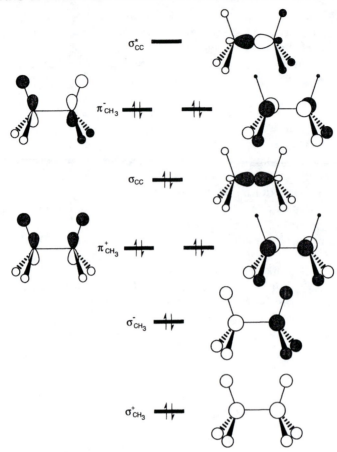

8-12

thus has one C—C bond of σ type, and of course none of π type. The connection to the Lewis structure is obvious.

One important difference between ethane and ethylene is the absence of a significant rotation barrier around the C—C bond in the saturated hydrocarbon. The energy difference between eclipsed and staggered forms (**8-13**) is only 12 kJ mol^{-1}. This result is simply understandable by consideration of the form of the MO which describes the C—C linkage (**8-14**). The overlap between the fragment orbitals leading

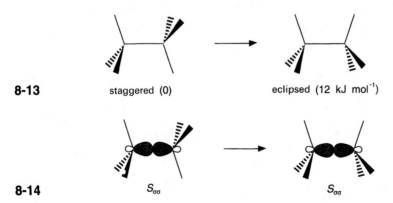

8-13   staggered (0)   eclipsed (12 kJ mol^{-1})

8-14 $S_{\sigma\sigma}$ $S_{\sigma\sigma}$

to σ_{CC} is axially symmetric (as in ethylene too) and so does not change on rotation of one methyl group relative to the other. One can in fact show that the small energy barrier is associated with interactions between the orbitals which characterize the C—H bonds.

8.2. Conjugated polyenes

The analogy between H_2 and ethylene which we established in Section 3.5 extends to conjugated polyenes having n carbon atoms as long as the σ and π orbitals are separable. This will only be the case if the molecule is planar. If xy is the molecular plane then the set of π orbitals, p_z on each carbon, antisymmetric with respect to this plane will lead to a set of π molecular orbitals. All other orbitals on carbon ($2s$, $2p_x$, $2p_y$) and hydrogen ($1s_H$) have no overlap with these AOs since they are symmetric with respect to this plane. The MOs which result are closely related to the MOs found for the model systems H_n in which a single hydrogen $1s$ orbital per atom generates a set of MOs. Really the only difference in the MO diagram is the replacement of $1s_H$ by carbon $2p_z$. In order to determine the number of π electrons for a given molecule we need to subtract from the valence electron sum two electrons for each C—C and C—H σ bond.

The form, (symmetry, orbital coefficients, etc.) and relative energies of the MOs of the 'linear' conjugated molecules, allyl (**8-15**) and butadiene (**8-16**) and the cyclic systems, cyclopropenyl (**8-17**), cyclobutadiene (**8-18**) and benzene (**8-19**) are then

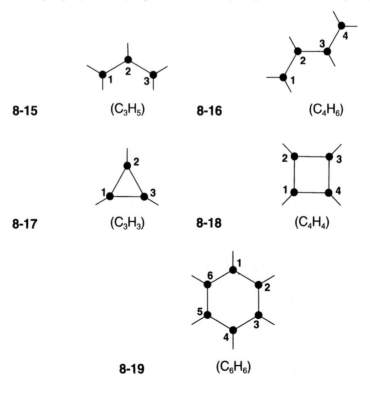

8-15 (C_3H_5) **8-16** (C_4H_6)

8-17 (C_3H_3) **8-18** (C_4H_4)

8-19 (C_6H_6)

just like those of the model linear and cyclic H_n systems of Chapter 4. For the cyclic systems the geometries correspond exactly. The carbon atoms in benzene form a regular hexagon just as they do in cyclic H_6. However for allyl (**8-15**) and butadiene (**8-16**) the carbon atoms do not lie in a straight line as they do in the all-hydrogen model compounds. The C—C—C bond angles are around 120°. This, in fact, is of little importance for the π MOs. The lateral overlap between p_z orbitals on adjacent carbon atoms depends only upon the C—C distance not upon whether the carbon skeleton is linear or not. Thus there is a one-to-one correspondence between the π levels of these 'linear' or straight-chain polyenes and the levels of H_n just as in the cyclic case.

8.2.1. MOs and electronic structure of allyl

The π MOs of this molecule (**8-20**) may be deduced from the MOs obtained for the linear H_3 molecule in Section 4.1.3. The π_1 orbital is bonding between the central carbon (C_2) and the terminal ones (C_1 and C_3). The π_2 orbital is nonbonding since it has no coefficient on C_2, and π_3 is antibonding between central and terminal atoms.

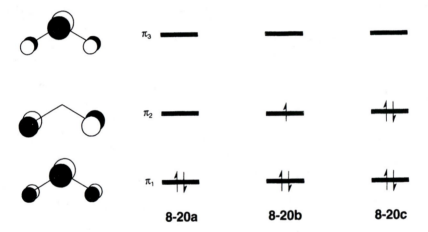

In the neutral allyl species there are a total of seventeen valence electrons (four from each carbon and one from each hydrogen). Of these, fourteen occupy deep-lying orbitals which characterize the five C—H and two C—C σ bonds. Thus three are left to fill the π manifold. The electronic configurations of the neutral species (**8-20b**) and the corresponding anion (**8-20c**) and cation (**8-20a**) are $\pi_1^2 \pi_2^1$, $\pi_1^2 \pi_2^2$ and π_1^2 respectively.

In the allyl cation only the orbital π_1 is occupied (**8-20a**). Since this orbital is bonding between C_1 and C_2 and between C_3 and C_2 there exists some double bond character between the carbon atoms. We may extend the idea of bond order from Section 7.1.5a to the polyatomic case. If n_b and n_a are respectively the number of electrons in bonding and antibonding orbitals and the orbitals concerned are spread over k linkages then the bond order *per linkage* is equal to

$$(n_b - n_a)/2k \tag{1}$$

For the present case since there is one π bonding electron pair but two C—C linkages the π bond order is $\frac{1}{2}$, i.e., there is only partial double bond character between the carbon atoms. This result is in accord with the idea of resonance between the two mesomeric structures of **8-21**. Here too, on average the C—C bond order is between

8-21

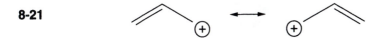

a single bond and a double one. These mesomeric structures also suggest that the positive charge is located on the end atoms, C_1 and C_3, of the molecule, and not on the central one. This is also in accord with the orbital picture. From **8.20** notice that the π electron density is preferentially located on C_2 in π_1, leaving the end atoms deficient in electron density.

In the allyl radical the extra electron occupies the π_2 orbital. Occupation of this nonbonding orbital does not change the bonding situation between the central atom and its neighbors. Indeed there is the same type of resonance between the two double-bond structures of **8-22**. The form of the molecular orbital tells us too that

8-22

the unpaired electron will be found on the terminal carbon atoms and not on the central one. Again this is indicated by the form of the mesomeric structures of **8-22**.

Finally, in the anion π_2 is doubly occupied (**8-20c**), and once again there is an increase in the electron density on the terminal atoms and no change in the C—C bond order. Just as in the neutral radical species and the cation the C—C π bond order is $\frac{1}{2}$. Just as in the radical the extra density residues on the terminal atoms (**8-23**).

8-23

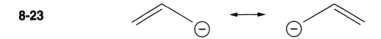

8.2.2. MOs and electronic structure of butadiene

The form and relative energies of the π MOs of butadiene may be deduced from those of the linear H_4 molecule (Section 4.1.4). Butadiene has a total of twenty-two valence electrons of which eighteen are involved with C—C and C—H σ bonds (3 C—C and 6 C—H). The π manifold thus contains four electrons leading to the ground electronic configuration $\pi_1^2\pi_2^2$ (**8-24**). The form of the occupied MOs tells us about the relative strengths of the central (C_2—C_3) and terminal (C_2—C_1 and C_3—C_4) CC bonds. Since π_1 is bonding between all three pairs of atoms, and π_2 bonding between C_2—C_1 and C_3—C_4 but antibonding between C_2 and C_3, we should expect to see shorter terminal than central CC bonds. This is borne out experimentally. The terminal CC distances are clearly short (134 pm) and appropriate for a C=C double bond. The central CC distance is 148 pm which is somewhere between the value expected for a C=C and a C—C (154 pm). In order to show this we really need to evaluate the bond order in a more sophisticated way. Not all bonding orbitals

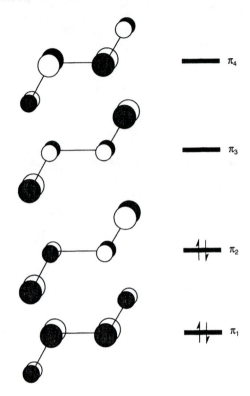

8-24

are equally bonding of course, neither are all antibonding orbitals equally anti-bonding. Thus equation (1) is useful only as a first approximation. We might expect that orbitals containing large coefficients on the two atoms on either side of the bond would be more bonding (or antibonding) than orbitals where the coefficients are small. A qualitative argument taking into account the relative sizes of the p_z coefficients on the carbon atoms in the π_1 and π_2 orbitals will suffice here. The contributions to the bond order of the terminal linkages occur both in π_1 and π_2 via the product of large and small coefficients on adjacent atoms. For the central linkage however, the bonding contribution in π_1 results via the product of two large coefficients, but the antibonding contribution in π_2 via the product of two small coefficients. Thus the orbital picture automatically leads to a reduction of double bond character between C_2 and C_3 rather than complete removal of the π component. This result can be obtained, though somewhat arbitrarily by consideration of the Lewis structures. While the lowest energy structure shows a double bond for the terminal linkages and a single bond for the central one (**8-25**), admixture of the two resonating mesomeric structures of **8-26** leads to a strengthening of the central

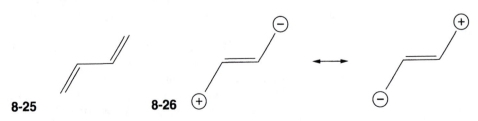

8-25   **8-26**

bond at the expense of the outer ones. The only problem with this argument is that, unlike the orbital picture we do not know *a priori* how important the structures of **8-26** actually are.

8.2.3. MOs and electronic structure of cyclopropenyl

The MOs of cyclopropenyl (**8-27**) are the π analogs of the triangular H_3 system of Section 4.1.5. In the cation there are fourteen valence electrons. After subtracting the twelve electrons which characterize the C—C and C—H σ bonds, there remain two electrons which occupy the lowest energy π orbital of **8-27**.

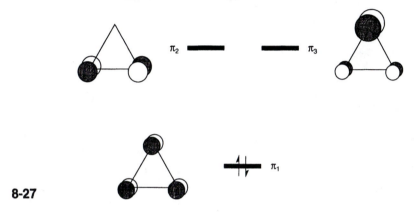

8-27

In this orbital, bonding between all three carbon atoms, the AO coefficients are equal such that the π bonding character is equal between all three centers. Since there is only one pair of bonding π electrons and three C—C linkages, the π bond order is equal to $\frac{1}{3}$. The C—C distance in the $C_3R_3^+$ unit is around 137 pm, shorter than a typical C—C single bond of 154 pm. The orbital picture meshes nicely with the set of three resonating mesomeric structures which are used to describe the ion (**8-28**).

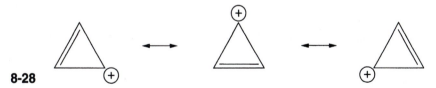

8-28

8.2.4. MOs and electronic structure of cyclobutadiene

The π orbitals of square cyclobutadiene (**8-29**) are the analogs of square H_4 described in Figure 4.1. The π_1 orbital is bonding between all carbon atoms, π_2 and π_3 are a degenerate pair of nonbonding orbitals and π_4 is antibonding. There are twenty valence electrons of which sixteen occupy C—C and C—H σ bonding orbitals, leaving four electrons for the π manifold. In the ground electronic configuration, $\pi_1^2 \pi_2^1 \pi_3^1$, two electrons occupy the bonding orbital while the other two are located, one each, in π_2 and π_3 with their spins parallel if Hund's rule were obeyed. Cyclobutadiene

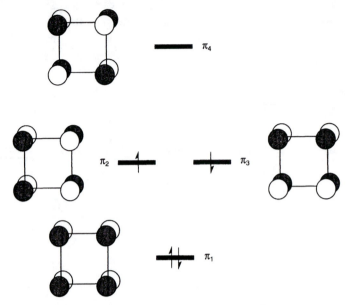

8-29

itself is not known in this geometry but its dianion $C_4H_4^{2-}$ is found as a square unit coordinated to transition metal atoms. In this species (as in the neutral molecule too) there is one pair of π bonding electrons for four C—C linkages and thus the C—C π bond order is $\frac{1}{4}$. The C—C distances found in such organometallic complexes are quite a bit shorter than a typical C—C single bond. The half occupation of the pair of degenerate orbitals π_2 and π_3 in the neutral molecule has some important geometric consequences which we will study later in exercises 9.2 and 9.3.

8.2.5. MOs and electronic structure of benzene

Benzene has thirty valence electrons of which twenty-four are associated with C—C and C—H σ bonds. The remaining six electrons occupy the lowest three orbitals of the π manifold (**8-30**), the analogs of the orbitals of cyclic H_6 as shown in **4-16**. In the deepest-lying orbital, π_1, the orbital coefficients are such that the bonding character between each pair of adjacent carbon atoms is equal. In π_2, however, bonding only occurs between the pairs of atoms C_2 and C_3 and between C_5 and C_6 since the coefficients on C_1 and C_4 are identically zero. In π_3 though, while C_1 is bonded to C_2 and C_6, C_4 is bonded to C_3 and C_5, there are antibonding interactions between C_2 and C_3 and between C_5 and C_6. This state of affairs reminds us of the situation in H_3 described in the Appendix to Chapter 4, where although the form of each of the degenerate orbitals suggests unequal interactions around the ring the equal occupation of both of them leads to completely equivalent interactions between all pairs of connected atoms. Thus if we consider the pair of orbitals π_2 and π_3 the contribution to C—C π bonding is equal for each bond. Since there are three occupied bonding orbitals and six CC linkages the π bond order is $\frac{1}{2}$. This description is in accord with the two resonating mesomeric forms (the Kekulé structures) of **8-31a** in which single and double bond characters alternate around the ring. Conventially we use the description of **8-31b** to show that the six π electrons are delocalized around

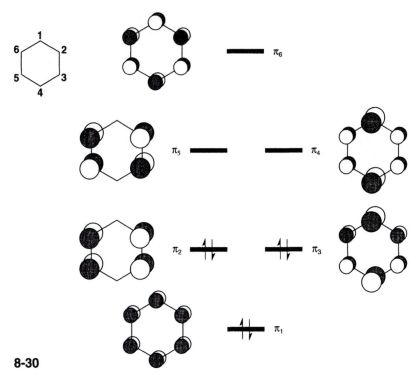

8-30

the ring as initially described in Chapter 1. The observed C—C distances of 140 pm are nicely in accord with this view, lying between typical single bond (154 pm) and double bond (134 pm) distances.

8-31a ⟷ **8-31b**

8.2.6. Aromatic and antiaromatic compounds and Hückel's rule

Consider the three cyclic hydrocarbons just described, C_3 (cyclopropenyl), C_4 (cyclobutadiene) and C_6 (benzene). Their orbital patterns are in fact quite similar. They each have an orbital, bonding everywhere, at lowest energy. This is followed at higher energy by pairs of orbitals. In the C_3 and C_4 systems there is just one pair, π_2 and π_3, but in the C_6 molecule there are two, π_2 and π_3, and π_4 and π_5. From the point of view of chemical stability, molecules prefer to have closed shells of electrons and we know from experience that the benzene molecule is quite a stable species. Its electronic configuration is such that the lowest three orbitals are occupied. Similarly the $C_3R_3^+$ ion is known (R is an organic group), both accompanied by an anion (e.g., SbF_6^-) or complexed to a transition metal atom. The square planar S_4^{2+} molecule is also known. It has the same number of valence electrons as $C_4H_4^{2-}$, a sulfur lone pair replacing the two electrons in each C—H bond. $C_3H_3^+$ has two π electrons which fill the deepest lying π orbital (π_1) and $C_4H_4^{2-}$ has six π electrons which fill the bonding π orbital (π_1) and the two nonbonding orbitals π_2 and π_3, thus satisfying our criterion. The general rule for stability of these cyclic conjugated hydrocarbons is that they possess $(4n + 2)\pi$ electrons (where $n = 0, 1$, etc.). This is *Hückel's rule*. The two electrons fill the lowest energy orbital (π_1) whereas the $4n$ electrons fill the n pairs of doubly degenerate π orbitals above it. The counter n is of course equal to zero for $C_3H_3^+$ but equal to 1 for $C_4H_4^{2-}$ and for benzene. Molecules which satisfy this rule are said to be *aromatic*, those which do not *antiaromatic*. Molecules of the latter type are often structurally unstable as we will see later for C_4H_4 itself in Chapter 9.

EXERCISES

8.1 *Electronic structure of cyclopentadienyl* (C_5H_5)

The cyclopentadienyl molecule, and its substituted derivatives, are very important as ligands in transition metal chemistry, leading to stable organometallic complexes.

(i) Give the form and relative energies of the π MO for this radical using the MOs found for the cyclic H_5 species in exercise 6.7.

(ii) How many π electrons are there? Write down the most stable electronic configurations.

(iii) In transition metal complexes cyclopentadienyl has a tendency to take an extra electron from the metal atom, leaving the metal in an oxidized state. Why is this?

8.2 *MOs of formaldehyde*

The form of the highest occupied and lowest unoccupied orbitals of formaldehyde (H_2CO) are given below.

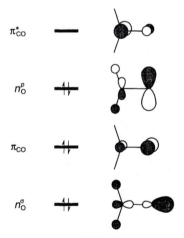

(i) Justify the names given to these orbitals.

(ii) The two excited states which are of chemical importance for this molecule result from the excitation of an electron from n_O^σ to π_{CO}^* ($n \to \pi^*$) and of an electron from π_{CO} to π_{CO}^* ($\pi \to \pi^*$). Describe how you would expect the CO bond length to change as a result of these excitations.

8.3 Find a qualitative explanation of the form and energetic ordering of the π MOs of benzene by assembling the MO diagram via interaction of two allyl units.

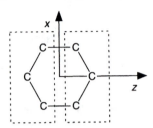

(i) Locate the energies of the different fragment π MOs on an energy diagram.

(ii) Which interactions are possible by symmetry? Select the single two-orbital interaction and thus deduce the form of two of the π orbitals of benzene.

(iii) To determine the four other π MOs of benzene, use the same method as used for the MOs of the A_2 molecule. First construct the provisional orbitals by only allowing degenerate MOs to interact. Then recover the final MOs by allowing mixing between the provisional orbitals of the same symmetry.

8.4 *MOs of cyclopropane*

Construct the three bonding MOs which characterize the C—C bonds of cyclopropane and their antibonding counterparts using the n_σ and n_p orbitals of the three methylene fragments which comprise the molecule. To simplify the problem assume that three orbitals are generated via interaction of the n_σ set and three others via interaction of the n_p set.

n_σ n_p

(i) Interaction of the n_σ orbitals. The orientation of the n_σ orbitals in the molecule is shown in the figure below. To construct the MOs consider the first fragment to be made up of the methylene groups 1 and 3 and the second to consist of the methylene group 2.

(a) Give the form and relative energies of the fragment orbitals.

(b) Which interactions are possible by symmetry?

(c) Give the form and relative energy of the three MOs obtained after interaction. (To help you make an analogy between this problem and that of triangular H_3 of Section 4.1.5.)

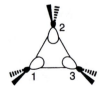

(ii) Interaction of the n_p orbitals. The orientation of the n_p orbitals in the molecule is shown in the figure below. Using the same fragmentation as before construct an orbital interaction diagram from the fragment orbitals. Give the form and energy of the MOs which you generate. (The two bonding levels are degenerate.)

(iii) Draw out on an energy diagram the three bonding orbitals which characterize the three C—C bonds.

8.5 H_3BNH_3

The molecule H_3BNH_3 consists of a pyramidal BH_3 fragment linked to a pyramidal NH_3 fragment.

  (i) Construct an MO diagram for the molecule.
 (ii) Identify the B—N bond order.
(iii) Indicate the direction of the dipole moment in the molecule.

8.6 Localize the occupied MOs of the BH_3 molecule to produce three localized functions which individually describe the three BH bonds.

8.7 Localize the occupied MOs of water to produce a pair of 'rabbit ear' lone pairs and two localized OH bonds.

APPENDIX

Bond localization

One of the obvious differences between the molecular orbital approach and the Lewis scheme is the very different way the bonds are represented in molecules such as BeH_2, BH_3 and CH_4. Lewis structures stress a localized approach where two electrons are assigned to a specific AH linkage. The molecular orbital approach describes the chemical bonds via occupation of bonding orbitals, delocalized over the entire molecule. One can readily see that the electronic description of a large molecule, such as cholesterol, will be very complex in molecular orbital terms, but will still be simply described by a Lewis structure. Although the MO description is the 'correct' one in the sense that there are two very different IPs for CH_4 rather than the single one expected on the localized approach, it is useful to search for a linkage between the two theoretical models.

Lewis' scheme works via the (arbitrary) division of the total number of bonding electrons equally between the bonds of the AH_n molecule. There is a way of similarly reapportioning the electron density obtained via the molecular orbital method to produce a similar but orbital-based description. This is by a process known as *bond localization*. The four electrons in BeH_2 lie in the pair of bonding molecular orbitals $1\sigma_g$ and $1\sigma_u$ (5-10) which have different energies. These energy levels have a well defined form and energy and are orthogonal. It is possible though to construct two new orbitals which while not having this property are localized between the Be and each hydrogen atom.

8-32 shows two new functions which can be created by adding together the $1\sigma_g$

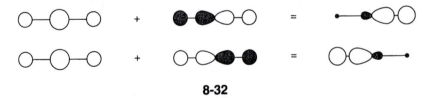

8-32

and $1\sigma_u$ orbitals or subtracting one from another. Notice that when added together with suitable coefficients the contribution from the left-hand hydrogen atom disappears completely and that the result is an orbital with mixed s and p character on the central atom and hydrogen $1s$ character on the right-hand atom. When the two MOs are subtracted one from the other an analogous orbital directed towards the left-hand hydrogen atom results. (The details of construction of the *hybrid* orbital on the central atom were described in more detail in Chapter 6.) These two new functions thus represent bonding orbitals localized between each of the two pairs of atoms. They are also of the same energy. This is a very general result. Given n doubly occupied bonding molecular orbitals delocalized over n 'bonds', a set of n localized orbitals may be generated, each lying between a bonded pair of atoms. If there are less occupied bonding orbitals than there are 'bonds' then this process is not applicable. An example is the π manifold of benzene. Here there are three bonding pairs but six close pairs of carbon atoms. Recall that the Lewis approach now requires the use of resonance structures to reproduce the observed equality of the six CC distances. The molecular orbital approach treats this problem no differently than in the BeH_2 case. The delocalized MO approach is applicable everywhere, the localized approach only under special conditions.

It is important to note that the new functions of **8-32** are not orthogonal and thus are not solutions of an eigenvalue equation of the Schrödinger type. Since they have the same energy, only one IP is predicted using them, a result we know is not true. However if we are interested in the *electron density* associated with the Be—H bonds then this is conceptually a useful way. This localization scheme is a very useful way to provide a link between the Lewis picture and that derived 'exactly' using the rules of quantum mechanics, but we must emphasize once again that the functions which result are not those which are allowed quantum mechanically and so must not be used in correlations with physical observables such as photoelectron spectra.

Part III
Introduction to the study of the geometry and reactivity of molecules

9 Orbital correlation diagrams: the model systems H_3^+ and H_3^-

As a rule a molecule adopts the geometry that minimizes its energy. There are often slight differences between the geometries of molecules studied in the gas phase and in solution or in crystals as the result of intermolecular interactions in the condensed phases. Sometimes the structures are completely different as a result of very strong interactions between the particles. Cesium forms a solid where each atom has twelve equidistant neighbors, whereas in the vapor above the boiling liquid Cs_2 molecules are prevalent. On the other hand the C—H bond length in gaseous methane is very close to that found in the crystalline solid. Most of what we will describe in the rest of the book applies to the structure of isolated molecules, as an approximation (often a very good one) to the structures of molecules under reaction conditions in solution.

The VSEPR model of Chapter 1 is based on the idea that the minimum energy geometry of a molecule is that which minimizes the repulsions between the valence electron pairs around the central atom. This is the structure where the pairs lie as far apart from each other as possible. The simplicity of this model is its great virtue and it is extremely successful in leading to a qualitative picture of the geometry just by counting electrons. The molecular orbital approach to the geometry problem is more difficult to use than the VSEPR scheme. It requires a detailed knowledge of the electronic structure of the molecule, relative energies, form of the molecular orbitals and of the electronic configuration. However the effort required to master this type of analysis is more than justified in the detailed information which results. It is not possible, for example, to deduce the structure of excited electronic states from VSEPR, but such information falls out naturally from the orbital picture. In addition, although we will not discuss such topics in this book, the molecular orbital picture is applicable to many other areas where VSEPR ideas are just not good enough. These include the structures of transition metal complexes and the realm of organometallic chemistry.

Two methods are largely used to study the geometries of molecules within the framework of molecular orbital theory, the study of correlation diagrams (Chapters 9 and 10) and interactions of fragments (Chapter 11). In this chapter we will illustrate the use of orbital correlation diagrams by studying the angular geometries of two very simple molecules, namely H_3^+ and H_3^-. H_3^- is not known and accurate calculations suggest it is unstable with respect to $H_2 + H^-$, but it is a prototypic system for several more complex molecules. H_3^+, on the other hand, is a molecule which exists in large quantities in interstellar space and, although a very reactive molecule, may be studied experimentally.

The orbital correlation method, which traces the energies of the molecular orbitals as a function of the molecular deformation (in this case θ, the H—H—H angle) is a method initially used by Mulliken and greatly extended by Walsh. The resulting diagram is often called a *Walsh diagram*. As always these may be constructed numerically *via* readily available computer programs. Here we use the approximate, but insightful methods developed in earlier chapters, rules based on symmetry and overlap ideas. Once the correlation diagram has been obtained then the variation in total energy for a given electron configuration may be studied and the lowest energy structure identified. For example, if there is an energetic stabilization on moving from a linear to bent geometry for a triatomic molecule then a bent geometry will be found. If there is an energetic destabilization on bending then the linear geometry will be found.

9.1. Rules for drawing orbital correlation diagrams

9.1.1. Stabilization or destabilization of the MOs

Construction of the orbital correlation, or Walsh, diagram consists of determining how each MO changes in energy on distorting the molecule. Bonding interactions lead to a stabilization, antibonding interactions to a destabilization. Geometrical changes result in changes in orbital overlaps which directly lead to changes in the orbital energy. Three simple rules are sufficient to understand them.

   (i) If bonding interactions increase and/or antibonding interactions decrease on distortion, the orbital is stabilized.

   (ii) If bonding interactions decrease and/or antibonding interactions increase on distortion, the orbital is destabilized.

   (iii) If the orbital character is such that on distortion there is no change in these interactions the orbital remains unchanged in energy.

A very simple example is the behavior of the MOs of the dihydrogen molecule as the molecule is stretched (**9-1**). Since σ_{H_2} is a bonding orbital (where the coefficients

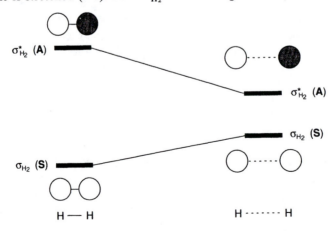

9-1

are of the same sign) the overlap integral decreases as the bond is stretched and this orbital is destabilized. The opposite is true in the antibonding $\sigma_{H_2}^*$ orbital (where the coefficients are of opposite sign) and this orbital is stabilized.

This type of analysis forms the basis for construction correlation diagrams. In addition two fundamental rules have to be obeyed. The first is the *conservation of orbital symmetry* during the process, and the second is that *orbitals of the same symmetry may not cross.*

9.1.2. Conservation of orbital symmetry

As a given molecule undergoes a structural change some elements of symmetry, present initially may disappear, but others are conserved during the geometrical distortion. It is useful to classify the symmetry properties of each MO with respect to this latter set of symmetry elements. An important rule is that the symmetry characteristics of all MOs are preserved along the distortion coordinate. In other words *an MO which is symmetric with respect to a given symmetry element remains so as the distortion proceeds.* In the same way if an orbital is antisymmetric with respect to a given element then it remains so.

To illustrate this rule let us return to the example of stretching the H—H bond in H_2 (**9-1**). This distortion preserves all of the symmetry elements present in 'normal' dihydrogen. In order to distinguish between the symmetry of σ_{H_2} and $\sigma_{H_2}^*$ we just need to keep the mirror plane which bisects the H—H bond. So the σ_{H_2} orbital, symmetric with respect to this plane in 'normal' H_2 is still symmetric in a considerably stretched molecule, and the σ_{H_2} orbital at the left-hand side of **9-1** correlates with the σ_{H_2} orbital at the right hand side of **9-1**. In the same way the pair of $\sigma_{H_2}^*$ orbitals (antisymmetric with respect to this symmetry element) are correlated with each other. We could imagine that if a symmetric orbital with its coefficients of the same sign did correlate with an antisymmetric one, with its coefficients of opposite sign, then somewhere along the distortion coordinate at the crossover point the coefficients would both have to be zero. This result makes no physical sense at all.

9.1.3. The non-crossing rule for MOs of the same symmetry

Suppose we have two MOs, S_1 and S_2 of the same symmetry (**9-2a**). During a distortion from geometry (I) to geometry (II) their energies will in general change. The non-crossing rule states that the energy curves associated with two MOs of the same symmetry may not cross. In other words, as in **9-2a** the lowest energy orbital (of a given symmetry) of structure I correlates with the lowest energy orbital of structure II. If the orbital composition of the two orbitals in I and II changes such that the two curves appear to cross (the dashed lines in **9-2b**) then one observes in practice an avoided crossing in the crossing region shown by the solid lines. This energetic avoidance gets stronger as the orbitals of the same symmetry come closer in energy. The two orbitals avoid each other for a very simple reason. Being of the same symmetry they can interact with each other. Thus the lower energy orbital is stabilized and represents an in-phase or bonding mixture of the two starting orbitals. Analogously the higher energy orbital is destabilized and represents the out-of-phase or antibonding mixture of these two orbitals. The result is a smooth energetic evolution of both orbitals. Of course, depending upon the size of the overlap, sometimes the energy separation at the crossing points will be large (a strongly avoided crossing) and sometimes it will be small (a weakly avoided crossing). We note finally that we should expect quite a change in the form of the MOs during the distortion as a result of such mixing.

If the symmetry of the two orbitals are different then they remain orthogonal all along the distortion coordinate the no avoided crossing occurs. Such an allowed crossing is shown in **9-2c** for the two orbitals labeled A and S.

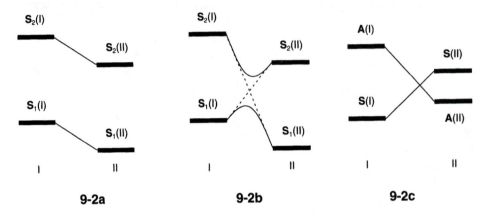

9.2. Orbital correlation diagram for bending H_3

Here we study the evolution of the MO diagram on bending an H_3 unit with special reference to the geometries expected for H_3^+ and H_3^-. The same diagram will apply in both cases; only the orbital occupation is different.

9.2.1. Geometrical model

This distortion is characterized by a single geometrical parameter, the angle θ between the H_a—H_b and H_b—H_c bonds (**9-3**). We assume that these H—H bond lengths remain equal and constant during the distortion as θ is varied. From a geometric point of view the distortion leads to a shortening of the $H_a \ldots H_c$ distance until $\theta = 60°$ all three H—H distances are equal.

9-3

P H_a H_b H_c $\longrightarrow$ P H_b H_a H_c

$\theta = 180°$ $\theta < 180°$

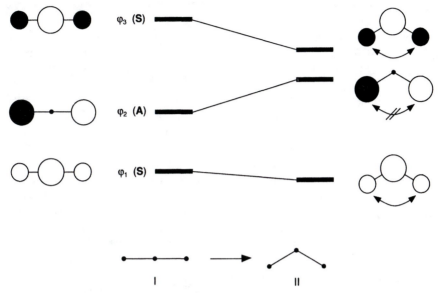

Figure 9.1. Energetic evolution of the MOs of an H_3 unit on bending.

9.2.2. Symmetry of the MOs

The starting point for the analysis of the orbital structure of H_3 is the linear geometry. The form and relative energy of the orbitals, established in Chapter 4, are shown at the left-hand side of Figure 9.1. The orbital ϕ_1 is bonding between neighboring atoms, ϕ_2 is nonbonding and ϕ_3 is antibonding.

We saw in the previous section the importance of a knowledge of the symmetry properties of MOs with respect to those symmetry elements which are preserved along the distortion coordinate. The motion from the linear geometry (I) to the bent geometry (II) is accompanied by the loss of several symmetry elements. For example the inversion center in I at H_b is absent in II. Such symmetry elements will not need to be considered. Of those that remain we will use the plane P, perpendicular to the plane of the molecule and bisecting the $H_a H_b H_c$ angle (**9-3**). With respect to this plane, ϕ_1 and ϕ_3 are symmetric (since the coefficients of the terminal hydrogen atom orbitals are of the same sign) and ϕ_2 is antisymmetric (since these coefficients are of opposite sign.) The rule of conservation of orbital symmetry requires that such labels are valid for all θ.

9.2.3. Energetic evolution of the MOs

We need to study the changes involved with the bonding and antibonding interaction within each MO on bending in order to analyze their energetic evolution.

First, we examine the interactions between the atoms which are bonded in the linear geometry, (H_a, H_b) and (H_b, H_c). Within our geometric model, since the distances between these atoms do not change, the bonding interactions in ϕ_1 and the antibonding ones in ϕ_3 do not change as θ is varied, the central $1s$ orbital being spherically symmetrical. Since the central atom coefficient in ϕ_2 remains equal to zero all along the distortion coordinate the interactions between the bonded atoms remain zero too. Thus from the point of view of interactions between the pairs of bonded atoms of the linear geometry the energy of all three orbitals remains invariant with θ.

The factor which controls the energetic evolution of the MOs on bending is the interaction which develops between the terminal atoms H_a and H_c as θ decreases. In ϕ_1, the coefficients on H_a and H_c are of the same sign (Figure 9.1) for all θ. As the molecule bends a bonding interaction develops between the two end atoms and the orbital is stabilized. The same argument can be applied to ϕ_3. Its energy drops on bending too as a stabilizing interaction develops between the end atoms. On the other hand since these coefficients are of the opposite sign in the antisymmetric orbital ϕ_2, an antibonding interaction develops here on bending, resulting in a destabilization of this orbital. Thus a simple examination of the relative signs of the coefficients on the terminal atoms, H_a and H_c allows one to predict the energetic evolution of the orbitals on going from linear to bent H_3.

We can ask what happens energetically to these levels as the molecule is severely bent to an equilateral triangle ($\theta = 60°$) and beyond ($\theta < 60°$). Certainly the orbital interactions we have described increase in magnitude as the angle decreases. An important point however concerns the possible crossing of the orbitals concerned. The non-crossing rule does not allow the two symmetric orbitals, ϕ_1 and ϕ_3, to cross but does allow a crossing of the levels ϕ_2 and ϕ_3 since these are of different symmetry. We have just shown how ϕ_3 drops in energy but ϕ_2 increases in energy as θ decreases. At some stage they will cross. This in fact takes place at the equilateral triangular geometry where $\theta = 60°$ (Figure 9.2). Indeed the orbital structure here, a pair of degenerate antibonding orbitals lying above a non-degenerate bonding one, is just what was found in Chapter 4 for this molecule. For $\theta < 60°$ ϕ_3 steadily proceeds downward in energy and the energies of ϕ_2 and ϕ_3 are inverted compared to those for $\theta > 60°$.

There is an interesting correspondence between the orbital correlation diagram of Figure 9.2 and the picture of Figure 4.8 which showed the result of substituting one H atom by an atom X with an electronegativity different from that of hydrogen. The resemblance is not completely fortuitous. The two perturbations of the equilateral triangular structure, making $\theta \neq 60°$ or changing the electronegativity of one of the atoms, both split the degeneracy. Since the degeneracy, as we have noted results from the presence of a three-fold rotation axis in the molecule, either perturbation is sufficient to remove it and hence lift the degeneracy of the pair of levels.

9.3. Geometry of H_3^+

The H_3^+ molecule possesses two valence electrons, which in the ground electronic state occupy the deepest lying MO, ϕ_1. From Figure 9.2 it is clear that H_3^+ must adopt a bent geometry since ϕ_1 is stabilized as θ becomes smaller.

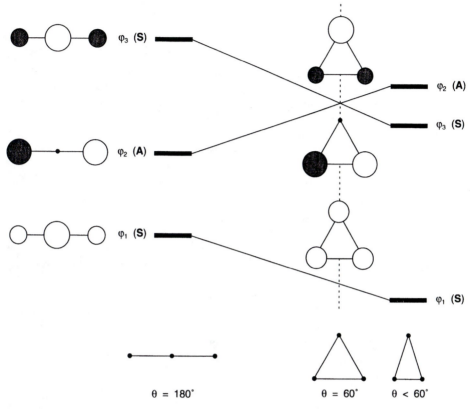

Figure 9.2. Orbital correlation diagram for the transformation H_3 (linear) to H_3 (triangular).

We must note though that this conclusion is based on the change in the electronic energy of H_3^+ associated with double occupation of ϕ_1. We should remember though that it is the total energy (electronic plus nuclear) which governs the geometries of all molecules. In fact the electronic energy, as we have defined it, actually follows the total energy quite well, except of course we have ignored the strong nuclear repulsion term between H_a and H_c that will become increasingly important as these two atoms come closer together. The limitations of the qualitative approach for H_3^+ are now apparent. The minimum in the total energy of the molecule as θ decreases is a balance of the electronic driving force which favors a bent geometry and the nuclear repulsions between H_a and H_c which increases with smaller angles. A quantitative treatment which takes into account both of these factors shows that H_3^+ indeed has an equilateral triangular geometry with $\theta = 60°$, a theoretical result which, as noted earlier, is in accord with experiment. (It is interesting to note that given a simple restriction the equilateral triangular geometry is one to be expected if the nuclear repulsions alone are taken into account. Using a model where the sum of the H—H distances remains constant on distortion, the smallest nuclear repulsion between the three nuclei occurs for $\theta = 60°$. We leave the reader to show this.)

9.4. Geometry of H_3^- and the rule of the highest occupied MO

H_3^- has four valence electrons so that in its lowest energy state the lowest two MOs are doubly occupied. From Figure 9.2 we can see that on bending away from the linear geometry ϕ_1 is stabilized but ϕ_2 is destabilized. Thus it will be difficult to make any predictions concerning the geometry of H_3^- since the two contributions to the electronic energy change in opposite directions. In general this will be quite a severe problem since most molecules of interest contain more than just a single pair of electrons. One way around the problem is to perform some numerical calculations and actually enumerate the two contributions. There are however some rules which will help enormously.

9.4.1. Rule of the highest occupied MO (HOMO)

This rule simply states that in many cases the molecule adopts the geometry which best stabilizes the *highest occupied MO (HOMO)*. It implies that variations in the energy, via the changes in the overlaps of the atomic orbitals which it comprises, are dominant in setting the lowest energy geometry. We may partially justify the rule in the following way. For the H_2 molecule we showed in Chapter 3 that there is a larger destabilization of the antibonding orbital than stabilization of its bonding analogue. On changing the H—H distance (which changes the overlap) then, as a result, the antibonding orbital changes in energy much more appreciably than the bonding orbital does. The same effect is found in general for the MOs of more complex molecules. As the energy increases and the stack of orbitals is climbed, their energetic variation during geometry changes increases. It turns out because of this, that the energetic variation of the highest occupied orbital of the set is extremely important in setting the equilibrium geometry.

It goes almost without saying that such a simple rule has some exceptions. Here we will mention two which will be discussed in the next chapter.

(i) When the HOMO is only occupied by one electron. This occurs in radical species and in excited electronic states. The problem here is easily understood. The energetic driving force associated with the HOMO is only half as large as for the case where it contains two electrons.

(ii) When the distortion is such that the energy of the HOMO does not change with geometry. In this case we need to consider the properties of the orbital immediately below it in energy, that is to say the highest occupied orbital whose energy *does* change with geometry.

9.4.2. Geometry of H_3^-

For linear H_3^- the HOMO is the antisymmetric orbital ϕ_2 (Figure 9.2). Since this orbital is destabilized on bending, our rule above tells us that the H_3^- molecule will remain linear. This is also the geometry which minimizes the nuclear repulsions between the hydrogen atoms using a model which keeps the H_a—H_b and H_b—H_c distances equal during the distortion.

It is useful to see how the electronic energy varies as the molecule is bent from $\theta = 180°$ through the equilateral triangle with $\theta = 60°$ to even smaller values of θ (**9-4**). In the linear geometry the electronic configuration is $\phi_1^2 \, \phi_2^2$. As the molecule bends ϕ_2 is destabilized, leading to an energetic preference for the linear geometry,

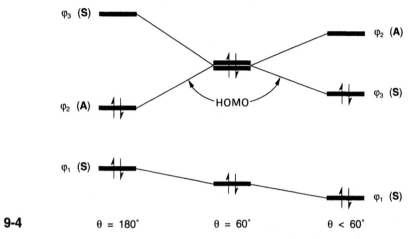

9-4 $\theta = 180°$ $\theta = 60°$ $\theta < 60°$

but at $\theta = 60°$, ϕ_2 and ϕ_3 are degenerate, and for $\theta < 60°$ ϕ_3 is now the HOMO. This orbital, ϕ_3, starts off at high energy at $\theta = 180°$ but is stabilized as θ decreases. Thus past $\theta = 60°$ the energy of the HOMO decreases and lower values of θ become energetically favored. The behavior of the electronic energy on bending, $E(\theta)$, is thus given by a curve of the form shown in Figure 9.3. The equilateral triangular structure lies at an energy maximum, a result exactly opposite to that found for H_3^+ with an energy minimum at this point. There are two energy minima for H_3^-, one at the linear geometry and one for an isosceles triangle. Judging the lower energy arrangement using a qualitative approach is difficult. Although quantitative calculations show that for reasonable H—H distances (around 100 pm) the linear structure is preferred (and shown this way in Figure 9.3), there is qualitative result which we may use to show that the isosceles triangular structure is unlikely in other ways. **9-5** shows the occupied MOs of H_3^- in this geometry. Both ϕ_1 and ϕ_3 are bonding

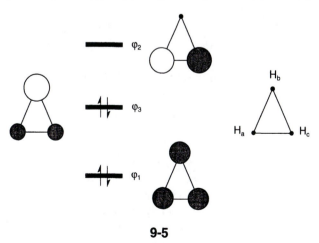

9-5

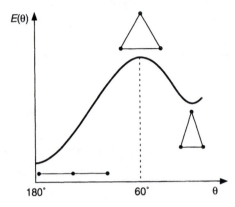

Figure 9.3. Energy of H_3^- as a function of the HHH angle θ.

between H_a and H_c leading overall to a strong bond between these two atoms. However ϕ_1 is bonding but ϕ_3 antibonding between H_b and H_a or H_c. Thus in this geometry although H_a and H_c are bonded together, H_b has no bond to either H_a or H_c. From the size of the atomic orbital coefficients in the two MOs we can see that the electron density lies preferentially on H_b. So a useful model, applicable to the isosceles triangular structure of H_3^- is of an H^- ion in contact with an H_2 molecule. H^- is a closed shell species, just like the He atom, and the H_2 molecule has one deep lying doubly occupied orbital. As a result the interaction between the two species is very much like the repulsive interaction between two closed shell He atoms. In other words if we made an H_3^- molecule with this geometry it would immediately dissociate to H^- and H_2. We return to this point in more detail in Section 11.2.2.

9.4.3. The Jahn–Teller effect

From **9-4** the instability of the equilateral triangular structure for H_3^- is directly related to the presence of two degenerate MOs which hold two electrons. Any variation of the geometry, whether to higher or lower θ lifts the degeneracy. As a result one orbital is stabilized and the other destabilized. With two electrons in the HOMO, either distorted structure is more stable than the equilateral one. This result, a very general one is known as the *Jahn–Teller effect*. Geometries where the electrons occupy the two components of a degenerate HOMO in an asymmetric way are always unstable with respect to some lower symmetry arrangement.

9.5. Conclusion

This preliminary study of the geometries of the H_3^+ and H_3^- species highlights at the same time both the predictive power of the qualitative orbital correlation method and its limitations. It dramatically illustrates the primary importance of the number of valence electrons on the geometry of a molecule. Thus the equilateral triangular geometry is the most favored one for H_3^+ with two electrons, but the least favored for H_3^- with four electrons. The same ideas may be generally applied to much more complex molecules. For example the geometrical change associated with coordination of oxygen to hemoglobin to give oxyhemoglobin, of immense biological importance, may be simply understood in this way. Since some orbitals go up in energy during a distortion, and some go down the overall energetic behavior will crucially depend upon the number of valence electrons and the nature of the HOMO.

EXERCISES

9.1 Construct an orbital correlation diagram for the MOs of H_4 undergoing the following distortion: tetrahedral (I) → butterfly (II) → square plane (III). During the first step the angle H_c—O—H_d changes from 109.5° to 180° while H_a—O—H_b remains fixed. (O is the origin of the geometrical coordinates appropriate to the problem.) In the second step H_a—O—H_b changes from 109.5° to 180° with H_c—O—H_d remaining fixed. The distances of the hydrogen atoms to the origin remain constant during both steps.

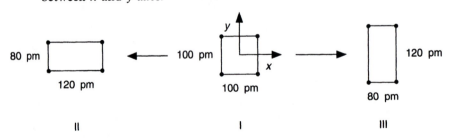

Start off with the MOs of tetrahedral H_4 and trace out their energetic changes during the two steps. Verify that the orbitals you end up with in structure III are those obtained in exercise 4.4, for the square plane using this axis system.

9.2 (i) Construct the orbital correlation diagram for the MOs of square planar H_4 (I) as it is distorted to a rectangular geometry (II or III). For the square planar structure, use the MOs given in Section 4.1 where the atoms lie between x and y axes.

80 pm [rectangle] ← 100 pm [square with y,x axes] → [rectangle] 120 pm

120 pm   100 pm   80 pm

II   I   III

(ii) Using the rule of the HOMO predict which geometry, square or rectangular, is most favorable for the electronic configuration where all the electrons are paired up in the lowest energy MOs.

(iii) What do you predict will be the geometry of cyclobutadiene? (See Section 8.2.4.)

9.3 Answer the same questions as in 9.2 except consider the distortion of the square plane (I) to a rhombus (II or III). For square planar H_4 use the MOs established in exercise 4.4 where the atoms lie along the x and y axes.

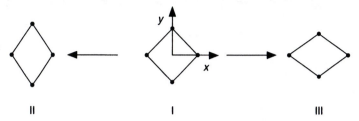

10 Geometry of AH_2 and AH_3 molecules

In this chapter we will apply the methods developed earlier for studying the geometries of H_3^+ and H_3^-, to molecules a little more complex. Specifically we will look for a way of predicting whether AH_2 molecules are linear or bent and whether AH_3 molecules will be planar or pyramidal. We will study those cases where A is an element from the second or third row of the periodic table. At the end of the chapter we will see how the model allows prediction of the geometries of AX_n molecules where X is also a non-transition element.

10.1. AH_2 molecules

Molecules of this type may be either linear or bent. If we assume that the A—H distances remain constant then the geometry problem is described by a single parameter, the angle θ between the two bonds (**10-1**). The analysis of this problem

10-1 $\theta = 180°$ $\theta < 180°$

uses the same strategy employed for the model systems H_3^+ and H_3^-. There are four steps.

  (i) Start off with the orbitals of linear AH_2 with $\theta = 180°$.
 (ii) Determine the symmetry properties of the MOs with respect to the symmetry elements preserved during the deformation.
(iii) Trace out the correlation diagram of the orbitals as the molecule is bent ($\theta < 180°$). To do this study the changes in the orbital overlaps in each orbital. Take into account the non-crossing rule for orbitals of the same symmetry.
 (iv) Finally place the relevant number of electrons into the collection of MOs to find the variation of the total energy with angular geometry.

10.1.1. MOs of linear AH₂

The MOs of the linear AH_2 molecule were constructed in Chapter 5. Briefly we recall that there are (**10-2**):

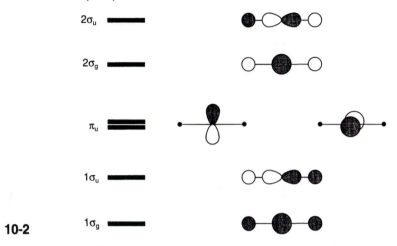

10-2

(i) Two MOs bonding between the central atom and the hydrogen atoms ($1\sigma_g$ and $1\sigma_u$).
(ii) Two antibonding MOs ($2\sigma_g$ and $2\sigma_u$).
(iii) At an intermediate energy a pair of nonbonding, degenerate orbitals (π_u) namely the p_x and p_z orbitals localized completely on the central A atom. (These orbital labels are appropriate for the choice of y as the internuclear axis of the linear molecule.)

10.1.2. Orbital correlation diagram, linear to bent AH₂

(a) Conserved symmetry elements

When the AH_2 molecule bends away from the linear geometry certain elements of symmetry disappear (**10-3**). These include the rotation axis y, the inversion center located at A and all symmetry planes which contain y (with the exception of yz). Three symmetry elements are preserved; (i) the two-fold axis around z, C_2^z, i.e., a rotation of 180° around z which interchanges the two equivalent hydrogen atoms, (ii) the plane xz which bisects the HAH angle and (iii) the yz plane which contains all three atoms.

10-3

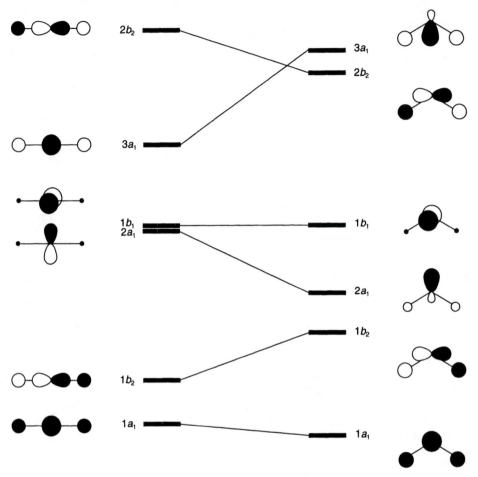

Figure 10.1. Orbital correlation diagram for an AH$_2$ molecule as it bends away from the linear geometry.

The bending of the molecule is accompanied therefore by a reduction in symmetry. Group theory has labels to describe this. We say that the symmetry has dropped from $D_{\infty h}$ to C_{2v}. We now need to redefine the symmetry properties of the MOs of linear AH$_2$ with respect to the new collection of symmetry elements. We will look in turn at each orbital at the left-hand side of Figure 10.1.

The orbitals which earlier we labeled $1\sigma_g$ and $2\sigma_g$ are symmetric with respect to all three elements (SSS). The same is true of the nonbonding p_z orbital, one of the degenerate π_u pair. Using group theoretical conventions we use the label a_1 to describe functions which are totally symmetric in the C_{2v} group. Thus $1\sigma_g$ (bonding) becomes $1a_1$, p_z (nonbonding) becomes $2a_1$ and $2\sigma_g$ (antibonding) becomes $3a_1$. The orbitals $1\sigma_u$ and $2\sigma_u$ are antisymmetric with respect to the two-fold rotation around z and the xz plane but symmetric with respect to the yz plane (AAS). They become respectively the $1b_2$ and $2b_2$ orbitals. Finally the p_x orbital, the other partner of π_u, is antisymmetric with respect to rotation around z and on reflection in the yz plane, but symmetric with respect to the plane xz (**ASA**). It is labeled $1b_1$.

(b) Correlation diagram

The orbital correlation diagram which shows the energy changes of the six MOs of AH$_2$ as a function of θ is given in Figure 10.1. In order to understand their energetic evolution we need to study the variation in atomic orbital overlap within each molecular orbital. We assume that the A—H distances remain constant.

The $1a_1$ orbital is bonding everywhere. A change in the angle θ does not change any of the overlaps between the central atom and the hydrogen $1s$ orbitals. The s orbital on the central A atom is spherically symmetrical and these overlap integrals only depend upon the AH distance. However a bonding interaction does develop between the two hydrogen atoms since their coefficients are in phase. This increases as the H . . . H distance (and thus θ) decreases. This orbital is thus weakly stabilized via this interaction between nonbonded atoms.

The $1b_2$ orbital is bonding between A and H, and antibonding between the two hydrogen atoms. When the molecule bends, two factors contribute to the destabilization of this orbital (**10-4**). First the bonding interaction between A and H decreases,

10-4

since the p_y orbital does not point directly at the $1s_H$ orbitals any more. Second the antibonding interaction between the two hydrogen atoms increases as θ decreases. The first is the more important of the two since it is associated with interactions between neighboring atoms. The $1b_2$ orbital is thus quite strongly destabilized.

The nonbonding orbital $1b_1$ $(=p_x)$ has no hydrogen atom contribution at the linear geometry. In addition since the two hydrogen atoms remain in the yz plane for all θ they may not mix at all with the p_x orbital. This orbital thus remains unchanged on bending (**10-5**), and is a purely A located nonbonding p_x orbital.

10-5

For the antibonding $2b_2$ orbital the distortion leads to the same two effects present for $1b_2$, except that here they operate against each other (**10-6**). The reduction in the antibonding interaction between p_y and $1s_H$ as the molecule bends leads to a stabilization. Opposing this is the increasing antibonding effect between the two

10-6

hydrogen atoms. Just as before, the first factor, operating between adjacent atoms is more important, and the orbital is stabilized.

The change in the energies of the $2a_1$ and $3a_1$ orbitals on bending are not quite so straightforward. The arguments used up until now relied on the fact that the atomic orbital composition of the MOs changed little during the distortion. Let us call $2a_1^o$ and $3a_1^o$ the orbitals which we would *expect* to find using this supposition and $2a_1$ and $3a_1$ the 'real' orbitals that we would get from a calculation, or by using the approach of Section 6.3. The energy of $2a_1^o$ $(=p_z)$ which has no hydrogen contribution in the linear geometry is therefore expected to remain unchanged on distortion. This is shown by the dashed line in **10-7**. On the other hand, in the $3a_1^o$

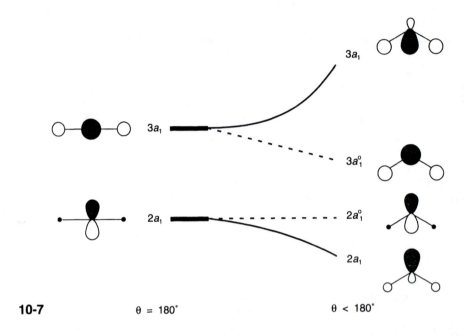

10-7 $\theta = 180°$ $\theta < 180°$

orbital, whereas the antibonding interactions between adjacent atoms remain unchanged on distortion (interactions between *s* orbitals at a fixed distance) the bonding interactions between the hydrogen atoms is strengthened on bending. This orbital will then be stabilized as shown by the dashed lines in **10-7**. In fact the situation for $2a_1$ and $3a_1$ is very different. As shown by the solid lines, $2a_1$ is stabilized and $3a_1$ destabilized on distortion. We can understand this behavior by simple recourse to symmetry arguments. Although these two orbitals have different symmetry in the linear molecule (π_u and $2\sigma_g$ in **10-2**), they become of the same symmetry (a_1) in the bent molecule, and will thus 'repel' each other. The lower energy orbital of the pair is stabilized $(2a_1^o \rightarrow 2a_1)$ and the higher energy orbital destabilized $(3a_1^o \rightarrow 3a_1)$ as the two mix together. Alternatively we could say that, although their overlap is zero at $\theta = 180°$, their overlap is non-zero for the bent case and gradually increases as the molecule bends away from the linear geometry. We may see this in a simple way. The $3a_1^o$ orbital is delocalized over both the A and $H_a \ldots H_b$ fragments as in equation (1).

$$3a_1^o = \lambda s - \mu(1s_a + 1s_b) \tag{1}$$

Its overlap with $2a_1^o$ $(=p_z)$ can therefore be broken down as the sum of two terms, the first between p_z and the s orbital part of $3a_1^\circ$, and the second

$$\langle 2a_1^\circ \mid 3a_1^\circ \rangle = \langle p_z \mid \lambda s - \mu(1s_a + 1s_b) \rangle \tag{2}$$

between p_z and the $(1s_a + 1s_b)$ component of $3a_1^\circ$. The first of these is identically zero for all θ since these two orbitals, located on the same atomic center, are orthogonal. The second term is, however, non-zero for all $\theta \neq 180°$, i.e., as soon as the hydrogen atoms leave the nodal plane of p_z. The closeness of the two energy levels $2a_1^\circ$ and $3a_1^\circ$, and the increasingly important overlap between them, leads to a strong interaction between the two on bending. The result is that $2a_1^\circ$ is depressed in energy and increasingly so as the molecule bends the antibonding analog being increasingly destabilized. For the latter this is so strong that $3a_1$ is destabilized on bending, in contrast to the stabilization afforded $3a_1^\circ$.

This orbital mixing leads, of course, to a significant change in the orbital description. The form of $2a_1$ is derived from the in-phase, bonding, combination of $2a_1^\circ$ and $3a_1^\circ$. To see how the mixing occurs in detail we need to look at the overlap of each component of $3a_1^\circ$ with the p_z orbital comprising $2a_1^\circ$. The in-phase mixture is found by adding the two orbitals together in a way which leads to a positive overlap integral. As shown in **10-8**, since the A atom s and p_z orbitals are orthogonal

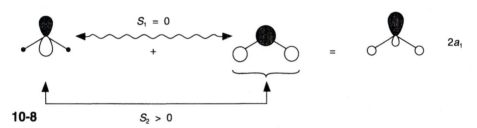

10-8

all we need to ensure is that the hydrogen atom contributions add in such a way that they make a positive overlap with p_z. The converse is true for $3a_1$ (**10-9**). In each case there is a mixing of the pure s and p orbitals on the central atom and the formation of an s/p hybrid, whose large lobe points outward in the case of $2a_1$ and inward, towards the hydrogen atoms in the case of $3a_1$. Thus $2a_1$, which starts off in the linear geometry as a nonbonding, purely p orbital, is stabilized partly because it has acquired some s character ($\varepsilon_s < \varepsilon_p$), and partly because of the bonding interaction between the hydrogen atoms (**10-8**). On the other hand, $3a_1$ is destabilized because of the development of an s/p hybrid whose large lobe is involved in an antibonding interaction with the hydrogen atoms (**10-9**). This contribution overwhelms the bonding interaction which develops between the hydrogen atoms.

10-9

We have therefore been able to take into account, in a qualitative way, the interaction between the levels $2a_1^o$ and $3a_1^o$ when the molecule bends. It is pertinent though to ask the question why do similar interactions not take place between $1a_1$ and $2a_1$ or between $1a_1$ and $3a_1$, or even between the two b_2 orbitals. Such interactions do in fact occur, but because of the large energy separation between these pairs of orbitals, they are sufficiently weak to be ignored on our simple scheme. However, inclusion of the $2a_1^o$–$3a_1^o$ mixing is essential to understand Figure 10.1. Finally we note that the orbital correlation diagram is just the same as that which would have been obtained by using the orbitals derived for bent AH_2 using the two fragments, A and $H_a \ldots H_b$.

10.1.3. Geometry of AH$_2$ molecules

Use of the orbital correlation diagram of Figure 10.1 allows us to understand the geometry changes observed for AH_2 molecules both as the number of electrons changes and when the molecule is in an excited electronic state. All that is needed is the filling of the relevant orbitals with electrons and asking how the total energy is expected to change on bending.

First, supposing that the molecules are in their electronic ground state, for species with only two valence electrons (H_3^+, LiH_2^+), only the $1a_1$ orbital is doubly occupied. Since this orbital is stabilized when the molecule bends, a nonlinear geometry is expected. The optimal value of θ varies from system to system, being $60°$ for H_3^+ and $\sim 22°$ for LiH_2^+. The latter is thus best described as a Li^+ cation coordinated to an H_2 molecule. For molecules with four valence electrons (BeH_2, BH_2^+) the occupied orbitals, $1a_1$ and $1b_2$ have opposite energetic behavior on bending. Using the rule that the energetics of the HOMO are dominant then the unfavorable behavior of $1b_2$ on bending indicates that a linear geometry is preferred. For molecules with five (BH_2, CH_2^+) or six (BH_2^-, CH_2, NH_2^+) electrons $2a_1$ is singly or doubly occupied respectively and bent geometries are found. Although we suggested in the previous chapter that the rule of the HOMO was often inapplicable for those cases where the HOMO was singly occupied, in the present case the stabilization of $2a_1$ is so large that its energetic behavior prevails even when only one electron is present. However, we should expect that the driving force for bending should be reduced for the case for one electron, compared to that for two. This is indeed true and is reflected in the values of the bond angles found in these molecules; $131°$ for BH_2 and $102°$ for BH_2^-. With seven (CH_2^-, NH_2, OH_2^+) or eight (NH_2^-, H_2O) electrons the $1b_1$ orbital is populated. Since this orbital does not change in energy on bending the geometries of these molecules are those expected by double occupancy of $2a_1$. Thus we find strongly bent structures for these molecules: $99°$ for CH_2^-, $103°$ for NH_2, $110°$ for H_2O^+, $104.5°$ for H_2O and $104°$ for NH_2^-.

Excitation of a molecule from its ground state to an excited electronic state usually involves a change in the way the electrons are arranged in the MOs. Often, as a result there is a change in angular geometry. Consider, for example, the five-electron molecule BH_2, bent in its electronic ground state. The excitation $2a_1 \rightarrow 1b_1$ depopulates the $2a_1$ orbital which favors a bent geometry, and populates the $1b_1$ orbital whose energy does not depend on θ. The geometry of the excited state species, BH_2^*, with the configuration $1a_1^2\ 1b_2^2\ 1b_1^1$, is thus determined by the energetic behavior of $1b_2$. Since this favors the linear geometry, the electronic excitation is associated with the geometrical change, bent → linear. We thus have an explanation of the geometrical change in **1-39**. With seven valence electrons the NH$_2$ molecule is strongly bent. The excitation $1a_1^2\,1b_2^2\,2a_1^2\,1b_1^1 \rightarrow 1a_1^2\,1b_2^2\,2a_1^1\,1b_1^2$ involves promotion of an electron from $2a_1$ to $1b_1$. This is not sufficient to create a linear molecule, since one electron remains in $2a_1$, but the angle opens up from 103° to 144°.

CH$_2$ is an interesting molecule. We can envisage two forms, one with the spins paired, strongly bent with the configuration $1a_1^2\,1b_2^2\,2a_1^2$, stabilized by the strong change in $2a_1$ on bending but energetically unfavorable as regards the exchange energy associated with parallel spins, and the other less bent with the configuration $1a_1^2\,1b_2^2\,2a_1^1\,1b_1^1$ but stabilized by the presence of the two electrons with parallel spins (in $2a_1$ and $1b_1$). The two forms are indeed found close in energy. The molecule with its spins paired (we call this the *singlet*) has a bond angle of 110°, and is less stable than the molecule with parallel spins (we call this the *triplet*) with a bond angle of 136°. The reactions of CH$_2$ may be divided into two types depending on which spin state is involved.

10.2. AH$_3$ molecules

For the tetraätomic AH$_3$ molecules several geometries are possible. The molecule could be trigonal planar, trigonal pyramidal, T-shaped or pyramidal but without a three-fold axis. We limit ourselves here to the comparison of the trigonal planar and trigonal pyramidal systems where all H—A—H angles are equal. If the A—H distances are assumed to remain constant, a single parameter, θ of **10-10** suffices to

10-10

describe the distortion. This angle, between the three-fold axis and one A—H bond varies from 90° (planar) to >90° (pyramidal). A value of θ of 109.5° corresponds to a fragment of a regular tetrahedron with A at its center, and values up to around 115° are found for molecules such as NH$_3$ and OH$_3{}^+$.

The following discussion is identical to that used for the AH$_2$ molecules. Starting off with the orbitals of the trigonal planar molecule, those of the pyramidal variant may be generated.

10.2.1. MOs of trigonal planar AH$_3$

The MOs of trigonal planar AH$_3$, where A is an element from the second or third row of the periodic table, were constructed in Chapter 5. Briefly they are (**10-11**):

   (i) Three MOs bonding between the central atom and the hydrogen atoms. They comprise the totally symmetric orbital $1a_1'$ and the degenerate pair of orbitals $1e'$ ($1e_x'$ and $1e_y'$).

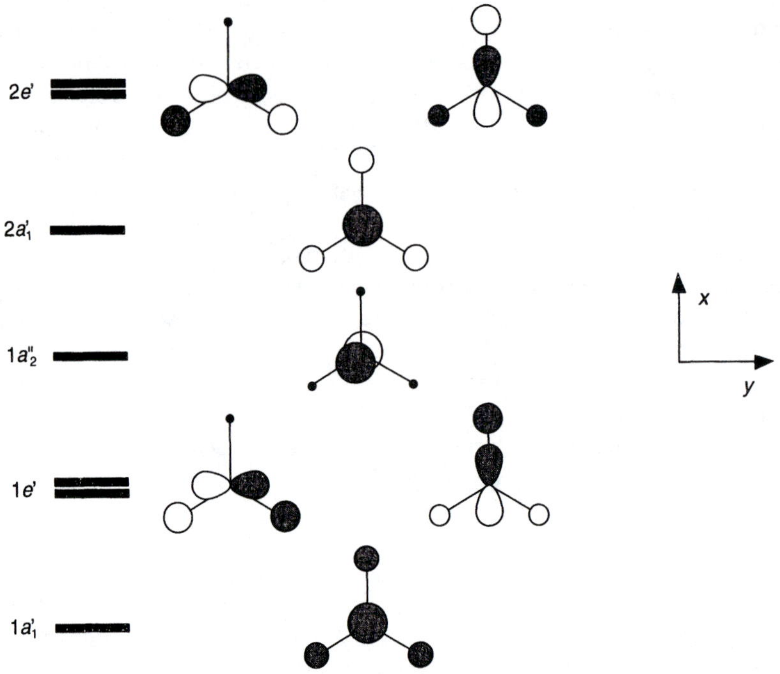

10-11

(ii) Three MOs antibonding between the central atom and the hydrogen atoms. These are the orbitals $2a_1'$ and $2e'$ ($2e_x'$ and $2e_y'$).

(iii) A nonbonding orbital a_2'' at an intermediate energy, a p_z orbital completely localized on A.

10.2.2. Orbital correlation diagram for trigonal planar to pyramidal AH_3

(a) Conserved symmetry elements

Several symmetry elements are lost as the trigonal planar AH_3 molecule distorts to a trigonal pyramid (10-10). Among these are the molecular plane, xy and the three two-fold rotation axes which contain an A—H bond. Others are conserved. These include the planes of symmetry which contain z and an A—H bond. Importantly, as we shall see, the three-fold rotation axis remains. Overall the symmetry of the molecule is lowered and, using the nomenclature of group theory, we say that the symmetry has been lowered from D_{3h} to C_{3v}. The new group theoretical labels for the orbitals are shown at the left-hand side of Figure 10.2.

The orbitals $1a_1'$ and $2a_1'$ are symmetric with respect to all of the elements of symmetry which are preserved. Since the xy plane of symmetry has disappeared the superscript is inappropriate and both of these orbitals are now labeled as a_1. The same is true of the p_z ($1a_2''$) orbital. It loses its double prime but also becomes an a_1 orbital, since it is symmetric with respect to the three-fold rotation and the three mirror planes. (Its subscript in the planar molecule told us that it was antisymmetric with respect to the two-fold rotation axes of the planar geometry, absent in the

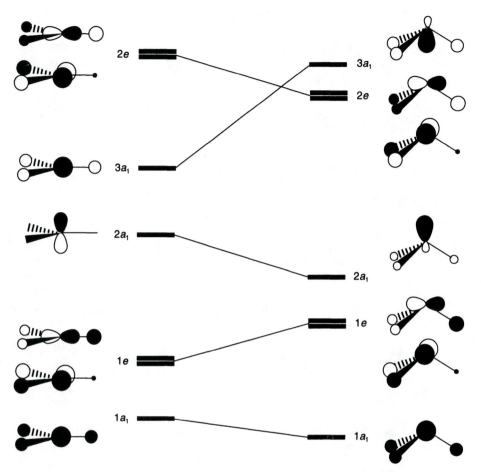

Figure 10.2. Orbital correlation diagram for an AH$_3$ as it bends away from a trigonal planar to trigonal pyramidal geometry.

non-planar one.) Thus $1a_1'$ becomes $1a_1$, $1a_2''$ becomes $2a_1$ and $2a_1'$ becomes $3a_1$, respectively bonding, nonbonding and antibonding orbitals. The orbitals $1e'$ and $2e'$ remain degenerate, a consequence of the retention of the three-fold axis, but just lose their primes. They then are labeled $1e$ ($1e_x$, $1e_y$) and $2e$ ($2e_x$, $2e_y$).

(b) Correlation diagram

The energetic evolution of the various MOs during the distortion is shown in Figure 10.2, and has some strong similarities to that for AH$_2$ in Figure 10.1.

For the $1a_1$ orbital pyramidalization does not change the overlaps between the central atom and hydrogen orbitals since the AH distances remain constant. This orbital is however, slightly stabilized since the hydrogen atoms come together, strengthening the bonding interaction between them. (Their coefficients are of the same sign.)

The pair of orbitals $1e_x$ and $1e_y$ are strongly destabilized for two reasons (**10-12**). The bonding interaction between the central atom A p orbitals and the hydrogen $1s$ orbitals decreases as the hydrogen atoms move away from the p orbital lobes. Also the hydrogen atoms come closer together resulting in an increase in the antibonding interactions between them. This is easy to see in $1e_y$ where the coefficients are of

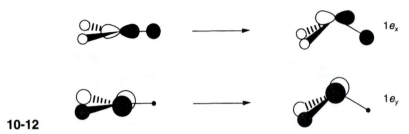

10-12

opposite sign on H_a and H_c. For $1e_x$ the small bonding interaction between H_a and H_c is outweighed by the large antibonding interactions between H_b and both H_a and H_c. Of the two effects the first is larger being an interaction between bonded atoms.

For the two antibonding $2e$ orbitals (**10-13**), pyramidalization leads to the same

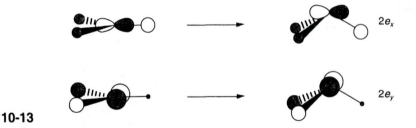

10-13

two contributions to the change in orbital energy, but here they oppose each other. The reduction in the antibonding interaction between the central atom and hydrogen leads to a drop in energy, but the increase in the antibonding interactions between the formally nonbonded hydrogen atoms leads to a destabilization. Since the first effect is the dominant one, the $2e$ orbitals are stabilized on bending.

It remains to look now at the two a_1 orbitals at intermediate energy namely $2a_1$ and $3a_1$. These look very similar indeed to the $2a_1$ and $3a_1$ orbitals of AH_2 (Figure 10.1). Just as before they will be strongly modified in character on bending. The behavior of the provisional orbitals, $2a_1^o$ and $3a_1^o$, on bending is shown by the dashed lines in **10-14**. $3a_1^o$ is stabilized via the increase in the bonding interaction between the hydrogen $1s$ orbitals, but $2a_1^o$ remains unchanged in energy since it is an orbital purely localized on A. However, these two orbitals, of different symmetry in the planar geometry, become of the same symmetry on pyramidalization, and may mix together, just as in AH_2. The result is shown by the full lines in **10-14**. As soon as the hydrogen atoms leave the nodal plane of p_z the overlap between $2a_1^o$ and $3a_1^o$ becomes non-zero. A strong interaction between the two leads to the two final MOs $2a_1$ and $3a_1$. The form of the lower energy partner is given by the bonding admixture of $2a_1^o$ and $3a_1^o$, set, as in **10-15**, by ensuring that the hydrogen contributions to

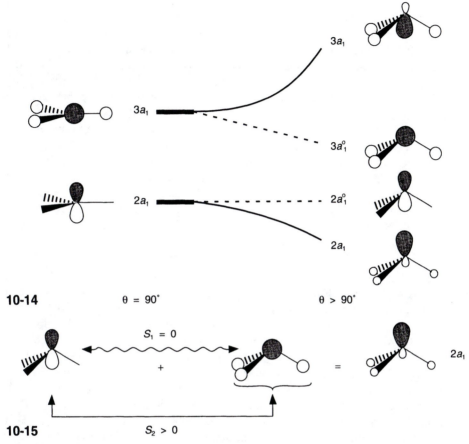

10-14   θ = 90°   θ > 90°

10-15 $S_2 > 0$

$3a_1^0$ have a positive overlap with p_z. The form of $3a_1$ is likewise determined by ensuring a negative overlap of this type (**10-16**). In each MO an s/p hybrid orbital is created whose large lobe points towards the hydrogen atoms in $3a_1$ and away from them in $2a_1$. The $2a_1$ orbital is stabilized via this admixture of s orbital character as for AH$_2$, and to a lesser extent by the direct in-phase interactions between the nonbonded

10-16 $S_2 < 0$

hydrogen atoms. On the other hand the $3a_1$ orbital is strongly destabilized by the presence of the large antibonding interaction between the large lobe of the hybrid and the hydrogen atom s orbitals. This is offset a little by the bonding interaction between the nonbonded hydrogen atoms. For strong distortions $3a_1$ becomes higher in energy than $2e$. This is an allowed crossing between orbitals of different symmetry.

We should finally note that this diagram is identical to that which would be generated by using the pyramidal AH_3 orbitals obtained by a different route in Chapter 6.

It is important at this stage to comment on the nature of these s/p hybrid orbitals which result from the mixing of the a_1 orbitals both in AH_2 and AH_3. The ratio of s to p character in the hybrid depends very much on the parameters describing the atomic orbitals of the problem as well as the size of θ. The extent of the mixing between $2a_1^o$ and $3a_1^o$ is controlled, as we have seen, by the overlap between the hydrogen $1s$ orbitals with the central atom p orbital. This will depend on the nature of A, the A—H distance and on the angle θ. For small distortions the mixing will be small but will increase with distortion. Dependence on the nature of A comes too via equation (1). The ratio of μ to λ will be set by the relative energetic location of the s orbitals of different types. Determination of a numerical value for the s/p mixture needs recourse to calculation for a specific case. The same comment applies to other molecules such as acetylene and ethylene where such mixing also occurs.

10.2.3. Planar or pyramidal geometries?

Here we look at the electronic ground states of molecules having between three and eight valence electrons. For molecules such as LiH_3^+ the $1a_1$ orbital is doubly occupied and one electron occupies the degenerate pair $1e$ (Figure 10.2). Their behavior on bending is different for the two orbitals. $1a_1$ is stabilized on bending but $1e$ is destabilized. Even though only one electron occupies $1e$, the HOMO governs the geometry and the molecule is planar. For four, five and six electrons $1e$ is progressively filled, and all these molecules are strongly planar. Examples include BeH_3^+, (four electrons), BeH_3 (five electrons) and BeH_3^-, BH_3, AlH_3, CH_3^+ and SiH_3^+ (six electrons). Although all of these molecules are planar, not all of them are trigonal. Those molecules which contain an incomplete $1e$ orbital are Jahn–Teller unstable in exactly the same way as described for H_3^- in Section 9.4.3. Recall that this molecule could distort either towards a linear geometry opening up one H—H—H angle towards 180° from 60°, or towards a geometry where this angle closes down from 60°. In a similar way Jahn–Teller unstable AH_3 molecules can distort either towards a T or Y shape as shown in **10-17** where one HAH angle

10-17

opens up from 120° or closes down from 120° respectively. During such a distortion one component of $1e$ is stabilized and the other destabilized.

For seven electron molecules, $1e$ is full and the last electron occupies $2a_1$ (Figure 10.2). Here the rule of the HOMO does not universally work. The single electron in $2a_1$, stabilized on bending, is often insufficiently stabilized so as to outweigh the destabilization of four electrons in $1e$. As a result some of these molecules, those containing a second row atom, such as BH_3^-, CH_3 and NH_3 are planar but others containing a third row atom, such as AlH_3^-, SiH_3 and PH_3^+, are pyramidal.

In all cases though the energy associated with bending is quite small. Molecules such as CH_3 are thus only tenuously planar. With two electrons in $2a_1$, all eight electron AH_3 molecules are pyramidal. Examples include CH_3^-, NH_3, OH_3^+, SiH_3^-, PH_3 and SH_3^+ with a valence electron configuration $1a_1^2\ 1e^4\ 2a_1^2$. The situation does change in excited states. For example, the first excited state of ammonia with the electronic configuration $1a_1^2\ 1e^4\ 2a_1^1\ 3a_1^1$ is strongly planar, a result of there being only one electron in $2a_1$ and one electron in $3a_1$, an orbital strongly destabilized on bending.

10.3. Extension to more complex molecules

One can ask whether the results which we have obtained in this chapter really justify the effort needed to master the techniques of analysis. Certainly if they were limited to discussion of the simple examples discussed so far our gain in understanding of the details of these molecules has taken considerable effort. However the ideas behind the molecular orbital approach we have used are applicable to a wide range of molecules and solids beyond the scope of this introductory text. Here we will describe an example which will indicate how more complex systems may be tackled.

10.3.1. The geometries of AX_2 and AX_3 molecules

In this chapter the tri- and tetraätomic molecules we have considered corresponded to the case where the central atom A, is a main group element from the second or higher row of the periodic table, and the ligands are hydrogen atoms. We would like to know whether similar ideas are applicable to molecules such as NF_3 (instead of NH_3) and OCl_2 (instead of OH_2). The difference of course lies in the fact that these molecules contain atoms, X, which carry both s and p orbitals. The details of the orbital picture have to be much more complex since each ligand atom X has four valence orbitals compared to the one for hydrogen. There are, however, two simplifications which we may use to understand these more complex systems. First we refer to the σ manifold of the orbitals for an A_2 diatomic such as N_2, shown in **7-22**. We notice that as a result of the mutual polarization of the two atoms this set of orbitals comprise a σ bonding orbital formed via overlap of two s/p hybrids and a corresponding σ antibonding orbital, along with two lone pair orbitals. The latter are characterized by large hybrid lobes pointing away from the bonding region. Such polarization occurs in AX_3 molecules. Thus we may draw out the orbital picture for the σ orbitals of an AX_3 system in the following very simple way. Instead of using a single hydrogen $1s$ orbital as in AH_3 we use the pair of orbitals shown in **10-18**. Since the interaction of the outward pointing hybrid ϕ_0 with the central atom orbitals is very small we ignore it when constructing the orbital diagram. It will become a

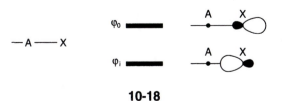

10-18

lone pair orbital on the ligand X. It is easy to see that the orbital diagrams constructed using the inward pointing orbital ϕ_i will be similar in very way to the orbitals for AH_n constructed using the hydrogen $1s$ orbital. The orbital ϕ_i has exactly the same symmetry properties as the ligand $1s$ orbitals. For AX_n molecules we often talk of σ-only molecular orbital diagrams which are identical to those for AH_n except that ϕ_i replaces $1s_H$ in their construction. The second approximation involves the remaining two p orbitals on atom X, lying perpendicular to the A—X axis (**10-19**).

10-19

These are of π type with respect to the AX bond and will interact with central atom orbitals of the correct symmetry to generate a π manifold of orbitals. An approximation, which varies from being very good to quite poor, depending on the system, is to ignore their interactions altogether. A part of the justification is that π interactions are invariably smaller than σ ones and may be quite small between elements which are not from the second row of the periodic table. These orbitals may be then regarded as lone pair orbitals on the atom X. Remember too the neglect of π bonds in the VSEPR scheme.

We can now see how an atom such as fluorine, with seven electrons can behave, in terms of chemical bonding, just like a hydrogen atom. One electron occupies the hybrid ϕ_i responsible for bonding to the central atom, and the other six electrons occupy nonbonding, lone pair orbitals on fluorine. Thus the geometrical preferences for NF_3 and OF_2 are just like those for NH_3 and OH_2.

As a general rule we may predict the geometries of AX_n systems by first filling all of the lone pair orbitals on the X atoms, which are usually of greater electronegativity than A and thus lie deep in energy, and then assign all the remaining electrons to the σ manifold. Either Figure 10.1 or 10.2 may then be used to predict the geometry. As example, consider the carbonate ion CO_3^{2-}. With a total of 24 electrons, $6 \times 3 = 18$ are assigned to oxygen lone pairs, leaving six for the σ manifold. These six electrons fill the $1a_1'$ and $1e'$ orbitals of Figure 10.2 and the ion is predicted to be planar, as it is. Such a treatment does not of course account for the double bond character in the CO bonds apparent in the Lewis structures of **1-20** since we have neglected π type interactions here. But it does give us an easy way of predicting the molecular geometry by concentrating on the energetically dominant σ interactions. Interactions of π type are in fact often important. In Part four Problem 1 they will be the focus of an interesting structural problem.

EXERCISES

10.1 *Geometries of the molecules AlH_2, SiH_2 and PH_2*

In their ground electronic states these molecules have the geometries shown.

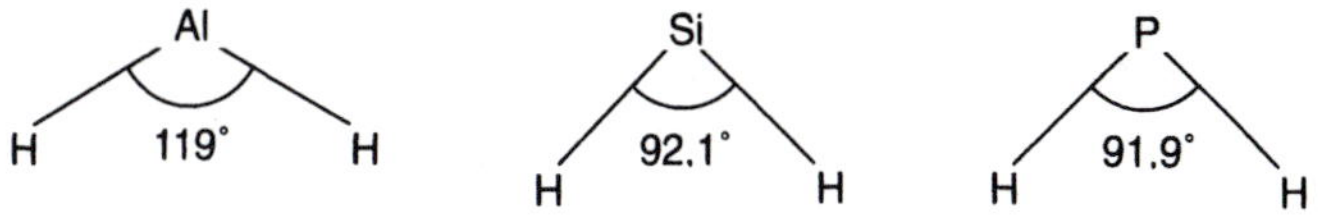

(i) Write down the ground electronic configuration for each molecule.
(ii) Explain why the molecules are bent.
(iii) Justify the trend $\theta_{Al} > \theta_{Si} \simeq \theta_P$

The electronic transition $2a_1 \rightarrow 1b_1$ leads to a geometry change for all three molecules.

(iv) Predict the geometry of AlH_2^*
(v) Account for the facts that SiH_2^* and PH_2^* are both bent but with larger angles ($\theta_{Si} = 122°$, $\theta_P = 123°$).

10.2 *The dications AH_2^{2+}*

(i) Predict the geometries of CH_2^{2+} and SiH_2^{2+}.
(ii) Why are the dications PH_2^{2+} and SH_2^{2+} bent? Explain why the bond angles are 124.8° and 96° respectively.

10.3 *Walsh diagram for BH_2*

The calculated energies (in eV) of the MOs of BH_2 as a function of the angle θ are given in the table.

MO/θ	180°	160°	140°	120°	100°
$2b_2$	21.14	19.59	15.72	11.16	7.13
$3a_1$	6.49	7.58	10.73	15.68	21.52
$1b_1$	−5.70	−5.70	−5.70	−5.70	−5.70
$2a_1$	−5.70	−5.95	−6.53	−7.21	−7.85
$1b_2$	−14.21	−14.17	−14.04	−13.77	−13.28
$1a_1$	−18.56	−18.57	−18.58	−18.62	−18.68

   (i) Draw out the energetic evolution of the MOs of BH_2 as a function of the angle θ to and thus generate a Walsh diagram. Verify that the picture is the same as that given in Figure 10.1.

   (ii) How many valence electrons are there in $BH_2{}^+$, BH_2 and $BH_2{}^-$?

   (iii) Using the rule of the HOMO, predict the structure (linear or bent) of each of these species.

   (iv) Compare the electronic configurations of the ground and first excited electronic states of BH_2. What do you expect to be the geometry of BH_2^*.

10.4 Use the appropriate figures of Chapter 10 and the method described in Section 10.3 to predict the geometry of SO_2 and BCl_3.

11 Molecular geometry using fragment molecular orbitals

Another method of tackling molecular geometry problems consists of studying the interactions between two fragments A and B. In this approach the molecule is divided into two fragments each characterized by a set of orbitals which are known. Each orbital on A can interact simultaneously with several orbitals on B via non-zero overlap integrals. The result can be complex both from the point of view of the energies and nature of the orbitals. In this chapter we will only consider some simple cases and will assume that the total interaction can be decomposed into a *series of interactions between two orbitals*, one on A and one on B, and that the different contributions are *additive*. The interactions which result between the orbitals of the two fragments will in general vary with their relative orientation. The energetic preferences as a function of electron count may then be used to specify the overall molecular geometry. The geometry change already described for ethylene in its ground and excited states in Section 3.5.2 is a simple example of this type.

11.1. Two- and four-electron interactions

Among the orbitals which characterize the two fragments A and B some are occupied and others are empty. The interactions which develop between pairs of fragment orbitals will lead in some cases to a stabilizing and in other cases a destabilizing effect depending upon the orbital occupancy. Three situations may be identified and are shown schematically in **11-1**.

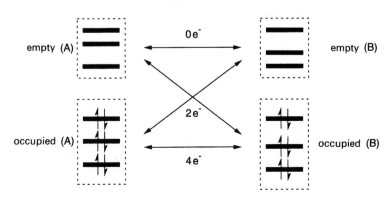

11-1

(i) Four-electron interactions ($4e^-$) between pairs of doubly occupied orbitals.

(ii) Two-electron interactions ($2e^-$) between a pair of orbitals, one of which is doubly occupied and the other empty, or between two singly occupied orbitals.

(iii) Zero-electron interactions ($0e^-$) between empty orbitals.

11.1.1. Energetic consequences

The crucial point of the analysis concerns the stabilizing or destabilizing nature of these interactions. As we saw in Chapter 3 this is determined by the number of electrons involved.

First consider the two-electron case. Two situations frequently encountered are shown in **11-2** and **11-3**. In the former, two orbitals, one doubly occupied lying deeper in energy than another empty orbital, give rise to a bonding and antibonding pair of orbitals. With two electrons in the bonding partner an electronic stabilization proportional to $S^2/\Delta\varepsilon$ results where S is the overlap integral and $\Delta\varepsilon$ the energy separation of the starting orbitals. Thus the stabilization energy decreases as the energy gap $\Delta\varepsilon$ increases. In the second case, **11-3**, the two orbitals are either of equal or similar energy and each contain a single electron. Here the stabilization energy is proportional to the overlap S alone. In the four-electron case (**11-4**) the bonding and

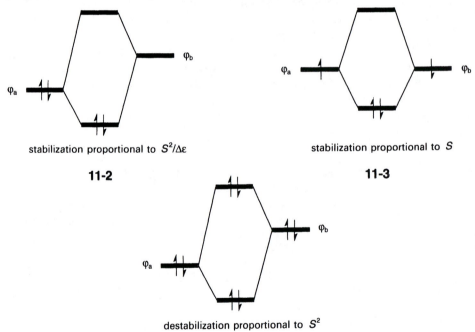

stabilization proportional to $S^2/\Delta\varepsilon$

11-2

stabilization proportional to S

11-3

destabilization proportional to S^2

11-4

antibonding orbitals are both doubly occupied. The destabilization of the antibonding orbital with respect to ϕ_b is larger than the stabilization of the bonding orbital with respect to ϕ_a. Overall then the total electronic energy balance is unfavorable and such four-electron interactions are destabilizing, the magnitude of the effect increasing as S^2. We have already seen the chemical results of such effects. Two atoms of helium or other inert gases repel each other. The case of the excited state of ethylene (Section 3.5.2) is the two-electron analog of this with one electron in the bonding and one in the antibonding orbital.

Finally we note that a zero-electron interaction has no energetic consequences since the orbitals involved are empty.

The geometry predictions we will make by identifying all the two- and four-electron interactions as a function of orientation will rely on the fact that they are additive.

11.1.2. Electron transfer

A second factor which differentiates the two- and four-electron interactions concerns the transfer of electronic charge from one fragment to another. Specifically we shall compare the case where one doubly occupied orbital interacts with another vacant orbital higher in energy (**11-2**), with the case where the two interacting orbitals are doubly occupied (**11-4**). Since ϕ_a lies deeper in energy than ϕ_b, the bonding MO (ϕ_+) is largely localized on ϕ_a, the orbital closest to it in energy. Analogously ϕ_- is largely localized on ϕ_b. These results are illustrated in **11-5** and **11-6**, where for simplicity we have used s orbitals on the two fragments A and B.

11-5 A B φ_+ $\qquad$ **11-6** A B φ_-

In the case of the two-electron interaction (**11-2**), before the orbitals interact two electrons are localized completely on ϕ_a. Afterwards, since they now occupy ϕ_+, an orbital with some ϕ_b character, there has been a partial electron transfer from A to B. This result was apparent in the assembly of the orbital interaction diagram for the molecule H_3BNH_3 of exercise 8.5 where the electron transfer is often indicated via the Lewis structure as $H_3B^--^+NH_3$. For the four-electron interaction (**11-4**), initially there are exactly two electrons on each fragment, two in ϕ_a and two in ϕ_b. After interaction, although the two electrons initially in ϕ_a are partially donated to ϕ_b in ϕ_+, the converse is true in ϕ_-. Here the two electrons initially in ϕ_b are partially donated to ϕ_a. Thus there is zero (or little) charge transfer associated with the four-electron interaction. We note finally that for the case shown in **11-3** there is no charge transfer between the fragments since the bonding MO is equally developed on both centers.

11.2. Model examples of H_3^+ and H_3^-

To illustrate the general approach we will study first the structures of the H_3^+ and H_3^- molecules studied at length in earlier chapters. In Chapter 9, using the orbital correlation method, we showed why their geometries are triangular and linear respectively. Here we explore these two possible geometries by considering the fragments H_2 and H^+ (for H_3^+) and H_2 and H^- (for H_3^-).

11.2.1. Geometry of H$_3^+$

First we consider the triangular geometry (**11-7**). The H$_2$ fragment possesses bonding (σ_{H_2}) and antibonding ($\sigma_{H_2}^*$) orbitals, and the fragment H$^+$ a single 1s_H orbital. Of these only σ_{H_2} is occupied, with two electrons. The orbitals σ_{H_2} and 1s_H are symmetric with respect to the plane P perpendicular to H—H, and passing through H$^+$. The $\sigma_{H_2}^*$ orbital is antisymmetric. The only interaction which can occur is between 1s_H and σ_{H_2} (**11-8**). Overall this a stabilizing two-electron interaction (**11-9**).

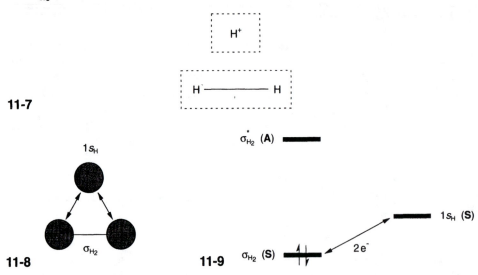

11-7

11-8

11-9

For the linear geometry (**11-10**) the 1s_H orbital can interact with both σ_{H_2} and $\sigma_{H_2}^*$ orbitals (**11-11** and **11-12**). This leads to a stabilizing two-electron interaction with σ_{H_2} and a zero-electron interaction with no energetic consequences with $\sigma_{H_2}^*$ (**11-13**).

To compare the relative stabilities of the two geometries, linear and triangular, it is necessary to compare the stabilization of the σ_{H_2} via its interaction with 1s_H in the two cases. The stabilization is proportional to $S^2/\Delta\varepsilon$. The energy denominator

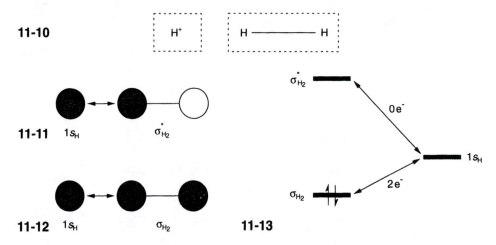

11-10

11-11   1s_H $\sigma_{H_2}^*$

11-12   1s_H σ_{H_2}

11-13

is the same in both cases so we must focus upon the variation in S. In the triangular geometry the $1s_H$ orbital overlaps with the two AOs which comprise σ_{H_2} (**11-8**) but in the linear form with only one. (The second is too far away to be involved.) Since the overlap is larger in the triangle, this is the geometry favored.

11.2.2. Geometry of $H_3{}^-$

For the two geometries, triangular (**11-14**) and linear (**11-15**), where H^- interacts with H_2, the fragment orbitals are the same as before, except their orbital occupation is different. ($1s_H$ is doubly occupied.) By symmetry the only interaction which can occur in the triangular arrangement is the destabilizing four-electron one, the analog of **11-9** and shown in **11-16**). In the linear geometry there is one four-electron destabilization but one two-electron stabilization (**11-17**). First let us compare the

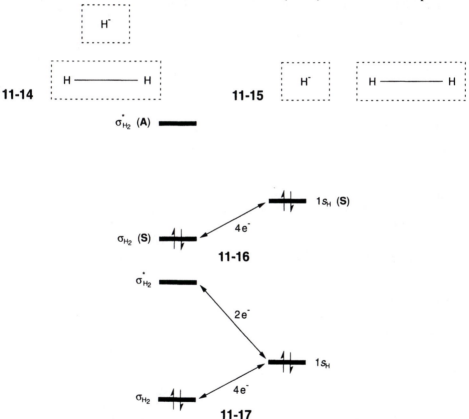

two four-electron interactions. There are the analogs of the two-electron interactions for $H_3{}^+$. Just as the two-electron (stabilizing) term in $H_3{}^+$ was larger for the triangle, so is the four-electron (destabilizing) term larger for $H_3{}^-$. Thus on the basis of this factor only the linear geometry is more stable than the triangular. Second, the linear geometry is the only one for $H_3{}^-$ with a stabilizing two-electron interaction (**11-17**). On both counts therefore the linear arrangement is more stable than the triangular one.

We will continue this type of analysis but using the methylene (CH_2), methyl (CH_3) and vinyl (CH_2=CH) groups as fragments.

11.3. Hyperconjugation

11.3.1. Conformation of the $C_2H_4^{2+}$ dication

Recalling the electronic structure of ethylene, C_2H_4 (**8-8**) we note that the highest energy orbital, π_{CC} is occupied by a pair of bonding electrons. Their presence is responsible for the planar structure of this molecule. By removing these two electrons to give the dication $C_2H_4^{2+}$, only a σ bond remains between the two carbon atoms. We might therefore expect a significant elongation of the C—C bond (from 134 to about 154 pm) and removal of the large rotational barrier present in ethylene. For the latter we would expect a value, around 12 kJ mol^{-1}, similar to that for ethane. The results of numerical calculations however, show a very different state of affairs (**11-18**).

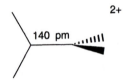

11-18

(i) The planar geometry is indeed not found to be the lowest energy arrangement. The perpendicular geometry is strongly favored by 117 kJ mol^{-1}.
(ii) The C—C distance does indeed increase but only to 140 pm. This is a value somewhere between those expected for typical single and double CC bonds and close to that found for benzene. Some π character seems to exist but the two CH_2 units are orthogonal.

Two structures may be envisaged for the $C_2H_4^{2+}$ dication, the planar (**11-19a**) and perpendicular (**11-19b**) forms. In each case the molecule may be decomposed into two CH_2 fragments which either lie in the same plane or in two orthogonal ones.

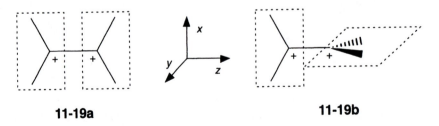

11-19a   **11-19b**

The deepest lying orbitals of CH_2^+ and their occupation are shown in **11-20** for the two possible fragment orientations. The antibonding orbitals are much higher in energy and are not shown. They will of course be involved in stabilizing interactions

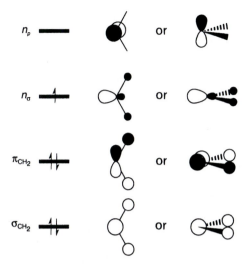

11-20

with the occupied orbitals of the other fragment, but these will be weak due to the large energy separations between them.

Irrespective of the conformation used (**11-19a** or **11-19b**) the σ_{CH_2} and n_σ orbitals are the only ones symmetric with respect to the two planes xz and yz. They cannot interact with the other orbitals (π_{CH_2} and n_p) which are antisymmetric with respect to one or other of these planes. The interactions between σ_{CH_2} and n_σ vary little from one conformation to the other. An example is shown in **11-21** for the σ_{CH_2} orbitals.

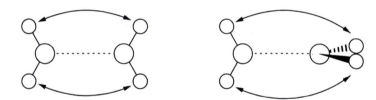

11-21

The overlap between the carbon $2s$ orbitals on each fragment is the same in both conformations but the small contributions arising from nonbonded $1s_H$ interactions are a little different. These are small since the hydrogen atoms are far apart. The same argument applies to the n_σ orbitals. The origin of the structural differences in the $C_2H_4^{2+}$ dication does not lie then with the n_σ and σ_{CH_2} orbitals.

Now consider the interactions between the 'π' type orbitals of the two fragments (π_{CH_2} and n_p). In the planar geometry (**11-19a**) the n_p orbitals are just the atomic p_y orbitals on each carbon whereas the π_{CH_2} orbitals involve a p_x component. These orbitals are labeled $\pi_{CH_2}^x$ in **11-22**. By symmetry the only interactions which may result are between the doubly occupied $\pi_{CH_2}^x$ orbitals (symmetric with respect to xz and antisymmetric with respect to yz) and between the empty n_p^y orbitals (antisymmetric with respect to xz and symmetric with respect to yz). The overall energy balance for this π orbital set is thus unfavorable being the sum of a destabilizing four-electron interaction and a zero-electron inteaction with no energetic consequences.

For the orthogonal geometry (**11-19b**) the relevant orbitals are labeled n_p^y and

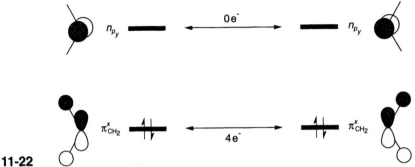

11-22

$\pi^x_{CH_2}$ on the left-hand fragment and in the right-hand fragment (rotated by 90°) $\pi^y_{CH_2}$ and n^x_p (**11-23**). The interactions allowed by symmetry are thus between $\pi^x_{CH_2}$ and n^x_p, and between $\pi^y_{CH_2}$ and n^y_p. In this geometry there are therefore two two-electron interactions which lead to a stabilization of this orthogonal geometry. This electronic stabilization of the orthogonal geometry over the planar one thus allows a clear understanding of the origin of the rotation barrier.

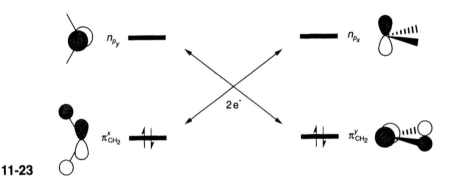

11-23

This analysis also allows understanding of other geometric properties of the $C_2H_4^{2+}$ dication. The σ bond between the two carbon atoms is assured via the bonding combination of the singly occupied n_σ orbitals (**11-20**). However the π-type interactions we have described augment the CC bonding above that of a single bond. As shown in **11-24**, each pair of interactions leads to a two-electron stabilization leading to two bonding orbitals. These MOs are in-phase combinations of the fragment orbitals π_{CH_2} and n_p and are largely localized on the CH_2 group, the deepest lying starting orbital. It is this π bonding character which, when added to that of the σ bond, explains the shortening of the C—C bond below the value expected for a single bond. The situation is similar to that shown earlier for π type interactions in heteronuclear AB diatomics, with the difference that the deeper-lying orbital is CH bonding. This type of chemical bonding is called *hyperconjugation* to describe the CC and CH aspects of the interaction.

One last point to consider concerns the electron transfer associated with the two-electron interactions in this orthogonal geometry. Before interaction the two electrons occupy π_{CH_2}, but afterwards they are partially delocalized onto the adjacent carbon atom. Overall such transfer is symmetric as far as the carbon atoms are concerned since there is transfer from left to right in the right-hand partner of **11-24**

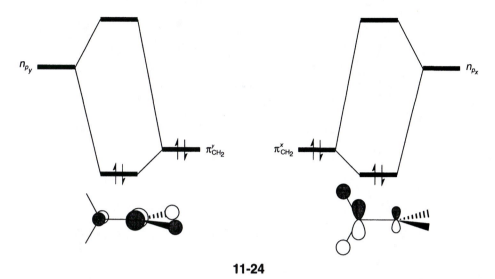

11-24

and from right to left in the left-hand one. However there is a net transfer of electrons from a C—H bonding orbital to a nonbonding one. This drop in the C—H bonding character in the new orbitals leads to a reduction in CH bond strength. As a result the CH distances increase from 108 pm in ethylene to 122 pm in its dication. Thus CC bonding has increased at the expense of CH bonding.

11.3.2. The ethyl cation $CH_3CH_2^+$

The natural fragmentation to use here consists of the methyl unit, CH_3, and the methylene cation, CH_2^+. The two conformations we will consider are those where the CH_2^+ fragment lies in the yz plane (**11-25a**) and that where it is rotated by 90° and lies in the xz plane (**11-25b**).

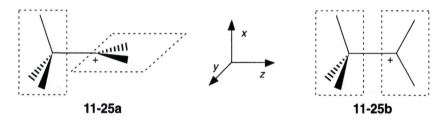

11-25a   **11-25b**

(a) Interactions between the fragment orbitals

The deepest lying orbitals of the two fragments are shown in **11-26**, those of CH_3 at the left and those of the two conformations of CH_2^+ at the right (**a** and **b**). Once again the antibonding orbitals, which lie high in energy, are not shown, since their energy differences relative to the bonding set are so large that their mutual interaction will be weak.

As in the previous example the interactions between σ_{CH_3} or σ_{CH_2} and n_σ are

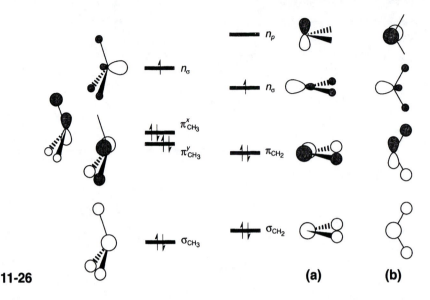

11-26   **(a)**   **(b)**

of no practical importance in determining the conformational preferences. Since their amplitude is directed along the C—C axis the overlap integrals involved do not depend upon the fragment orientation. We must then analyze the interactions which develop between the π_{CH_3} orbital on CH_3 and the π_{CH_2} and n_p orbitals of $CH_2{}^+$.

In conformation **a**, the carbon p_y component of the π_{CH_2} orbital and the carbon p_x orbital which comprises n_p^x can interact with symmetry-appropriate orbitals on CH_3, namely $\pi_{CH_3}^y$ and $\pi_{CH_3}^x$. The result is **(11-27)** one two-electron stabilization and

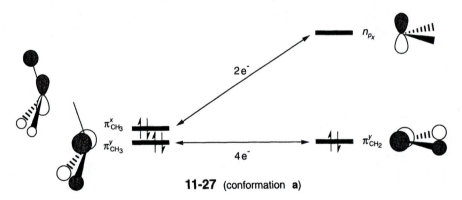

11-27 (conformation **a**)

one four-electron destabilization. The first of the two is a hyperconjugation interaction, similar to that described for the case of the ethylene dication $C_2H_4{}^{2+}$. **11-28** shows the analogous result for conformation **b**, where the $CH_2{}^+$ group is rotated around the C—C axis by 90°. All that has changed are the x, y labels of the $CH_2{}^+$ orbitals, but each of them finds a symmetry match with the pair of π_{CH_3} orbitals on CH_3. The picture is identical to that shown in **11-27** for conformation **a**, and results from the degeneracy of the π_{CH_3} orbitals on CH_3, the relevant overlap integrals being identical in the two cases. The hypercongugative interaction, identical for the

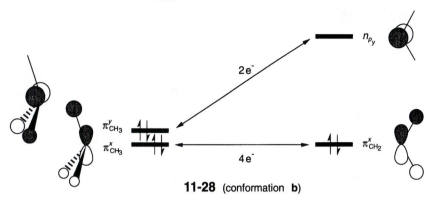

11-28 (conformation **b**)

two conformers, leads to free rotation of the CH_2 unit around the C—C axis in the ethyl cation.

An interesting point concerns the stabilization we have described at geometries intermediate between the two conformational extremes. It is easy to show that in fact it does not change with rotational angle at all, and thus our statement concerning free rotation is generally valid. For some arbitrary geometry, since the $\pi^x_{CH_3}$ and $\pi^y_{CH_3}$ orbitals are degenerate, all that needs to be done is to write a new pair of degenerate orbitals as a linear combination of the old. These two new orbitals can be chosen to have the same orientation with respect to the CH_2^+ n_p and π_{CH_2} orbitals as in one or other of the conformers **a** and **b**. The orbital interaction picture is thus identical to those just described.

(b) Consequences of hyperconjugation

The hyperconjugation interaction leads to a number of geometrical consequences. Since the two conformations are electronically equivalent we can choose one (**a**) for study. In the methyl fragment two electrons occupy the $\pi^x_{CH_3}$ orbital, bonding between the carbon and all three hydrogen atoms. In the ethyl cation these two electrons lie in an orbital bonding between $\pi^x_{CH_3}$ and n_p (**11-29**). There is thus a strengthening of

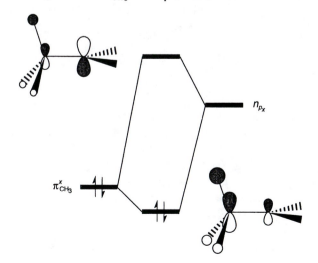

11-29

the CC bond which translates into a shortening of the CC distance. Accurate calculations show the distance to be 144 pm, significantly shorter than a typical single bond (154 pm). On the other hand the hyperconjugative interaction leads to the transfer of electrons from $\pi^x_{CH_3}$ to n^x_p. The depopulation of the $\pi^x_{CH_3}$ orbital, bonding between the carbon and hydrogen atoms leads to a weakening of the CH bonds. This point will be analyzed in more detail in exercise 11.3.

As in the case for many conjugated molecules (Section 8.2) the representation of the electronic structure by a single Lewis structure is not very good. For example the structure of **11-30a** does not show a strengthening of the CC and weakening of the CH bonds and electron transfer toward the CH_2 unit. It is possible to improve the description by including contributions from the three 'mesomeric' forms **11-30b, c, d**

11-30a   **11-30b**   **11-30c**   **11-30d**

to account for the hyperconjugation. We note though that these structures are not in accord with the rules usually used for resonance structures (Section 1.3.1) since each contains a nonbonded proton. These 'no bond' resonance structures are generally used to describe hyperconjugation.

(c) Application to the stability and electron affinity of carbocations

In addition to leading to the strong effect on the geometry of a molecule we have just described, hyperconjugation in the ethyl cation also influences the stability of the ion. The existence of such an interaction renders such ions intrinsically more stable than those where it does not occur, as in the methyl cation CH_3^+. This relative stability shows up in several chemical situations. Consider, for example, the gas phase reaction of these cations with hydride ion.

$$R^+ + H^- \rightarrow RH$$

This reaction is very exothermic and the variation in the size of the change in enthalpy can be used as a measure of the stability of the cation. When R^+ is more stable the reaction is less exothermic. Experimentally the exothermicity decreases from 1312 kJ mol^{-1} for the methyl cation to 1145 kJ mol^{-1} for the ethyl cation*. This trend can be generalized to the cases where the hydrogen atoms of CH_3^+ are gradually replaced by methyl groups. Hyperconjugation increases with the number of methyl groups attached to the cationic center and, as a result, the stability of the

---

* These values come from; F. A. Houle and J. L. Beauchamp, *J. Amer. Chem. Soc.*, **101**, 4067 (1980).

series increases in the order:

CH_3^+	<	$CH_3CH_2^+$	<	$(CH_3)_2CH^+$	<	$(CH_3)_3C^+$
methyl		ethyl		isopropyl		tertiarybutyl
$1312\ kJ\ mol^{-1}$		$1145\ kJ\ mol^{-1}$		$1032\ kJ\ mol^{-1}$		$961\ kJ\ mol^{-1}$

This is reflected in the exothermicity of the reaction with hydride. The more methyl groups which may be involved in a hyperconjugative interaction the higher the cation stability.

The hyperconjugation effect also shows up in the electron affinity of these carbocations. In CH_3^+ the lowest unoccupied orbital, the one which will receive the added electron is a pure carbon located $2p$ orbital (see the electronic structure of the isoelectronic molecule BH_3 in Section 5.2.3). In the ethyl cation this empty orbital is destabilized via an antibonding interaction with π_{CH_3} (**11-29**). As a result the electron affinity (9.83 eV for CH_3^+) drops to 8.4 eV for $C_2H_5^+$. This trend continues for the isopropyl cation (7.5 eV) and the tertiarybutyl cation (7.4 eV).

11.4. The *s-cis* and *s-trans* conformations of butadiene

The *s-trans* conformer of butadiene (**11-31a**) is slightly more stable than the *s-cis* (**11-31b**), by about $15\ kJ\ mol^{-1}$. Among the factors which may be used to explain this preference we will examine the role played by the π electrons. For this the molecule will be divided into two ethylenic fragments each carrying π bonding

11-31a   **11-31b**

and antibonding MOs (π, π^*). Since there are a total of four bonding π electrons two are placed in each unit such that π is doubly occupied and π^* empty. In this fragmentation the only element of symmetry common to the two fragments is the molecular plane. Both π and π^* on each unit are antisymmetric with respect to this plane so that interactions between both pairs are non-zero by symmetry. For each conformation there are therefore (**11-32**):

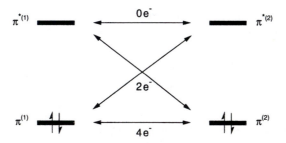

11-32

(i) A four-electron destabilization between the occupied orbitals $\pi(1)$ and $\pi(2)$.
(ii) Two two-electron stabilizing interactions between $\pi(1)$ and $\pi^*(2)$ and between $\pi^*(1)$ and $\pi(2)$.
(iii) A zero-electron interaction between $\pi^*(1)$ and $\pi^*(2)$ with no energetic consequences.

First, consider the four-electron interaction (**11-33**). In the two conformations the

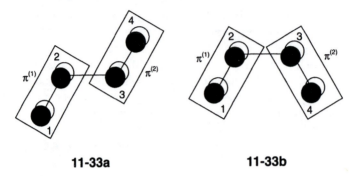

11-33a   **11-33b**

major overlap is between carbon atoms 2 and 3 which are geometrically the closest. This overlap is identical in the two conformations **a** and **b**. In this same way the overlaps 1–3 and 2–4 are equal, the distances between the two carbons being the same for both conformers. The only difference lies in the 1–4 interaction which is associated with a positive overlap. It is quite weak since these carbon atoms are far apart, but larger in the case of the *cis* conformation (**11-33b**) than in the *trans* (**11-33a**). This four-electron destabilization thus favors the *s-trans* geometry. Now consider the interaction of one π orbital with a π^* orbital located on the other fragment. The pair $\pi(1)$ and $\pi^*(2)$ are shown in **11-34a** for *s-trans* and **11-34b** for *s-cis*. Since the

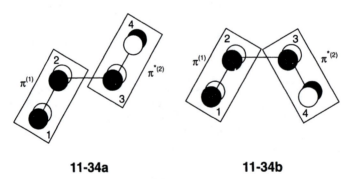

11-34a   **11-34b**

fragments are identical in the two conformations the energy difference between the two is the same irrespective of the geometry under consideration. The only factor which is different in the two conformations is the overlap between them. The largest contribution comes from the 2–3 overlap but this is the same in the two conformations. The 1–3 and 2–4 overlaps have opposite signs and exactly cancel (**11-34a** or **b**). Finally the small 1–4 overlap is of opposite sign to that between the carbon atoms 2 and 3, and leads to a reduction in the total overlap. This reduction is larger for *s-cis* (**11-34b**) than for *s-trans*, leading to a stronger overall stabilization for the latter.

A similar picture applies to the case of the interaction of $\pi(2)$ with $\pi^*(1)$. This larger two-electron stabilization contribution for the *s-trans* conformation and the smaller four-electron destabilization here are responsible for the increased stability of this conformer over the *s-cis*.

The energy difference between the two conformers is small, a direct result of the fact that the interactions responsible for the difference come from nonbonded atoms which are separated from each other by three chemical bonds. A similar analysis may be used to rationalize the energetic preference for the staggered conformation of ethane and the eclipsed conformation of propene (exercise 11.4). The four-electron destabilizing interactions which formed a part of this study are important in many molecules. For example, although the structure of CCl_4 is easily understood in terms of carbon–chlorine interactions which lead to CCl bonds, four-electron destabilizing interactions between the chlorine atoms tend to keep them apart too. In most molecules there are 'nonbonded repulsions' of this type which decrease in importance as the relevant internuclear separation increases.

References

$C_2H_4{}^{2+}$ *dication*: K. Lammertsma, M. Barzaghi, G. A. Olah, J. A. Pople, A. J. Kos, P. v. R. Schleyer, *J. Amer. Chem. Soc.*, **105**, 5252 (1983); R.H. Nobes, M.W. Wong, L. Radom, *Chem. Phys. Lett.*, **136**, 299 (1987).

$CH_3CH_2{}^+$ *cation*: R. Hoffmann, L. Radom, J. A. Pople, P. v. R. Schleyer, W. J. Hehre, L. Salem, *J. Amer. Chem. Soc.*, **94**, 6221 (1972).

Conformation of Butadiene: A. J. P. Devaquet, R. E. Townshend, W. J. Hehre, *J. Amer. Chem. Soc.*, **98**, 4068 (1976).

General Study of Hyperconjugation: L. Libit, R. Hoffmann, *J. Amer. Chem. Soc.*, **96**, 1370 (1974).

EXERCISES

11.1 Consider the methylene cyclopropene molecule whose Lewis structure is shown below. Although the molecule is only composed of carbon and hydrogen it possesses quite a large dipole moment (1.9 D) which comes from the polarization of the π electrons. This may be studied using the fragment orbital method.

(i) Suggest a fragmentation.
(ii) Analyze the interactions which develop between the fragment orbitals but do not go as far as drawing out the form of the MOs of the complete molecule.
(iii) Show the origin of the dipole moment and determine its direction.
(iv) Suggest some other Lewis structures for this molecule.

REFERENCE: T. N. Norden, S. W. Staley, W. H. Taylor, M. D. Harmony, *J. Amer. Chem. Soc.*, **108**, 7912 (1986).

11.2 Here we will study the $C_2H_4^{2+}$ dication in the geometry shown below.

$$2+$$

$$H-C-C\overset{H}{\underset{H}{\diagdown}}$$

The molecule may be decomposed into the two fragments CH^{2+} and CH_3.

(i) Analyze the interactions which take place between the orbitals of the two fragments.
(ii) What is the interaction responsible for the formation of the σ_{CC} bond?
(iii) The calculated CC distance (130 pm) is much shorter than that for a CC single bond (154 pm). Explain this result.

(Note: This geometry for the dication is less stable than that studied in Section 11.3.1).

REFERENCE: K. Lammertsma, M. Barzaghi, G. A. Olah, J. A. Pople, A. J. Kos, P. v. R. Schleyer, *J. Amer. Chem. Soc.*, **105**, 5252 (1983); R. H. Nobes, M. W. Wong, L. Radom, *Chem. Phys. Lett.*, **136**, 299 (1987).

11.3   Accurate molecular orbital calculations have shown that one of the consequences of hyperconjugation in the ethyl cation is a geometrical distortion of the methyl group.

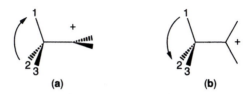

(a)   (b)

(i) The methyl fragment pivots slightly about the CC bond as shown. Show the orbital basis behind such a geometrical change and justify the result that it takes place in a different direction depending on the conformer.

(ii) In conformation (a) the CH_1 bond length increases (from 108.8 pm to 111.0 pm) much more markedly than do the CH_2 or CH_3 distances (108.8 pm to 109.1 pm). Show how this has a simple orbital explanation and predict how the CH distances will change in conformation (b).

(iii) With respect to the tetrahedral angle of 109.5°, the H_2—C—H_3 angle increases in conformation (a) to 113.6° but decreases in conformation (b) to 105.9°. Show how this comes about.

REFERENCE: W. A. Lathan, W. J. Hehre, J. A. Pople, *J. Amer. Chem. Soc.*, **93**, 808 (1971).

11.4   *Conformation of propene*

For propene ($CH_3CH=CH_2$) the most stable conformation is that in which a CH bond of the methyl group eclipses the C=C double bond (a) and not the one where the two are staggered (b). This result is surprising since staggered conformations are in general more stable since they minimize steric repulsions. This conformational preference will be analyzed by consideration of the 'π' type orbitals of the molecule, that is to say the orbitals which are antisymmetric with respect to the molecular plane in both (a) and (b).

(a) eclipsed   (b) staggered

(i) Choose a fragmentation.

(ii) List the π type orbitals on each fragment and specify their electron occupation.

(iii) Account for all of the interactions which may develop in each conformation and identify their stabilizing or destabilizing behavior.

(iv) Compare the interactions for the two cases (a) and (b) and thus show why
conformation (a) is favored.

Hint: The analysis is similar to that described for *s-cis* and *s-trans* butadiene in Section
11.4..

REFERENCE: W. J. Hehre, L. Salem, *Chem. Comm.*, 754 (1973).

11.5   The benzyl cation has the geometry given below. The CC bond lengths are
different from those found in benzene (140 pm). A useful model comes from
consideration of the π orbitals of the molecule.

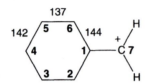

(i) Suggest a fragmentation.

(ii) Start off by analyzing the two-electron interactions between the fragment
orbitals to account for the differences between these CC distances and
those in benzene. (Use the π MOs of benzene of Section 8.2.5.)

(iii) Recover the same result by constructing the principal mesomeric structures of the molecule.

REFERENCE: W. J. Hehre, *Accts. Chem. Res.*, **8**, 369 (1975).

12 An introduction to the study of chemical reactivity

We have seen how a knowledge of the orbital structure of a molecule allows geometrical predictions. Another objective of molecular orbital theory is an understanding of chemical reactivity, that is to say the facility with which a given molecule transforms or reacts with others. Before discussing the use of the MO model we must briefly recall some features of the chemical reaction.

12.1. Description of a chemical reaction

Writing a chemical reaction in the form R → P represents the outcome of the reaction between a mixture of reactants (R) in some given proportions to give a set of products (P). Writing it this way has the advantage of simplicity but it gives no information as to how the transformation proceeds. In particular the participation of intermediates with short lifetimes are not included in this picture.

12.1.1. The reaction scheme and elementary processes

Writing down a reaction scheme consists of decomposing the reaction into a series of elementary processes which naturally include any intermediates which may be formed. Consider, for example, the addition of HBr to propene in solution. The overall scheme for the reaction is

$$HBr + CH_3CH{=}CH_2 \rightarrow CH_3{-}CHBr{-}CH_3$$

Three steps are necessary for the reaction to take place. The first is the fission of HBr.

$$HBr \rightarrow H^+ + Br^-$$

During the second the H^+ cation formed adds to the ethylenic double bond to give the isopropyl cation.

$$H^+ + CH_3CH{=}CH_2 \rightarrow CH_3{-}CH^+{-}CH_3$$

Finally in the last step the isopropyl cation combines with the bromide anion.

$$CH_3{-}CH^+{-}CH_3 + Br^- \rightarrow CH_3{-}CHBr{-}CH_3$$

The validity of a reaction scheme may be verified using several routes. One can trap the intermediates physically or chemically, or observe them directly using fast spectroscopy. On the other hand one can compare the experimental reaction rate with that deduced from the reaction scheme. A theoretical reaction route may be derived by knowing the rates of the individual steps of the reaction and the concentration of the reactants.

12.1.2. Reaction mechanism

'Ideally, knowledge of a reaction mechanism implies that we know the exact positions of all of the atoms in the molecules, including solvent molecules, at all times during the reaction. Additionally, we should know the nature of the interactions and/or bonds between these atoms, the energy of the system at all stages, and the rate at which all changes occur."*

The study of the reaction mechanism to give the information required by this definition is obviously extremely involved and there are only a few reactions for which such a mechanism has been elucidated. For the majority of reactions the problem reduces to the unraveling of the series of elementary processes which comprise the overall reaction. Thus the protonation of propene, used as an example above, leads to the formation of a carbocation sure enough but its detailed geometry, whether it is a 'classical' or 'bridged' species, is still open to question.

12.1.3. Reaction coordinate

To describe each elementary process it is common to use a parameter which follows the course of the reaction from reactants to products and which is associated with the geometrical changes which occur. This, often ill-defined, parameter is called the reaction coordinate. Take as an example the combination of two hydrogen atoms to give the hydrogen molecule. Here the reaction coordinate is well-defined and from a geometric point of view is just the H...H distance which varies from infinity (isolated atoms) to 74 pm (the distance between the two atoms in the molecule).

For more complex reactions the definition of the reaction coordinate is more difficult since it does not in general reduce to a single geometric parameter. Consider, for example, the nucleophilic substitution reaction $RX + Y^- \rightarrow X^- + RY$ where R is an alkyl group and X and Y are electronegative atoms (e.g. Cl) or groups of atoms (e.g. CCl_3). For certain combinations of R, X and Y the reaction proceeds by the S_N2 mechanism where the reaction is characterized by a single step, namely the attack of the anion Y^- on the carbon atom directly opposite the C—X linkage. At the mid-point of reaction the system is an anion which formally has five bonds (12-1). Three geometric parameters vary considerably during the reaction:

(i) The C—X distance d_X changes from its equilibrium value in the RX reactant to an infinite value in the products.

___

* N. L. Allinger, M. P. Cava, D. C. De Jongh, C. R. Johnson, N. A. Lebel, C. L. Stevens *Organic Chemistry.* McGraw-Hill (1976) p. 288.

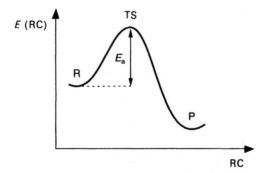

12-1

    (ii) The reverse is true for the C—Y distance d_Y. It changes from infinity to its equilibrium value in the RY product.

    (iii) The angle θ changes from 109.5° (pyramidalization towards the left) to 70.5° (pyramidalization towards the right).

The reaction coordinate (RC) is essentially some weighted combination of all three of these parameters which vary during the course of the reaction. Initially it is simply characterized by the approach of Y^- to RX (variation in d_Y) and at the end of the reaction by the departure of X^- (variation in d_X). Around the mid-point of the reaction all three parameters contribute to this coordinate.

12.1.4. Energy profiles, transition states and reaction intermediates

By drawing out the energy changes of the system as a function of the reaction coordinate we get the energy profile associated with the reaction. The form of this curve $E(RC)$ depends of course on the reaction studied. For example that for the combination $H + H \rightarrow H_2$ is one we have already encountered in Section 3.3.3, namely a Morse curve. For more complex reactions the reaction profile is not quite as simple. **12-2** shows a profile which is frequently found and allows us to point out

12-2

some important points. The reactants (R) and products (P) are stable species since they are located in energy minima. The highest point of the curve is called the transition state (TS) of the reaction. The molecular arrangement of this point is unstable with respect to both reactants and products and in general is not observable experimentally. The difference in energy, E_a, between reactants and the transition state is called the activation energy. It represents the energetic cost of the nuclear and electronic reorganization energy which takes place during the course of the reaction.

    An energy profile analogous to **12-2** is associated with each elementary process, where the reactants and products are associated with a single transition state. For a reaction that takes place in several steps the energy profile is obtained by linking together those appropriate for each of the elementary processes. As a result secondary minima appear (12-3) which correspond to different reaction intermediates (RI).

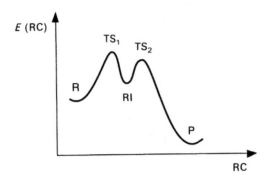

12-3

Contrary to a transition state, a reaction intermediate has a lifetime which is sufficiently long to be observed experimentally. These intermediates are separated from the reactants and products (and if present other intermediates) by a transition state.

It is during the elementary processes that the electronic structure of the species implicated in the reaction (reactants, intermediates or products) is considerably reorganized. These are therefore the steps which we will study in this chapter using molecular orbital theory. It is important to note that the rate of a chemical reaction is proportional to $\exp(-E_a/kT)$ where k is Boltzmann's constant and T the absolute temperature. Thus the rate of reaction associated with each elementary process depends critically on E_a and decreases as this energy barrier increases. In searching for the most favorable reaction pathway among several possibilities the problem then reduces to finding the path with the *smallest* E_a.

12.2. The frontier orbital approximation

Consider two molecules, A and B which react with each other to give a product, C. The electronic structure of each reactant can be described by a collection of MOs which are either occupied or vacant. For present purposes we shall assume that all of the occupied MOs are doubly occupied. As A and B come together three types of interactions develop between the two:

(i) Zero-electron interactions between empty orbitals.
(ii) Two-electron interactions between empty orbitals on A and filled ones on B, and vice-versa.
(iii) Four-electron interactions between filled orbitals on A and B.

Recall that the energetics associated with the three types of interaction are none, stabilizing and destabilizing respectively. If we can envisage two different mechanisms for a particular reaction (corresponding perhaps to two different geometrical approaches) then we can in principle compare the energetics associated with each.

For this it is necessary to compare the importance of the stabilizing and destabilizing interactions which result in each case, eventually leading to the identification of the most favorable mechanism. It is clear that such an approach is identical to that used in the preceding chapter to study the geometries of molecules starting off with fragment orbitals.

However, if the method is simple in principle, the situation can get very complicated due to the large number of orbitals which are involved. The problem is often simplified by using the *frontier orbital approximation*.

12.2.1. The method

The goal of the approximation is to judiciously select out, from all of the possible interactions, those which are of greatest importance.

In general, although the size of the two-electron stabilization is proportional to $S^2/\Delta\varepsilon$ (where S is the overlap integral and $\Delta\varepsilon$ the energy separation between two orbitals) it is difficult to give a general rule which will allow identification of those orbitals whose overlap is most important. The overlap part of this expression will have to be evaluated case by case. The $\Delta\varepsilon$ term is much easier to analyze; the closer the orbitals are in energy, the larger the stabilization. The important result is that the dominant interactions are those between the highest occupied orbital (HOMO) on A and the lowest unoccupied orbital (LUMO) on B, and vice-versa (**12-4**).

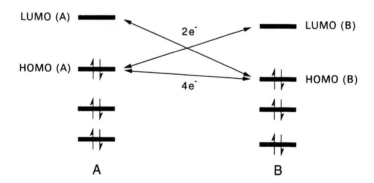

The set of repulsive interactions of the four-electron type between filled pairs of MOs are, as we have described earlier (see for example Section 11.4) associated with steric interactions. These four-electron destabilization terms only depend upon the overlap and increase as S^2. In addition, as discussed for the case of H_2 in Section 3.3.3, nuclear repulsions help in keeping a collection of atoms from collapsing to a point. In the simplest version of frontier orbital theory we neglect the role of such interactions. This does not mean that they are zero (in which case there would often be a predicted zero activation energy) but they are probably close enough for the different mechanisms that we will study. In any case we should avoid mechanisms that bring formally nonbonded atoms, which lie away from the reaction center, close together. When we do wish to improve the model to take into account these effects, often it is sufficient to consider only the interaction between the two HOMOs (**12-4**).

In summary we will find that consideration of the HOMO—LUMO interactions between a pair of interacting molecules, A and B, will often enable valuable insights into the way the two react.

12.2.2. Electrophilic and nucleophilic reactants

The molecules A and B are in general different and are thus characterized by two different sets of orbitals. An important consequence is that the two HOMO—LUMO interactions are of different sizes since the appropriate values of $\Delta\varepsilon$ may be quite different (12-5). If the difference between the two is large then a reliable picture is often found if the energetically dominant interaction only is considered.

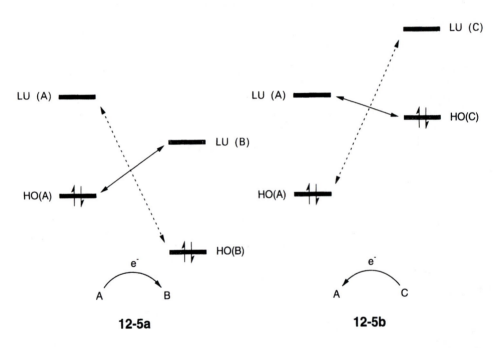

12-5a   **12-5b**

In the example shown in **12-5a** the two-electron interaction between the HOMO of A and LUMO of B is the preponderant one and results in electron transfer from A to B (Section 11.1.2). The reactant A, with its tendency to lose electrons is thus called a *nucleophile* and reactant B, with its tendency to accept electrons is called an *electrophile*. *In general a nucleophile is characterized by a high-lying HOMO and an electrophile by a low-lying LUMO.*

Most often the nucleophilic or electrophilic character of a reactant is not an absolute one, but depends on the partner with which the molecule is associated. Thus in the A, B pair of **12-5a**, A plays the role of the nucleophile, but in the A, C pair of **12-5b**, it plays that of the electrophile. An interesting example is carbon monoxide, CO, which possesses a high-lying HOMO and a low-lying LUMO (Section 7.2.3) such that it can behave at the same time both as a nucleophile and as an electrophile with suitable partners. A transition metal is one such example.

12.2.3. The validity of the approximation

(a) Neglect of other orbitals

The frontier orbital approximation is most valid when the HOMO is well separated from the other occupied MOs in the nucleophile and the LUMO is well separated from the other unoccupied MOs in the electrophile. In this case the interactions which are ignored are weak, controlled by large energy separations, $\Delta\varepsilon$. One should use the method with care therefore in those molecules where one or the other of the reactants has MOs close to those of the two frontier orbitals.

(b) Description of the energy profile of a reaction

During the course of the reaction $A + B \rightarrow C$ the geometries of the species A and B were progressively modified in such a way that they end up close to that of the product C. But, when using the frontier orbital method for this type of reaction the orbitals of the molecules A and B are those appropriate for their geometries in the isolated state. Thus the best description of the interactions are those which develop at the beginning of the reaction when the geometrical deformations are small. As the reaction progresses further this description becomes less and less valid since a geometrical distortion is necessary to reach the product (Figure 12.1). The frontier orbital approximation shows itself to be most defective around the transition state especially if the geometry resembles that of the product.

(c) Comparison of two reaction mechanisms

We wish to see whether it is possible to use these ideas to distinguish between two possible reaction mechanisms for a given chemical reaction. The most favorable route will be the one with the smaller activation energy, i.e., the one with the lowest-lying

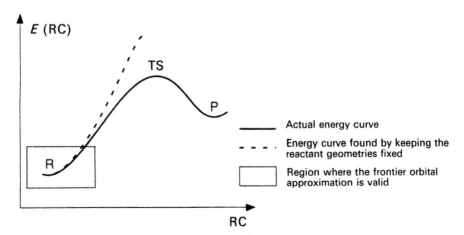

Figure 12.1. Energy profile and the frontier orbital approximation.

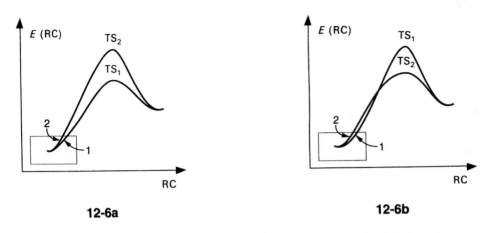

12-6a   **12-6b**

transition state. We can imagine two situations. In the first (**12-6a**) there is not a crossing between the two curves corresponding to the two mechanisms. This means that the route involving the lowest energy transition state (TS_1) is also the one which is of lower energy early in the reaction. In this case analysis of the energetics early on in the reaction leads to determination of the favored route. One can also envisage the situation shown in **12-6b** where a crossing takes place between the two energy curves. In this case since TS_2 lies lower than TS_1, route 2 is the favored one but the opposite (incorrect) conclusion would be drawn by considering the early behavior of the two curves. In fact an analysis of a large number of chemical processes has shown that the situation of **12-6a** (the absence of a crossing) is practically always observed. The study of the interactions which develop at the beginning of a chemical reaction are therefore very frequently a good indication of the relative energy of the transition state which is reached later in the reaction.

(d) Charge and frontier orbital control

Some chemical interactions are not usually studied using orbital models. An example is the interaction between a charged atom and a permanent dipole moment. Thus the stability of the F^- ... HF hydrogen bond is usually described via such a classical picture rather than an orbital one. When the principal interaction which develops during a chemical reaction is electrostatic in nature we say that the reaction proceeds under *charge control*. Contrarily, when the overlap between the frontier orbitals of the reactants governs the course of the reaction, then the reaction is under *frontier orbital control*. In the majority of cases both types of interaction are present but one is often dominant. It is simple to see that reactions under charge control are those between charged species where the orbitals are very contracted and thus overlap little, whereas those under frontier orbital control are either between neutral molecules or charged species if the latter have large overlaps. This is the case if the orbitals are diffuse. By way of examples, ions such as F^- and Li^+ favor charge control but species such as H^- or C_2H_4 favor frontier orbital control. In the following examples we shall limit ourselves to reactions under frontier orbital control, electrostatic factors being briefly discussed where needed.

12.3. Cycloaddition reactions

During a cycloaddition reaction the π systems of two molecules are used to form a cyclic product. For example the cycloaddition of butadiene and ethylene (called the *Diels–Alder reaction*) leads to the formation of cyclohexene, and the dimerization of ethylene to cyclobutane (**12-7**).

12-7

The shorthand used to describe those reactions highlights the number of π electrons in each molecule. Since in the Diels–Alder reaction, butadiene has four and ethylene two such electrons, such a reaction is called a [4 + 2] cycloaddition. In the same way the dimerization of ethylene is a [2 + 2] cycloaddition.

12.3.1. The [4s + 2s] thermal cycloaddition: the Diels–Alder reaction

This reaction is frequently used in synthetic organic chemistry with substituted ethylenes and butadienes. The reaction is stereospecific, the addition of a Z ethylene leading to a *syn* product, and the addition of an E ethylene to an *anti* product (**12-8**).

These results suggest that the reaction is concerted, that is to say that the two new CC bonds are formed simultaneously. We can see this in the following way. For the moment assume that one bond is in fact formed before the other. In this case we could obtain a non-cyclic biradical or a zwitterion (**12-9**). (The species with a separation of charge are called *zwitterions*.) Irrespective of which form we use in **12-9** there is free rotation about one of the CC bonds, shown in **12-10** for the biradical, which leads in general to a mixture of *syn* and *anti* isomers. Thus to be stereospecific the two new bonds have to be made at the same time.

12-9

12-10

The simplest mechanism which takes account of the stereospecificity of the reaction is the approach of the two molecules in parallel planes as shown in **12-11**. For both butadiene and ethylene the two new bonds are formed on the same side of the molecular plane. This reaction is said to be *supra-supra* and using the shorthand notation, [4s + 2s]. In a symmetric approach of the two molecules the plane P passes through the middle of the C_2—C_3 and C_5—C_6 bonds and lies perpendicular to the two molecular planes. P is a plane of symmetry for the whole system.

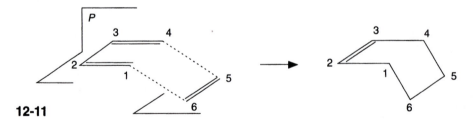

12-11

The frontier orbitals of ethylene are the π (HOMO) and π^* (LUMO) orbitals, and those of butadiene are the orbitals π_2 (HOMO) and π_3 (LUMO) of Section 8.2.2. With respect to the plane of symmetry P, the orbitals π and π_3 are symmetric, but the orbitals π^* and π_2 are antisymmetric (**12-12**). The symmetry properties of the frontier orbitals are such that the only interactions allowed are the stabilizing two-electron ones between the HOMO of one molecule and the LUMO of the other. It is also important to note that the large lobes of π_2 and π_3 of butadiene lie on the terminal carbon atoms of the molecule, C_1 and C_4, which are the carbon atoms which will form the new bonds to ethylene (**12-12**). Thus the overlaps between the two sets of orbitals are large. It is this pair of stabilizing interactions which are switched on

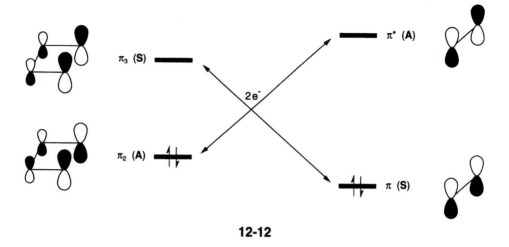

π_3 (S)

π^* (A)

2e⁻

π_2 (A)

π (S)

12-12

as the two molecules come together which are responsible for the ease with which this reaction takes place experimentally. We say that the Diels–Alder reaction is *allowed by symmetry*. On the other hand since the two molecules are in their lowest energy configuration to begin with, the energy necessary to get over the activation energy barrier comes from simple heating. Thus the [4s + 2s] cycloaddition is *thermally allowed*. We must note finally that in the simple scheme of **12-12** only the stabilizing interactions are shown. We must not forget that repulsive interactions, including four-electron interactions involving deep-lying orbitals and nuclear repulsions, have to be added to this picture to get the overall energetics.

12.3.2. Thermal [2s + 2s] cycloaddition: the dimerization of ethylene

Consider, just as before, the approach of the two molecules in two parallel planes to give a [2s + 2s] cycloaddition. The plane P which bisects each of the ethylenic C=C bonds and lies perpendicular to these molecular planes is a symmetry element for the system (**12-13**).

The frontier orbitals of each molecule are the π and π^* orbitals. With respect to the plane P the π orbitals are symmetric and the π^* orbitals antisymmetric (**12-14**). The only interactions which may take place between the orbitals of the two

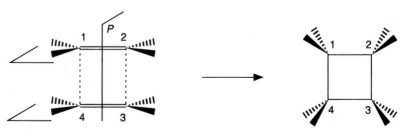

12-13

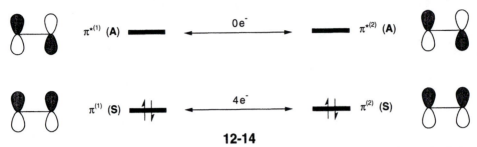

12-14

molecules are thus a destabilizing four-electron interaction between the occupied π orbitals and a zero-electron interaction, with no energetic consequences between the π^* orbitals (**12-14**). The HOMO–LUMO interaction is identically zero by symmetry. The overall interaction on bringing together the two molecules is thus a destabilizing one and so the dimerization as envisaged here is forbidden by symmetry. *In general thermal [2s + 2s] cycloadditions are forbidden by symmetry.*

12.3.3. Generalization to [ms + ns] cycloadditions

The two examples we have used show quite dramatically that thermal *supra-supra* cycloaddition reactions are strongly controlled by the number of electrons present. [2s + 2s] is thermally forbidden but [4s + 2s] is thermally allowed. These results may be generalized. It is quite easy to show in fact that [ms + ns] cycloadditions are thermally allowed if $(m + n)/2$ is odd but forbidden if $(m + n)/2$ is even. Another way of putting this is that the reaction is thermally allowed if $m + n = 4q + 2$, but forbidden if $m + n = 4q$, where q is an integer.

The important role of the number of electrons in controlling the mechanism of a chemical reaction has been the focus of much contemporary thought in theoretical chemistry. A collection of a set of rules which describe many organic reactions in these terms are known as the *Woodward–Hoffmann rules* after the two chemists who discovered them. They were established using the ideas of conservation of orbital symmetry along a reaction coordinate or using the frontier orbital approach developed principally by K. Fukui.

12.4. Further aspects of the [2 + 2] cycloaddition

The thermal [2s + 2s] cycloaddition reaction is forbidden by symmetry. Nevertheless, this reaction can lead to products under certain conditions. Several routes can be envisaged depending upon whether the mechanism is a concerted one or not.

12.4.1. Concerted mechanisms

It is the existence of the four-electron repulsion between two occupied π MOs and the absence of any two-electron stabilizing interactions which leads to the [2s + 2s] mechanism being thermally unfavorable. One way of changing this state of affairs is to change the electronic configuration of one of the reactants. Consider then the first excited state of ethylene where an electron has been promoted from the π to π^*

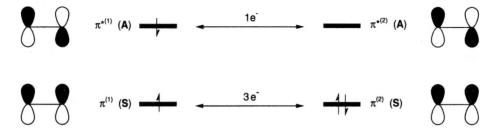

12-15

orbital. The interaction between the two molecules is now clearly less unfavorable
(**12-15**). The four-electron interaction has been replaced by a three-electron one which,
as discussed in Section 3.3.4, may be either weakly attractive or weakly repulsive. In
any case it is less destabilizing than the four-electron repulsion between the molecules
in their electronic ground states. In addition there is a stabilizing interaction coming
from the presence of one electron in the orbitals resulting from interaction of the
two π^* levels. The overall energy balance provided by this arrangement of electrons
is thus energetically favorable as the two molecules come together. This excited state
of ethylene may be generated by absorption of a photon and we may usually describe
this by saying that the [2s + 2s] cycloaddition reaction is *photochemically allowed*.

Just like the thermal pathway for the Diels–Alder reaction, the photochemical
dimerization of alkenes is stereoselective. If a solution of *cis*-2 butene is irradiated,
the only two molecules produced are the two isomers (**12-16**) expected via the
[2s + 2s] cycloaddition route.

12-16

Another way to get around the forbidden nature of the thermal reaction is to
change the relative orientation of the reactants, but still keeping them in their ground
electronic states. Consider the approach of two ethylene molecules, arranged such
that their molecular planes are perpendicular (**12-17**). The first is in a plane
perpendicular to the page (P_1) and the second (P_2) lies in a plane behind and parallel
to the page. The new CC bonds are thus formed on the same side of the

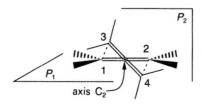

12-17

molecular plane for one molecule (a *supra* linkage) and on either side of the plane for the other (an *antara* linkage). The reaction is thus described as [2s + 2a]. The axis joining the mid-points of the two ethylenic C=C bonds is thus a two-fold rotation axis (C_2) for the system. With respect to this symmetry element the MO π_1 is antisymmetric but π_1^* symmetric. Conversely π_2 is symmetric but π_2^* antisymmetric.

The interaction diagram is now completely different from that found for the [2s + 2s] mechanism (12-18). Here the only non-zero interactions by symmetry are stabilizing two-electron ones between the HOMO of one molecule and the LUMO of the other. One can easily show (12-19) that the overlaps associated with these

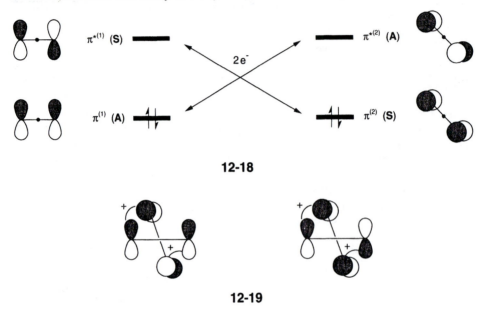

12-18

12-19

two interactions are quite different from zero. When studying this diagram do not forget that one molecule of ethylene lies above the other; they do not lie in the same plane.

This thermal cycloaddition is thus allowed by symmetry. Nevertheless this mechanism is not a very probable one for ethylene dimerization because it leads to unfavorable torsional constraints. However this mode of addition is observed in polycyclic systems where the double bonds, constrained by the presence of the ring systems are oriented in the proper way. The torsional constraints now do not arise since these repulsive forces have already been overcome in the course of the synthesis of the starting molecules.

12.4.2. Non-concerted mechanisms

The most favorable mechanism for ethylene dimerization thus has to be the one which reconciles at the same time a strong interaction between the frontier orbitals with these geometrical constraints. A good compromise is found when the two molecules approach in parallel planes, but in such a way that only two carbon atoms

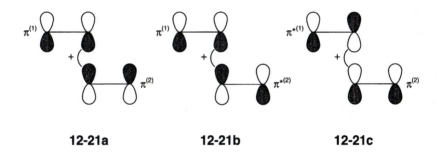

12-20

can interact with each other. A biradical or zwitterionic intermediate is formed which can readily rearrange to cyclobutane. This is shown in **12-20** for the biradical formulation.

In the course of the approach of the two molecules, a four-electron repulsive interaction develops between the two occupied π orbitals of the reactants in the same way as in the [2s + 2s] mechanism. However here the repulsion is somewhat weaker since the overlap only occurs between the orbitals located on carbon atoms 2 and 3 (**12-21a**). On the other hand there are two two-electron stabilizations between π and π^* orbitals which for the same reasons are weaker than in the [2s + 2a] case (**12-21b, c**). From the point of view of the orbital interactions this mechanism lies between those of the two concerted mechanisms. It is not as energetically favorable as [2s + 2a] but certainly more favorable than in [2s + 2s]. It is the absence of the geometrical constraints which make this mechanism more favorable than the concerted [2s + 2a] mechanism.

12-21a   **12-21b**   **12-21c**

The [2 + 2] cycloaddition via a thermal pathway is indeed observed experimentally, for example, in the case of enol ether with tetracyanoethylene. However, the reaction is not stereospecific. Starting from a *cis* alkene a mixture of *cis* and *trans* cyclobutanes is obtained, in variable proportions which depend on the nature of the enol and the reaction medium (**12-22**).

12-22

12.5. Examples of ionic reactions

This last section will be concerned with the study of some reactions where ions play an important part. As noted in Section 12.2 the interaction between the frontier orbitals is not the only factor responsible for determining energetic variations as the reaction proceeds. Electrostatic interactions can be the dominant features of reactions which are under charge control. This was not the case for the cycloaddition reactions just described since these involve neutral species. The importance of electrostatic effects will be briefly described for a series of frequently encountered organic reactions.

12.5.1. The S_N2 mechanism

In the S_N2 mechanism, the attack of the nucleophile Y^- on the side of the molecule opposite to the C—X bond (backside or antipodal attack) leads to an inversion of geometric configuration at the carbon atom. This is the Walden inversion shown in **12-23a**. It is possible to imagine attack of the nucleophile Y^- on the C—X bond in a way which leads to retention of configuration (**12-23b**). A study of the two mechanisms within the framework of the frontier orbital approximation allows an understanding of why the first mechanism is favored over the second. For reasons of simplicity we shall illustrate the arguments using a methyl halide CH_3X.

12-23a

12-23b

Since the ion Y^- is a reactive nucleophile, the principal interaction which occurs is between the HOMO of Y^- and the LUMO of CH_3X. For pictorial purposes we will represent the HOMO by a doubly occupied s orbital. (Of course it may be something else such as a p orbital which becomes an s/p hybrid orbital as the reaction proceeds.) For the cases when X is either Cl, Br or I the LUMO of CH_3X is an antibonding σ combination between two s/p hybrids on carbon and the halogen which we label σ^*_{CX}. We described this type of orbital in Section 10.3.1. Since X is more electronegative than C the orbital is largely localized on carbon (**12-24**). A somewhat different state of affairs holds for X = F. The C—F bond is very short in

12-24

CH_3F which means that the σ^*_{CF} orbital lies very high in energy and the LUMO of the molecule is $\pi^*_{CH_3}$ rather than σ^*_{CF}. This characteristic of fluorine derivatives means that they are not very reactive in substitution reactions.

Now we can study the interactions which develop during the two types of approach of **12-23**. In that of **12-23a**, which leads to inversion at carbon, a positive overlap develops between the HOMO of the nucleophile and the small carbon lobe of the σ^*_{CX} orbital (**12-25a**). Contrarily the attack of Y^- on the C—X bond leads to an overlap between the frontier orbitals that is close to zero, since the nucleophile is located close to the nodal surface of σ^*_{CX} (**12-25b**). The first type of approach is thus

the one where the two-electron stabilization is largest, leading to an explanation of the observation that it is the pathway with inversion, and not that with retention at carbon which is then one actually found.

Besides the energetic stabilization of this interaction of **12-25a** there is also an electronic reorganization inside the molecule. The bonding character between Y and carbon eventually develops into a real CY bond in the product. In addition electronic charge is transferred from Y^- to the σ^*_{CX} orbital which weakens the CX bond and facilitates its rupture.

This orbital analysis has similarities to the results found for the $H_3{}^-$ molecule. In effect, in the $Y^- + CH_3X$ system only four electrons play an important role. These are the two electrons in the HOMO of the nucleophile and the two electrons of the CX bond which is broken. In addition since the hydrogen atoms are 'spectators' during the reaction, these four electrons are localized over the three centers X, Y and the central carbon atom. Just as in $H_3{}^-$ with its four electrons and three centers, the system prefers a linear geometry over a triangular one.

Now consider the interactions between the charges on the various atoms. Since the atom X is more electronegative than carbon the CX bond is polarized such that negative charge accumulates on X. From the electrostatic point of view the first mechanism is favored over the second by avoiding the close $X^{\delta-}-Y^-$ repulsion which would occur in **12-25b**. There is however, an example which shows that this electrostatic interaction is not as important as the orbital one. The S_N2 reaction of a negatively charged nucleophile with a sulfonium salt, shown in **12-26**, leads to inversion at carbon even though, since the charges of $N_3{}^-$ and the SMe_2 groups are

12-26

of opposite sign, an electrostatic stabilization via the second pathway (**12-23b**) should occur. Such a result shows that the reaction is under orbital and not charge control.

12.5.2. Markovnikov's rule

In principle the addition of hydrogen bromide to propene can lead to two products, 2-bromopropane where the bromine is attached to the α carbon atom and 1-bromopropane where it is attached to the β carbon. The hydrogen atom is thus attached to the β and α sites respectively. Experimentally bromine adds preferentially to the most substituted carbon atom (*Markovnikov's rule*) leading to 2-bromopropane (**12-27**). Such a reaction is called *regioselective*.

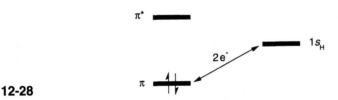

12-27

As described in Section 12.1.1 this reaction takes place in three steps. The nature of the product is determined in the second step, namely the attack of the H^+ on the double bond. It is this result which we will study using the frontier orbital concept.

(a) Frontier orbitals and the approach of the reactants

The H^+ cation is a reactive electrophile whose only valence orbital (1s) is empty. The frontier orbitals of propene are just the π and π^* orbitals associated with the double bond. The only interaction we need to consider is thus between the filled π MO of propene and the empty 1s orbital of H^+. This is a stabilizing two-electron interaction (**12-28**).

12-28

In propene the C=C π system is perturbed by the presence of the methyl group. In order to appreciate this we need to assemble an orbital interaction diagram using the ethylene and methyl fragments as in **12-29**.

12-29

The π_1 and π_1^* orbitals of the first fragment are antisymmetric with respect to the xz plane, and on the methyl group both the bonding $\pi_{CH_3}^y$ and antibonding $\pi_{CH_3}^{*y}$ orbitals of Section 6.4 are antisymmetric with respect to this plane. Calculations show that the $\pi_{CH_3}^{*y}$ orbital lies high in energy so that its influence on the π_1 and π_1^* levels of ethylene may be ignored. The orbital problem thus reduces to a three-orbital problem (12-30) which may be studied in the way described in Section 6.1. The

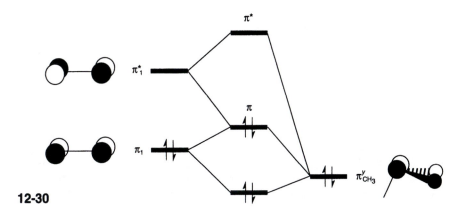

12-30

highest occupied orbital of the complete molecule is largely derived from the π_1 orbital of the ethylenic fragment mixed out of phase with the $\pi_{CH_3}^y$ orbital of methyl and in phase with the π_1^* orbital of ethylene (12-31). This polarization by the π^* level leads to an asymmetry in the π orbital coefficients. The p orbital contributions add

12-31

together on the β carbon (large coefficient) and subtract on the α carbon (small coefficient). The sense of the admixture of π^* and $\pi_{CH_3}^y$ is completely determined by the bonding and antibonding nature respectively of the two interactions. As a result the HOMO of propene is largely developed on the unsubstituted carbon atom, C_β (12-31). Consequently the overlap between the HOMO of propene and the H^+ $1s$ orbital is larger during the attack on the β than on the α carbon atom. This difference shows up directly in the energy profiles associated with each mode of approach as the reactants, $CH_2=CH-CH_3$ and H^+ approach from infinity. The lower energy plot is associated with attack on the β carbon (12-32).

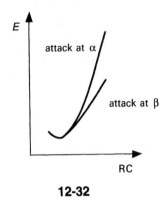

12-32

(b) Stability of the carbocations formed

The frontier orbital approximation rests on the analysis of the interactions between the MOs of the isolated reactants. It is thus most applicable when the geometrical deformations which take place during the reaction are small, that is to say at the beginning of the reaction coordinate. It is possible to study another part of the reaction by studying the stability of the carbocations formed via the two reaction routes. The attack of H^+ on the β carbon leads to the generation of a cation substituted by two methyl groups (CH_3—CH^+—CH_3), whereas attack at the α carbon leads to a cation substituted by a single ethyl group, $CH_2{}^+$—CH_2—CH_3.

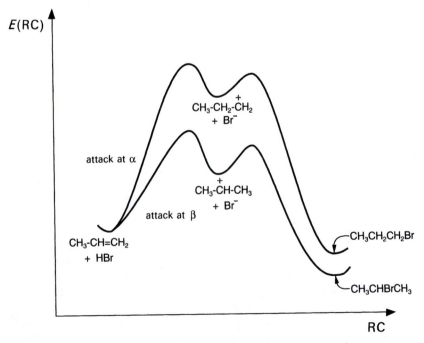

Figure 12.2. Reaction profiles for the two modes of attack of H^+ on propene.

Now, we have already seen in Section 11.3.2 that secondary carbocations are more stable than primary ones, and so the carbocation formed via β attack of H^+ is more stable than that formed via α attack. Combining this result with that of the previous section we can construct the energy profile associated with the formation of the carbocation. This is shown in the left-hand side of Figure 12.2. Attack on the β carbon, initially the most favored energetically also leads to the most favored cationic intermediate. There is no crossing of the two energy curves for the two mechanisms, and so one can conclude that the stabilities of the two transition states reflect the relative stability of the carbocation intermediates. This result is in accord with what is called *Hammond's postulate*, that in an endothermic reaction the transition state is closest in character and energy to the reaction product.

The last step of the reaction scheme involves the combination of the carbocation with the bromide ion. This is a strongly exothermic process with a rather small activation energy. It is dominated initially by the strong electrostatic interaction between the positively charged carbon atom and the negatively charged bromide ion. The most stable carbocation leads to the most stable product, and there is once again no crossing of the reaction profiles. This leads to the complete energy picture of Figure 12.2.

References

Frontier Orbital Approximation: K. Fukui, *Accts. Chem. Res.*, **4**, 57 (1971).

Woodward-Hoffmann Rules: R. B. Woodward, R. Hoffmann, *The Conservation of Orbital Symmetry*, Verlag Chemie Weiheim (1970).

The S_N2 Reaction: L. Salem, *Chem. Brit.*, **5**, 449 (1969).

Polarization of the π-system of propene: L. Libit, R. Hoffmann, *J. Amer. Chem. Soc.*, **96**, 1370 (1974).

EXERCISES

12.1 *Interaction of two methylene fragments*

    (i) Recall the form and relative energies of the MOs of bent AH_2 molecules.

    (ii) Give the electronic configuration of methylene, CH_2 where all the electrons are paired up in the deepest lying set of orbitals.

    (iii) When imagining the approach of two methylenes in this electronic configuration two extreme possibilities are apparent, a coplanar structure (**1**) which directly gives (geometrically) the reaction product (ethylene), and an orthogonal structure (**2**) in which the methylenes are located in two perpendicular planes. Analyze the interactions between the frontier orbitals for each arrangement and so deduce the energetically most favored route.

                 1   **2**

    (iv) Construct the molecular orbitals of structure **1** for the two cases of (a) a weak interaction between the fragments (appropriate for a large separation) and (b) a strong interaction between them (appropriate for the short distance in ethylene).

    (v) Put electrons in the molecular orbitals you have constructed and show why structure **1** is energetically disfavored at long C—C distances but favored at short distances.

    (vi) Describe in qualitative terms the course of the reaction to give ethylene starting off with two methylene units in the electronic configurations you chose in (ii).

REFERENCE: R. Hoffmann, R. Gleiter, F. B. Mallory, *J. Amer. Chem. Soc.*, **92**, 1460 (1970).

12.2 *Interaction between methylene and ethylene*

Energetically compare the two structures **3** and **4** for the initial attack of a methylene fragment on an ethylene molecule (i.e., a large distance between the two units). Consider only the case where the two reactants are in an electronic configuration where all electrons are paired.

3   **4**

REFERENCE: R. Hoffmann, *J. Amer. Chem. Soc.*, **90**, 1475 (1968).

12.3 *Bridged non-classical ions*

Here consider the interaction of an electrophile (modeled by H^+) or a nucleophile (modeled by H^-) on the ethylene molecule, the reactant being located in the plane which bisects the C=C double bond (**5** or **6**). The result is the formation of a non-classical, or bridged species in contrast to the classical species obtained by attack at one or the other of the two carbon atoms.

5   **6**

(i) Determine the stabilizing or destabilizing nature of the interactions which develop between the frontier orbitals of the reactants, and hence show that the bridged structure is quite unfavorable for one of the two systems. Finally correlate your result with Hückel's aromaticity rule.

(ii) It is interesting to ask how the formation of a non-classical ion is more or less favored depending on the nature of the electrophile. To do this compare the two cases $E^+ = H^+$ and $E^+ = Cl^+$. In each case construct the interaction diagram between the frontier orbitals and derive the correct electronic configuration. For Cl^+ only use the $3p$ AOs, assuming that the $3s$ orbital lies too deep in energy to be important. Hence elucidate the identity of E which leads to the lowest energy structure.

To simplify the construction of the energy level diagram use the following values of the orbital energies; $\pi(C_2H_4) = -13.21$ eV; $\pi^*(C_2H_4) = -8.24$ eV; $\varepsilon_{1s}(H) = -13.6$ eV; $\varepsilon_{3p}(Cl) = -13.7$ eV.

REFERENCES: W. J. Hehre, P. C. Hiberty, *J. Amer. Chem. Soc.*, **96**, 2665 (1974); S. Yamabe, T. Tsuji, *Chem. Phys. Lett.*, **146**, 236 (1988).

12.4 *Attack of an electrophile E^+ on H_2O and H_2S*

(i) Recall the electronic structure of water.

(ii) Consider the attack of an electrophile, E^+, on H_2O to give H_2O—E^+. What are the two important stabilizing interactions to be considered? Use your result to explain why the attack of an electrophile takes place with an approach angle, θ close to 45°.

(iii) On moving from H_2O to H_2S the energy separation between the two lone pair orbitals increases. Explain how you would expect the value of θ to be different from that for attack on water.

REFERENCE: P. Kollman, J. McKelvey, A. Johansson, S. Rothenberg, *J. Amer. Chem. Soc.*, **97**, 955 (1975).

12.5 *Carbonyl ylid*
(i) Give the form and the energetic ordering of the π MOs of planar carbonyl ylid (7) by suitable modification of those of allyl given in Section 8.2.1.

7

(ii) How many π electrons are there? Use the electronic structure of water to help you.
(iii) Experimentally carbonyl ylid is a reaction intermediate obtained by ring opening of an oxirane. It adds to alkenes to give a five-membered ring. Starting from an analysis of the frontier orbital interactions, show how the approach of the two molecules in planes parallel to each other is the favored route.
(iv) Hence deduce the product of the addition reaction below.

REFERENCES: R. Huisgen, *Angew. Chem.* (*Int. Ed. Eng.*), **16**, 572 (1977); K. N. Houk, N. G. Rondan, C. Santiago, C. J. Gallo, R. W. Gandour, G. W. Griffin, *J. Amer. Chem. Soc.*, **102**, 1504 (1980); F. Volatron, Nguyen Trong Anh, Y. Jean, *J. Amer. Chem. Soc.*, **105**, 2359 (1983).

Part IV
Problems

Problem 1. **Stabilization of a planar tetravalent carbon atom via π effects**

1. MOs of the square planar H_4 fragment

The four hydrogen atoms are located at the corners of a square lying in the xOy plane where O is the origin of the coordinate system. Construct the MOs of square-planar H_4 via the interaction of two fragments of your choice. Give the form and relative energies of these MOs.

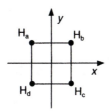

2. Square planar CH_4

(i) Construct the MOs of a square planar CH_4 molecule in which the carbon atom is located at the center of the square defined by the four hydrogen atoms. Use the symmetry properties of the two fragments, C and H_4, and the following energies for the relevant orbitals $\varepsilon_{2s}(C) = -19.4$ eV, $\varepsilon_{2p}(C) = -10.7$ eV, and -16.6 eV, -13.6 eV (twice) and -9.6 eV for H_4. We also specify that all of the antibonding orbitals between carbon and hydrogen lie above -9.6 eV.
(ii) Give the ground state electronic configuration.
(iii) How many bonding electrons are there? Why is this structure less stable than the tetrahedral geometry?

3. Stabilization of square planar tetravalent carbon via π effects

(i) One can try to stabilize this geometry by the introduction of appropriate substituents X.
   (a) Using the results from question 2 above identify the MO of CH_4 that it is important to stabilize. In the following only consider the interactions which occur between this orbital and the orbitals of the appropriate symmetry of the substituent(s) X.

(b) Replace one hydrogen by a lithium atom ($\varepsilon_{2p}(\text{Li}) = -3.5$ eV). Does this increase the stability of the planar structure?

(c) Answer the same question for X = F ($\varepsilon_{2p}(\text{F}) = -18.6$ eV).

(d) Recall the electronic structure of the trigonal planar BH_3 molecule. Draw out the form and indicate the energy of the lowest lying vacant orbital. $\varepsilon_{2p}(\text{B}) = -5.7$ eV. With X = BH_2 calculations show that the coplanar structure (1) is more stable than the perpendicular one (2) by 150 kJ mol^{-1}. Explain why.

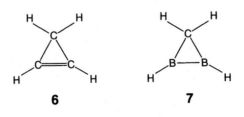

1   **2**

(e) For X = $CH_2{}^+$ the same structural preference is found as in (d) except that now the energy difference is 400 kJ mol^{-1}. Contrarily for X = NH_2 the perpendicular structure is more stable by 67 kJ mol^{-1}. Show the origin of these two results.

(ii) Calculations have been performed on several conformers of $C(BH_2)_4$.

   (a) Based on the results to the preceeding question which conformer out of **3, 4** and **5** should be most stable?

3   **4**   **5**

(b) By calculation, **4** is in fact more stable. **3** and **5** are 226 and 330 kJ mol^{-1} higher in energy. Provide an interpretation.

(iii) One can look to stabilize a planar tetravalent carbon in a small ring. Is it preferable to use cyclopropene (**6**) or diboracyclopropene (**7**)?

6   **7**

(iv) Construct a Walsh diagram for pyramidalizing square planar CH_4. For what electron count is a square planar geometry favored? Is this in accord with VSEPR?

PRINCIPAL THEORETICAL REFERENCES

R. Hoffmann, R. W. Alder, C. F. Wilcox, *J. Amer. Chem. Soc.*, **92**, 4992 (1970).

J. B. Collins, J. D. Dill, E. D. Jemmis, Y. Apeloig, P. v. R. Schleyer, R. Seeger, J. A. Pople, *J. Amer. Chem. Soc.*, **98**, 5419 (1976).

Problem 2. **Nucleophilic attack on a carbonyl group**

1. Electronic structure of the carbonyl group

(i) Recall the form of the π_{CO} and π^*_{CO} orbitals and their energy relative to the isolated atomic orbitals $2p(C)$ and $2p(O)$.

(ii) What is the polarity of the C=O bond?

(iii) Compare this with the case of the ethylenic double bond (C=C).

2. Nucleophilic attack on the carbonyl group

Consider the attack of the simplest nucleophile (H^-) on the simplest carbonyl-containing molecule (formaldehyde, H_2C=O)

(i) Two cases can immediately be distinguished, attack on the carbon or attack on the oxygen atom. What are the products of each mode of attack and which is thermodynamically more stable?

(ii) The transition states for each pathway can be represented by the structures **1** and **2**. In each model the attack of hydride proceeds perpendicularly to the plane of the carbonyl group.

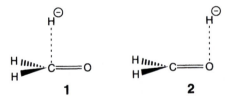

1   **2**

(a) Which is the structure favored electrostatically?

(b) Analyze the nature of the stabilizing and destabilizing interactions between the π and π^* orbitals of the C=O group and the hydride 1s orbital. Which is the approach favored on orbital grounds?

(c) Thus deduce the product formed on kinetic grounds.

(d) Which orbital interaction leads to charge transfer during nucleophilic attack?

(iii) A study using X-ray crystallography of a series of molecules containing both a carbonyl and an amine group (Figure 1) led to the geometrical results collected in Table 1.

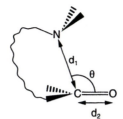

Figure 1

Table 1

d_1 (pm)	d_2 (pm)	$\theta(^\circ)$
291	121	105.0
256	122	101.6
199	126	110.2
164	138	110.9

(a) Recall the Lewis structure of an amine. Describe the electronic reasons for it being a nucleophile.

(b) Using the orbital results of question ii, explain why the distance d_2 increases as d_1 decreases.

(iv) The experimental results show that the attack is not perpendicular but that $\theta > 90^\circ$. By analyzing the variation in the overlaps between $1s(H)$ and the π_{CO} and π^*_{CO} orbitals as a function of geometry (Table 2) explain why the attack proceeds with $\theta > 90^\circ$.

Table 2

$\theta(^\circ)$	70	80	90	100	110	120
$\langle 1s \mid \pi \rangle$	0.13	0.10	0.09	0.08	0.07	0.06
$\langle 1s \mid \pi^* \rangle$	0.11	0.13	0.14	0.15	0.15	0.15

(v) Two carbonyl functionalities may be attacked during the reduction of succinimides (**3**) by $NaBH_4$. During this reaction BH_4^- behaves as a nucleophile and transfers a hydride. After reduction with $NaBH_4$, followed by hydrolysis, alcohols are formed. The orientation, **a** or **b**, depends on the nature of the substituents (Table 3).

Table 3

R_1	R_2	% on a	% on b
C_6H_5	C_6H_5	100	0
CH_3	CH_3	79	21

3

attack on **b** attack on **a**

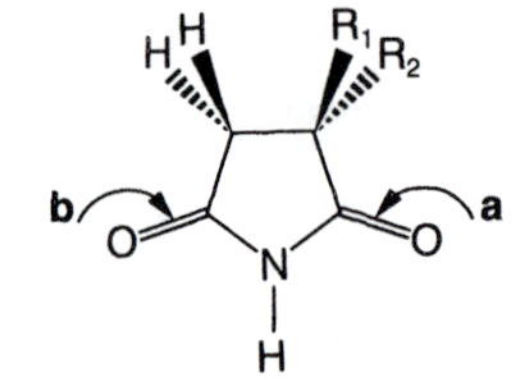

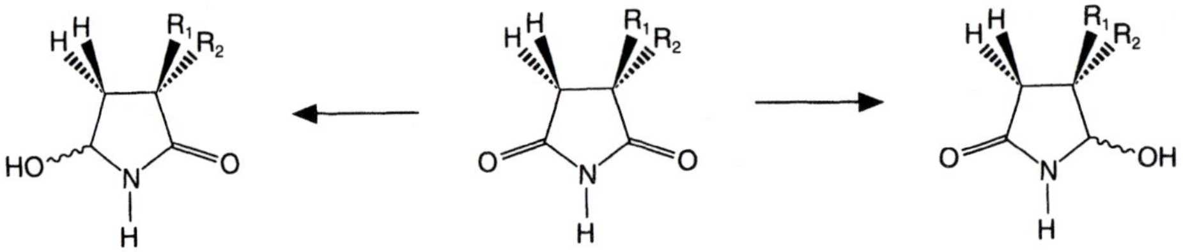

(a) When $R_1 = R_2 =$ phenyl, attack is found only at **a**, that is to say at the most sterically encumbered carbonyl group. Explain this result.
(b) Explain the difference in the product ratio for $R_1 = R_2 = CH_3$, compared to that for $R_1 = R_2 =$ phenyl.
(c) Predict the number of isomers expected after reduction ($NaBH_4$ followed by water) of the compounds **4** and **5**.

4

5

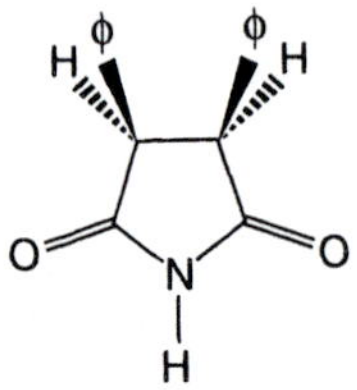

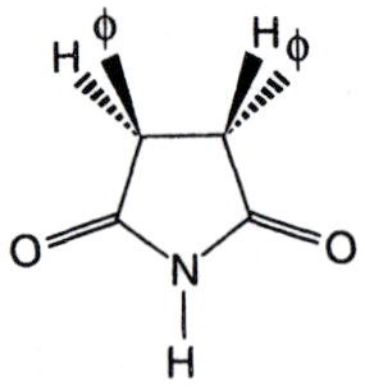

(vi) Explain the difference in reactivity found experimentally for the carbonyl and ethylenic groups. (Hint: use your answers to question 1.)

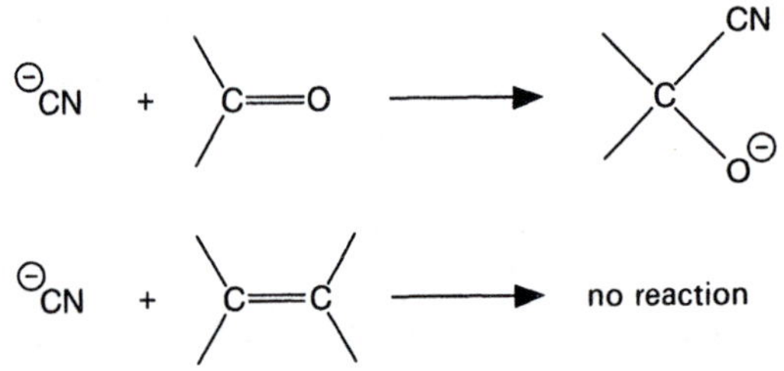

PRINCIPAL REFERENCES

H.-B. Bürgi, J. D. Dunitz, J. M. Lehn, G. Wipff, *Tetrahedron*, **30**, 1563 (1974).

H.-B. Bürgi, J. D. Dunitz, E. Shefter, *J. Amer. Chem. Soc.*, **95**, 5065 (1973).

J. D. Dunitz in *X-ray analysis and the structure of organic molecules*, Cornell University Press (1979).

J. B. P. A. Wijnberg, H. E. Schoemaker, W. N. Speckamp, *Tetrahedron*, **34**, 179 (1978).

N. T. Anh, O. Eisenstein, *Nouv. J. Chim.*, **1**, 61 (1977).

Problem 3. **Structure and reactivity of substituted cyclopropanes**

1. Molecular orbitals of cyclopropane

The form of the frontier orbitals of unsubstituted cyclopropane are given in Figure 1. These MOs will be used in the problem to study the structure and reactivity of substituted cyclopropanes.

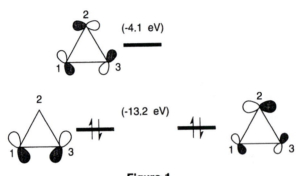

Figure 1

(i) Identify the symmetry properties of these MOs with respect to a plane which bisects the C_1—C_3 linkage, passes through C_2 and lies perpendicular to the paper.

(ii) Identify the bonding, nonbonding or antibonding character of these orbitals.

2. Structure of substituted cyclopropanes

Consider the carbinyl cyclopropane, **1** where a methylene cation replaces one of the hydrogen atoms of cyclopropane.

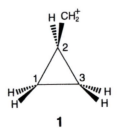

1

(i) One can envisage two conformations for **1**, the bisected structure (**2-a**) and the structure where the methylene group is rotated by 90° (**2-b**). Compare the interaction of the methylene 2p orbital with those of Figure 1.

2-a   **2-b**

(a) Set up an interaction scheme to obtain the orbital picture in conformations **2-a** and **2-b**. To do this make use of the symmetry properties of the orbitals. ($\varepsilon_{2p}(C) = -10.7$ eV.)

(b) Hence deduce that the CH_2^+ group behaves as an acceptor with respect to the cyclic system.

(c) Which conformation is more stable?

(d) Examine the electron transfer which occurs between the substituent and the ring in this conformation. What effect do you predict it will have on the CC distances?

(ii) Consider the molecule cyanocyclopropane, **3**.

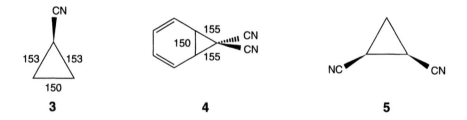

3   **4**   **5**

(a) Recall the form and relative energy of the π_{CN} and π_{CN}^* levels of the CN group. These are calculated by the extended Hückel method to lie at -14.72 eV and -8.27 eV respectively.

(b) Use symmetry arguments to list the two- and four-electron interactions between the ring frontier orbitals and those of the cyano group.

(c) Using overlap arguments, which are the two major interactions involving charge transfer between the two fragments? Which one is larger? Hence deduce whether the cyano group is a π donor or acceptor with respect to the ring.

(d) Provide a model to understand the bond lengths found experimentally in cyanocyclopropane (**3**). In the unsubstituted molecule all the CC distances are equal (151 pm).

(e) Assuming that the effect of two cyano groups is additive, show the origin of the bond length differences in the three-membered ring part of norcaradiene whose experimentally determined geometry is given in **4**. Predict the changes in the CC bond lengths in dicyanocyclopropane, **5**.

3. Reactivity

(i) Consider the equilibrium between norcaradiene (**6**) and cycloheptatriene (**7**). The equilibrium constant depends upon the nature of the substituents, R. Which side of the equilibrium is favored by a strong π-acceptor, R?

6   **7**

(ii) The Cope rearrangement is characterized by migration of a σ_{CC} bond (a sigmatropic rearrangement), **8**. The same type of process is observed in semibullvalene, **9**.

8   **9**

Where does the equilibrium lie in the schemes, **10, 11** and **12**? A is a π-acceptor substituent.

10

11

12

PRINCIPAL THEORETICAL REFERENCES

A. D. Walsh, *Trans. Faraday Soc.*, **45**, 179 (1949).

R. Hoffmann, *Tetrahedron Lett.*, 2907 (1970).

R. Hoffmann, *23rd Int. Cong. Pure Appl. Chem.*, **2**, 233 Butterworth, London (1971).

R. Hoffmann, W.-D. Stohrer, *J. Amer. Chem. Soc.*, **93**, 6941 (1971).

Problem 4. **Conformational consequences of hyperconjugation**

1. The fragments H_3 and H_2X

Figure 1 shows the form and relative energies of the MOs of the triangular H_3 unit. The coefficients shown are those evaluated by ignoring overlap between the orbitals as in Section 4.1.5b.

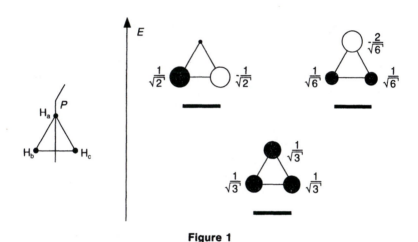

Figure 1

On replacing a hydrogen atom (H_a) by a more electronegative atom X, the symmetry plane P is conserved. Assume that the atom X only carries an s-type orbital.

  (i) Construct the MOs of H_2X by interaction of the H_2 fragment orbitals with that of X.
  (ii) Compare the picture with that for the H_3 reference molecule.
  (iii) Give a general qualitative description of the way the MOs of a molecule are modified when an atom is replaced by one which is more electronegative.

2. The CH_3 and CH_2X fragments

Recall (Figure 2) the form of the $1e$ (π_x and π_y) and $2e$ (π_x^* and π_y^*) orbitals of the CH_3 fragment.

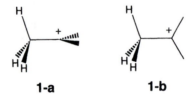

Figure 2

Replace H_a with a more electronegative atom X. Predict how the energies and form of the 1e and 2e molecular orbitals change on substitution, focusing especially on the contribution from the carbon 2p orbital. Use the results of question 1 to help you.

3. Conformation of the substituted ethyl cation ($CH_2XCH_2^+$)

(i) Consider the two conformations of the ethyl cation (**1-a** and **1-b**). In each case decompose the molecule into the two fragments CH_3 and CH_2^+. Only consider the interactions between the nonbonding 2p orbital on CH_2^+ with the relevant orbitals on the CH_3 fragment.

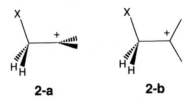

1-a   **1-b**

   (a) What are the orbitals which may interact in the conformation **1-a**? Which interaction(s) play an energetic role?
   (b) Answer the same questions for conformation **1-b**.
   (c) Do you expect a large energy difference between the two conformers?
(ii) Consider the two conformations **2-a** and **2-b** for the substituted cation. Using the results of the analyses above predict the more stable conformer when X is more electronegative than hydrogen.

2-a   **2-b**

4. Conformation of the substituted ethyl anion, $CH_2XCH_2^-$

In what follows, assume that the anionic center has a trigonal planar geometry.

(i) Using the analysis already undertaken for the cation, compare the energies of the two conformations **3-a** and **3-b**.

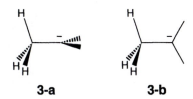

3-a   **3-b**

(ii) Predict which conformation will be more stable, **4-a** or **4-b**, for the case when X is more electronegative than hydrogen.

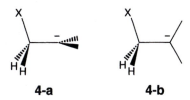

4-a   **4-b**

REFERENCE

R. Hoffmann, L. Radom, J. A. Pople, P. v. R. Schleyer, W. J. Hehre, L. Salem, *J. Amer. Chem. Soc.*, **94**, 6221 (1972).

Bibliography

This book has been designed to be an introduction to molecular orbital theory and its applications. There are several other books, written at a somewhat higher level which cover much larger areas of chemistry but using the same qualitative approach used here. In particular reference 1 traverses the field of chemistry from small fragments to solids and includes discussion of transition metal complexes. These books, which have been of valuable assistance to us in our efforts, will be useful to the reader who wishes a deeper level of understanding or a more exact approach to qualitative molecular orbital ideas in chemistry.

General

T. A. Albright, J. K. Burdett, M.-H. Whangbo, *Orbital Interactions in Chemistry*, John Wiley, New York (1985).

Shapes of molecules

J. K. Burdett, *Molecular Shapes*, John Wiley, New York (1980).
B. M. Gimarc, *Molecular Structure and Bonding*, Academic Press, New York (1979).

Reactivity

I. Fleming, *Frontier Orbitals and Organic Chemical Reactions*, John Wiley, London (1976).
R. G. Pearson, *Symmetry Rules for Chemical Reactions*, John Wiley, New York (1976).
R. B. Woodward, R. Hoffmann, *The Conservation of Orbital Symmetry*, Verlag Chemie, Weinheim (1970).

Structures of solids

R. Hoffmann, *A Chemist's View of Bonding in Extended Structures*, Verlag Chemie, Weinheim (1989).

Other books which may be useful:

Quantum mechanics

C. Cohen-Tannoudji, B. Diu, F. Laloe, *Quantum Mechanics*, Wiley, New York (1977) two volumes.
M. Karplus, R. N. Porter, *Atoms and Molecules*, Benjamin, Menlo Park, California (1970).
J. C. Slater, *Quantum Theory of Atomic Structure*, McGraw-Hill, New York (1960) volume 1.

The structure of complex molecules

W. J. Hehre, L. Radom, P. v. R. Schleyer, J. A. Pople, *Ab initio Molecular Orbital Theory*, John Wiley, New York (1986).

W. L. Jorgensen, L. Salem, *The Organic Chemist's Book of Orbitals*, Academic Press, New York (1973).

R. S. Mulliken, W. C. Ermler, *Diatomic Molecules*, Academic Press, New York (1977).

L. Salem, *The Molecular Orbital Theory of Conjugated Systems*, Benjamin, New York (1966).

L. Salem, *Electrons in Chemical Reactions: First Principles*, John Wiley, New York (1982).

Organic chemistry

N. L. Allinger, M. P. Cava, D. C. de Jongh, C. R. Johnson, N. A. Lebel, C. L. Stevens, *Organic Chemistry*, McGraw-Hill, Kogakusha (1977).

J. March, *Advanced Organic Chemistry*, McGraw-Hill, Kogakusha (1977).

The reader interested in the application of group theory to chemistry should consult:

Group theory

D. M. Bishop, *Group Theory and Chemistry*, Clarendon Press, Oxford (1973).

F. A. Cotton, *Chemical Applications of Group Theory*, 3rd Edition, Wiley-Interscience, New York (1990).

Answers to Exercises

Chapter 1

1.1

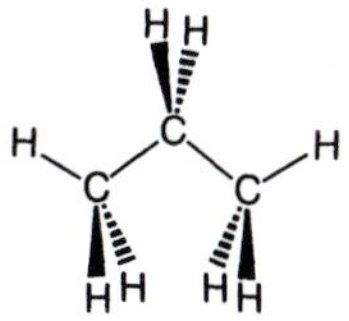

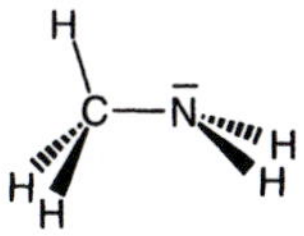

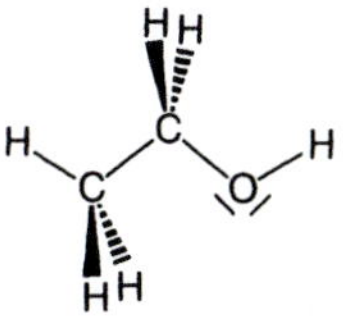

COH ≈ 109°

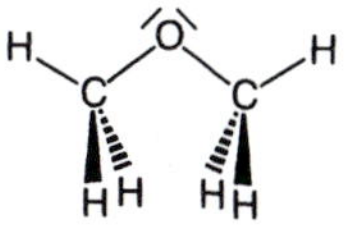

COC ≈ 109° 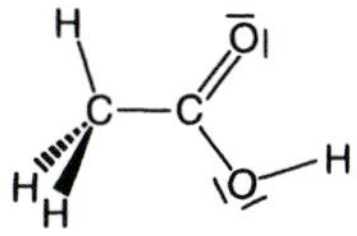   OCO ≈ 120° COH ≈ 109° 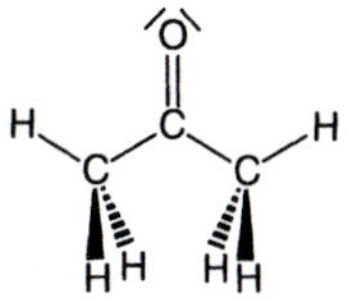   OCC ≈ CCC ≈ 120°

1.2 (i) (ii) (iii)

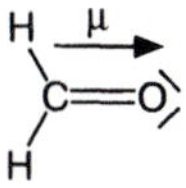

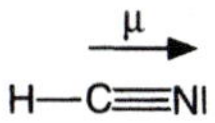

(iv) (a)

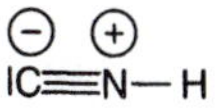

IC≡N—H   IC≡N—H   IC≡N—H
(linear) 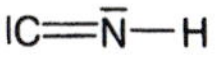   (bent)

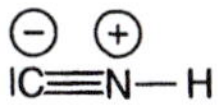

1.3 μ(*cis*) ≠ 0; μ(*trans*) = 0 (the dipole moments of the two C—Cl bonds cancel).

1.4 (i) (ii)

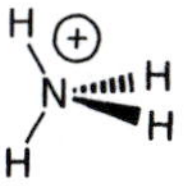

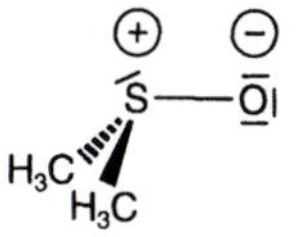

(iii) μ(NH$_4^+$) = 0

1.5 (i)

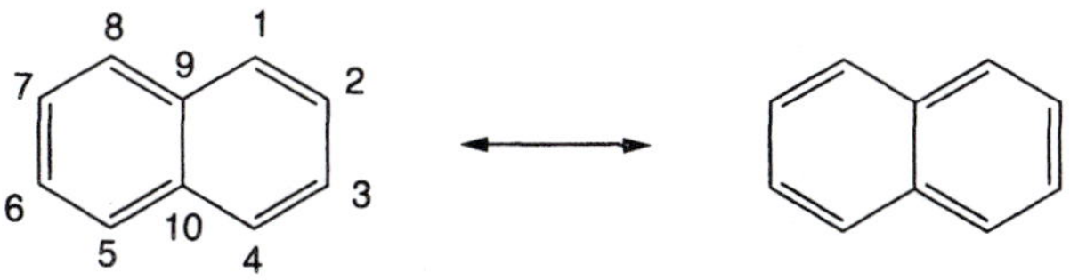

(ii) The C_1—C_2 linkage is represented twice by double bonds and once by a single bond in the resonance structures. It is then shorter than the C_9—C_{10}, C_2—C_3 and C_1—C_9 linkages which are represented twice by single bonds and once by a double bond.

1.6 (i)

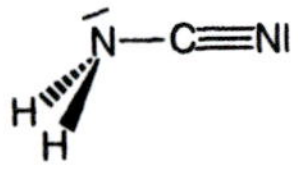

(ii)

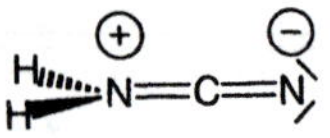

The molecule is planar

(iii) The important contribution to the structure is given in (ii). The situation is identical to that described for aniline.

1.7 (i)

(ii)

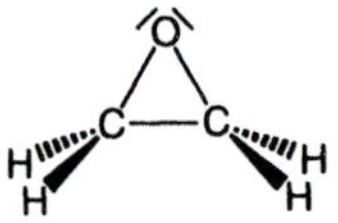

(iii)

1.8 (i) (a) 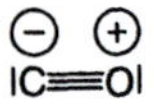 (b)

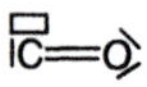

(ii) (a)

(b)

Chapter 2

2.1 (i) $3p \rightarrow 2s$ or $2p$; $3p \rightarrow 1s$. (ii) 656 nm; 102.5 nm. (iii) 1.51 eV or 145.6 kJ mol^{-1}.

2.2 (i) 54.4 eV (He$^+$) and 489.6 eV (C^{5+}). (ii) 9 (one s, three p and five d). 13.6 eV.

2.3 (i) $1s^2 2s^2 2p^6 3s^2 3p^4$. (ii) $Z^*(1s) = 15.70$; $\rho(1s) = 3$ pm; $Z^*(2s, 2p) = 11.85$; $\rho(2s, 2p) = 18$ pm; $Z^*(3s, 3p) = 5.45$; $\rho(3s, 3p) = 87$ pm.

2.4 41 pm (F); 78 pm (Cl); 111 pm (Br) taking $n = 4$. Using $n = 3.7$ leads to the value given in Table 2.3 (95 pm). The order of polarizability: F < Cl < Br.

2.5 $[\text{Xe}]6s^2 4f^{14} 5d^8$; $[\text{Xe}]4f^{14} 6s^1 5d^9$; $[\text{Xe}]4f^{14} 6s^0 5d^{10}$.

2.6 (i) He, Be, Ne, Mg, Ar, Ca. (ii) H, Li, B, F, Na, Al, Cl, K. (iii) C, O, Si, S.

2.7 $1s$: $\dfrac{d^2 S}{dr^2} = \dfrac{4}{a_0^3}\left(2r - \dfrac{2r^2}{a_0}\right)\exp\left(-\dfrac{2r}{a_0}\right)$; $\dfrac{d^2 S}{dr^2} = 0 \Rightarrow r = 0$ (minimum); $r = a_0$ (maximum such that $dS/dr = 0.541\, a_0^{-1}$).

$2s$: $\dfrac{d^2 S}{dr^2} = \dfrac{1}{8a_0^3}\left(2 - \dfrac{r}{a_0}\right)r\left(\dfrac{r^2}{a_0^2} - 6\dfrac{r}{a_0} + 4\right)\exp\left(-\dfrac{r}{a_0}\right)$; $\dfrac{d^2 S}{dr^2} = 0 \Rightarrow r = 0$ (minimum); $r = (3 - \sqrt{5})a_0$ (secondary maximum; $dS/dr = 0.052\, a_0^{-1}$); $r = 2a_0$ (minimum corresponds to a spherical node); $r = (3 + \sqrt{5})a_0$ (absolute maximum $dS/dr = 0.191 a_0^{-1}$).

$2p$: $\dfrac{d^2 S}{dr^2} = \dfrac{1}{24a_0^3}\dfrac{r^3}{a_0^2}\left(4 - \dfrac{r}{a_0}\right)\exp\left(-\dfrac{r}{a_0}\right)$; $\dfrac{d^2 S}{dr^2} = 0 \Rightarrow r = 0$ (minimum); $r = 4a_0$ (maximum; $dS/dr = 0.195\, a_0^{-1}$).

2.8 $\dfrac{d^2 S}{dr^2} = N^2 \dfrac{2r^{2n-1}}{a_0^{2n-2}}\left(n - \dfrac{Z^* r}{na_0}\right)\exp\left(-\dfrac{2Z^* r}{na_0}\right)$; $\dfrac{d^2 S}{dr^2} = 0 \Rightarrow \rho = \dfrac{n^2 a_0}{Z^*}$

2.9 $\langle 1s \mid 1s \rangle = \displaystyle\int_0^\infty \dfrac{4}{a_0^3} r^2 \exp\left(-\dfrac{2r}{a_0}\right) dr \int_0^\pi \sin\theta\, d\theta \int_0^{2\pi} \dfrac{1}{4\pi} d\phi$

$= \left(\dfrac{4}{a_0^3} 2 \dfrac{a_0^3}{8}\right)\left(2\dfrac{1}{4\pi} 2\pi\right) = 1$

2.10 (i) In the integrals $\langle 1s \mid 2p \rangle$ (or $\langle 2s \mid 2p \rangle$), we are only interested in the angular parts. For the orbital given in equation (30a) we get

$$\int_{\theta,\phi} YY' \sin\theta\, d\theta\, d\phi = \dfrac{1}{\sqrt{4\pi}}\sqrt{\dfrac{3}{4\pi}}\int_0^\pi \sin\theta\, d\theta \int_0^{2\pi} \cos\phi\, d\phi$$

$$\int_0^{2\pi} \cos\phi\, d\phi = [\sin\phi]_0^{2\pi} = 0 \text{ so } \langle 1s \mid 2p_x \rangle = 0$$

In the same way, for equation (30b): $\displaystyle\int_0^{2\pi} \sin\phi\, d\phi = 0$ so $\langle 1s \mid 2p_y \rangle = 0$. For

(30c): $\displaystyle\int_0^\pi \sin\theta\cos\theta\, d\theta = [\tfrac{1}{2}\sin^2\theta]_0^\pi = 0$ so $\langle 1s \mid 2p_z \rangle = 0$.

(ii) In $\langle 1s \mid 2s \rangle$, we are only interested in the radial part of the integral:

$$\int_r RR'r^2\, dr = \frac{1}{\sqrt{2}\, a_0^3} \int_0^\infty \left(2 - \frac{r}{a_0}\right) r^2 \exp\left(-\frac{3r}{2a_0}\right) dr$$

$$= \frac{1}{\sqrt{2}\, a_0^3}\left[4\,\frac{8a_0^3}{27} - \frac{6}{a_0}\,\frac{16a_0^4}{81}\right] = 0$$

Chapter 3

3.1 $0.627(\pi)$ and $0.828(\pi^*)$; 0.707 (π and π^*).

3.2   (i) $\varepsilon(1s_{He}) < \varepsilon(1s_H)$. (ii) 0.536. (iii) $\langle 1\sigma \mid 2\sigma \rangle = 10^{-3}$ (one does not get exactly zero because of rounding errors).

3.3   The overlap integral is zero at infinity, becomes negative as the nuclei approach each other and takes on the value of 1 (if the two orbitals are identical) when R is equal to zero. Schematically we have:

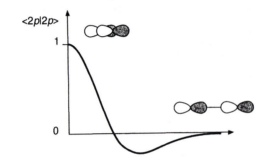

3.4   (i), (ii)

(iii) CH_3Cl: C_3 axis. 3 planes; NH_3: the same; CH_2CO: C_2 axis, 2 planes; *cis*: C_2 axis, 2 planes; *trans*: C_2 axis (perpendicular to the molecular plane and passing through the middle of the C—C bond), the molecular plane, inversion center (middle of the C=C bond).

3.5   (i) See Sections 3.5.2 and 3.5.5. (ii) The overlaps are larger in acetylene than in ethylene (the C—C distance is shorter in the former) and so the interactions are stronger. So $\varepsilon_\pi(\text{acetylene}) < \varepsilon_\pi(\text{ethylene})$; $\varepsilon_\pi^*(\text{acetylene}) > \varepsilon_\pi^*(\text{ethylene})$.

3.6 (i) 1 is more stable since there is no $p\pi$–$p\pi$ overlap in 2 to give rise to the formation of a π bond. (ii) 2. (iii) Both π_{NN} and π_{NN}^* will be full in geometry 1 which leads overall to a repulsion between the nitrogen atoms. Geometry 2 is preferred where there is no such overlap and thus no such repulsion. (iv) With four electron pairs at nitrogen (AX_3E) a pyramidal, rather than planar, geometry at N is preferred.

Chapter 4

4.1 (i) $\varphi_1 = \dfrac{1}{2\sqrt{1+3S}}(1s_a + 1s_b + 1s_c + 1s_d)$;

$\varphi_2 = \dfrac{1}{2\sqrt{1-S}}(1s_a + 1s_b - 1s_c - 1s_d)$;

$\varphi_3 = \dfrac{1}{\sqrt{2(1-S)}}(1s_a - 1s_b)$; $\varphi_4 = \dfrac{1}{\sqrt{2(1-S)}}(1s_c - 1s_d)$.

(ii) (a) $N_1 = N_2 = 0.5$; $N_3 = N_4 = 0.707$.
   (b) $N_1 = 0.322$; $N_2 = 0.686$; $N_3 = N_4 = 0.970$.

4.2 (i) $\varphi_1 = \dfrac{1}{2\sqrt{1+S+S'+S''}}(1s_a + 1s_b + 1s_c + 1s_d)$;

$\varphi_2 = \dfrac{1}{2\sqrt{1+S-S'-S''}}(1s_a + 1s_b - 1s_c - 1s_d)$;

$\varphi_3 = \dfrac{1}{2\sqrt{1-S+S'-S''}}(1s_a - 1s_b - 1s_c + 1s_d)$;

$\varphi_4 = \dfrac{1}{2\sqrt{1-S-S'+S''}}(1s_a - 1s_b + 1s_c - 1s_d)$.

(ii) (a) $N_1 = N_2 = N_3 = N_4 = 0.5$
   (b) $N_1 = 0.368$; $N_2 = 0.478$; $N_3 = 0.636$; $N_4 = 0.751$.

4.3 (i) H_a—H_c and H_b. (ii) A plane bisecting H_a—H_c and passing through H_b. (iii) $\sigma_{H_2}(S)$; $\sigma_{H_2}^*(A)$; $1s_b(S)$. Interaction between the symmetric orbitals is weaker (the distance is longer and so the overlap is smaller). (iv) φ_3 is less high in energy so $\varepsilon(\varphi_3) < \varepsilon(\varphi_2)$.

4.4 (i) (ii) (iii) (iv)

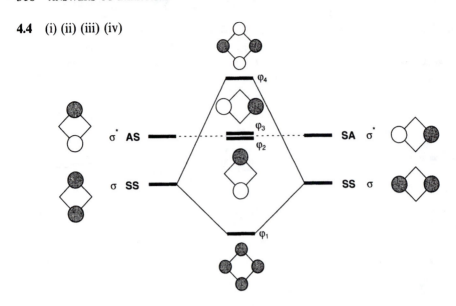

(v) The orbitals φ_1 and φ_4 are identical; orbitals φ_2 and φ_3 (degenerate) which are the two independent linear combinations of the orbitals φ_2 and φ_3 given in Figure 4.1. The result is equivalent (see Appendix).

4.5

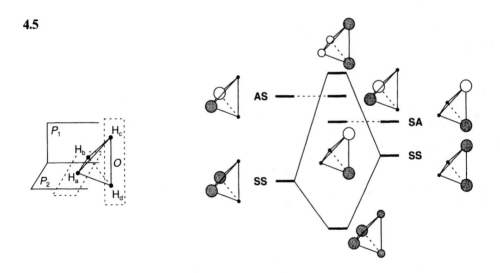

4.6 (i) H ... H and X ... X. (ii) xOz and xOy. (iii) (a) $\varepsilon(\sigma_{X_2}) < \varepsilon(\sigma_{H_2})$; $\varepsilon(\sigma_{X_2}^*) < \varepsilon(\sigma_{H_2}^*)$; $\sigma_{H_2}(SS)$; $\sigma_{X_2}(SS)$; $\sigma_{H_2}^*(AS)$; $\sigma_{X_2}^*(SA)$. (b) The interaction between σ_{H_2} and σ_{X_2}. In order of increasing energy, we find, φ_1 stabilized, located more on X_2 than on H_2, φ_4 stabilized and localized on X_2, φ_2 stabilized, largely localized on H_2, φ_3 unchanged, localized on H_2. (iv) The opposite trends.

Chapter 5

5.1 $\langle \varphi_2 \mid p_y \rangle = \dfrac{1}{\sqrt{2(1 - S)}} (\langle p_y \mid 1s_c \rangle - \langle p_y \mid 1s_a \rangle)$

$= \dfrac{1}{\sqrt{2(1 - S)}} (S_0 \cos 30° - S_0 \cos 150°) = \sqrt{\dfrac{3}{2(1 - S)}} S_0$

$\langle \varphi_3 \mid p_x \rangle = \dfrac{1}{\sqrt{6(1 - S)}} (2\langle p_x \mid 1s_b \rangle - \langle p_x \mid 1s_a \rangle - \langle p_x \mid 1s_c \rangle)$

$= \dfrac{1}{\sqrt{6(1 - S)}} (2S_0 - 2S_0 \cos 120°) = \dfrac{3}{\sqrt{6(1 - S)}} S_0 = \sqrt{\dfrac{3}{2(1 - S)}} S_0$

5.2 $\langle p_z \mid \varphi_4 \rangle = \dfrac{1}{\sqrt{2(1 - S)}} \langle p_z \mid 1s_c - 1s_d \rangle = \dfrac{1}{\sqrt{2(1 - S)}} 2S_0 \cos\left(90° - \dfrac{\alpha}{2}\right)$

$= \dfrac{2}{\sqrt{3(1 - S)}} S_0$

$\langle p_y \mid \varphi_3 \rangle = \dfrac{1}{\sqrt{2(1 - S)}} \langle p_y \mid 1s_a - 1s_b \rangle = \dfrac{1}{\sqrt{2(1 - S)}} 2S_0 \cos\left(90° - \dfrac{\alpha}{2}\right)$

$= \dfrac{2}{\sqrt{3(1 - S)}} S_0$

$\langle p_x \mid \varphi_2 \rangle = \dfrac{1}{2\sqrt{1 - S}} \langle p_x \mid 1s_c + 1s_d - 1s_a - 1s_b \rangle = 4 \dfrac{1}{2\sqrt{1 - S}} S_0 \cos\dfrac{\alpha}{2}$

$= \dfrac{2}{\sqrt{3(1 - S)}} S_0$

5.3 (i) Fragment H_a: $1s_a(S)$; fragment H_3: $\varphi_1(S)$, $\varphi_2(A)$, $\varphi_3(S)$: one MO $= \varphi_2$. (ii) $\langle 1s_a \mid \varphi_1 \rangle \neq 0$; $\langle 1s_a \mid \varphi_3 \rangle = 0$: φ_3 is a second MO. (iii) Interaction of $1s_a$ with φ_1 which gives a bonding MO (the deepest lying of all) and an antibonding MO (the highest lying).

5.4 (i) on $H_a H_b H_c$: φ_1, φ_2, φ_3; on $H_d \ldots H_e$: φ_4 (bonding), φ_5 (antibonding); H_d and H_e are far apart so φ_4 and φ_5 are at an energy intermediate between φ_1 and (φ_2, φ_3). (ii) $/xy$: $\varphi_1(S)$, $\varphi_2(S)$, $\varphi_3(S)$, $\varphi_4(S)$, $\varphi_5(A)$. So φ_5 is an MO. (iii) $/xz$: $\varphi_1(S)$, $\varphi_2(A)$, $\varphi_3(S)$, $\varphi_4(S)$: φ_2 is a second MO. (iv) $\langle \varphi_1 \mid \varphi_4 \rangle \neq 0$; $\langle \varphi_3 \mid \varphi_4 \rangle = 0$: φ_3 is a third MO. (v) A single interaction between φ_1 and φ_4 which leads to a bonding MO (the lowest in energy) and an antibonding MO (the highest in energy).

5.5 (i) $\varphi_1 =$ pseudo-s; $\varphi_2 =$ pseudo-p_x (with the opposite sign); $\varphi_3 =$ pseudo-p_y; no analogs for φ_4 and p_z. (ii) Same answers.

Chapter 6

6.1 (i) CH possesses five valence electrons so one must be unpaired. (ii) $|\dot{C}$—H. (iii)
(b) $1\sigma^2 2\sigma^2 \pi^1$ (c) $1\sigma^2$; C—H bond; $2\sigma^2$: lone pair on carbon; π^1: unpaired electron
on carbon (d) Because the unpaired electron is in a π-type orbital. (iv) The
ejected electron comes from a nonbonding orbital.

6.2 (i) · Be—H; $\overset{\oplus}{Be}$—H. (ii) $1\sigma^2 2\sigma^1$; $1\sigma^2$. The unpaired electron is located in an
orbital of σ type. (iii) 2σ is practically nonbonding. (iv) an electron moves from a
bonding orbital (1σ) to a nonbonding orbital (2σ) resulting in a weakening (and
thus an elongation) of the bond.

6.3 $|\overset{\ominus}{\underline{O}}$—H. (ii) $1\sigma^2 2\sigma^2 \pi^4$. $1\sigma^2$: O—H bond; $2\sigma^2$: σ lone pair on oxygen; π^4: two π
lone pairs on oxygen.

6.4 and 6.5   See the relevant diagrams in Sections 6.3.2 and 6.4.2.

6.6 (i) σ_{H_2} and $\sigma^*_{H_2}$ for H_a—H_b and $1s_c$ for H_c. (ii) None of the overlaps is zero;
we have a three-orbital interaction scheme (in place of the two-orbital one used
in Chapter 4).

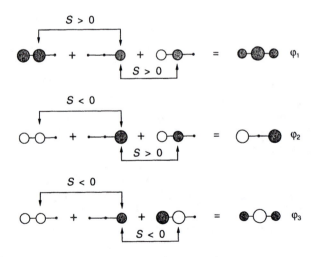

6.7 (i) (ii) (iii) Use the plane of symmetry which passes through H_a and bisects the
line connecting H_c—H_d.

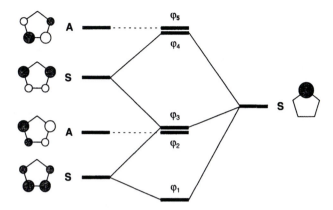

(iv) Antisymmetric orbitals

Symmetric orbitals

Chapter 7

7.1 (i) Identical to that for F_2. Valence electron configuration: $1\sigma_g^2 1\sigma_u^2 2\sigma_g^2 \pi_u^4 \pi_g^4$. (ii) An antibonding electron (in one of the π_g MOs) has been ejected; the bond is strengthened and thus becomes shorter. (iii) An electron moves from one of the π_g orbitals to the $2\sigma_u$ MO which is more antibonding. The bond is weakened and thus the internuclear separation increases.

7.2 S_2 has a valence electron configuration identical to that for O_2. $1\sigma_g^2 1\sigma_u^2 2\sigma_g^2 \pi_u^4 \pi_g^{x1} \pi_g^{y1}$. On removing one ($S_2^+$) or two ($S_2^{2+}$) electrons the bond becomes increasingly strengthened (π_g is an antibonding orbital). Contrarily adding one (S_2^-) or two (S_2^{2-}) electrons leads to a weakening of the bond since antibonding electrons are added.

7.3 (i) All of the antibonding orbitals are occupied and so the molecule is unstable (bond order = 0). (ii) One electron is removed from $2\sigma_u$ and the bond order becomes 1/2. Thus the Ne_2^+ ion is bound.

7.4   An electron is removed from the 3σ orbital which is close to being nonbonding. The bond length hardly changes.

7.5   (i) The number of valence electrons (11) is odd. (ii) (a) $1\sigma^2 2\sigma^2 1\pi^4 3\sigma^2 2\pi^1$. (b) $d(NO^+) < d(NO) < d(NO^-)$ since the 2π orbital is antibonding. One finds from experiment: $d(NO^+) = 106$ pm; $d(NO) = 115$ pm; $d(NO^-) = 127$ pm.

Chapter 8

8.1   (i)

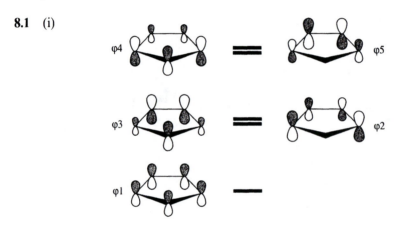

(ii) 5, $\varphi_1^2 \varphi_2^2 \varphi_3^1$ or $\varphi_1^2 \varphi_2^1 \varphi_3^2$. (iii) $C_5H_5^-$ possesses six π-electrons and so is aromatic.

8.2   (i) n_o^σ, weakly bonding and essentially localized on oxygen, approximately represents one of the oxygen lone pairs. The same is true of n_o^p which is weakly antibonding between C and O. The π_{CO} and π_{CO}^* orbitals are the π-orbitals of the system (see Section 3.5.3). (ii) In the excited configuration $n \rightarrow \pi_{CO}^*$ the π system has two bonding electrons and one antibonding one. In the configuration $\pi_{CO} \rightarrow \pi_{CO}^*$ there is a single bonding electron and a single antibonding one. With respect to the ground state ($r_{CO} = 121$ pm, two bonding π-electrons) the CO distance is longer in both of these configurations, and more so in the second ($r_{CO} = 148$ pm) than in the first ($r_{CO} = 138$ pm).

8.3   (i) (ii) The yz plane is a symmetry element for each fragment a and b. The following interaction diagram may then be constructed:

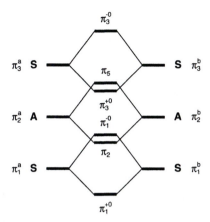

The interaction of the antisymmetric orbitals leads to the two molecular orbitals π_2 and π_5 of benzene. (iii) (b) One combines the zeroth order orbitals which have the same symmetry with respect to the xy plane. The combination of the π_1^{+0} orbital with π_3^{+0} (symmetric) allows generation of the MOs π_1 and π_4:

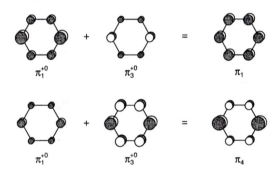

In the same way the combination of π_1^{-0} with π_3^{-0} (antisymmetric) leads to the MOs π_3 and π_6.

8.4   (i) (ii)

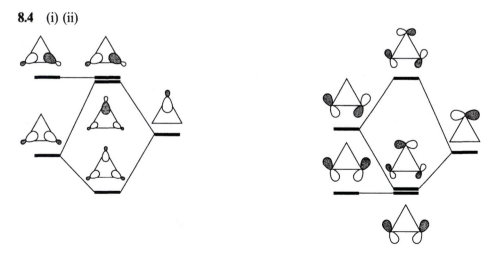

(iii)

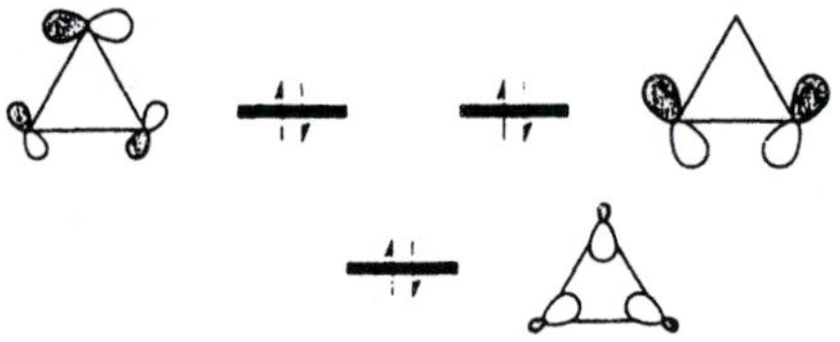

In the deepest-lying orbital there are three bonding interactions.

8.5 (i) The diagram will look very similar to that for the isoelectronic molecule ethane, H_3CCH_3, except that the energy levels for the NH_3 fragment will lie deeper than those for CH_3 and those for BH_3 will lie higher in energy. (ii) The B—N bond order, 1, will be the same as in ethane. (iii) The bonding orbitals will be more strongly localized on the more electronegative atom (here N) and so the dipole moment will have its negative end located on the nitrogen end of the molecule. However, electron donation is formally from nitrogen to boron, so that the dipole moment will be small.

8.6 From **5-22**

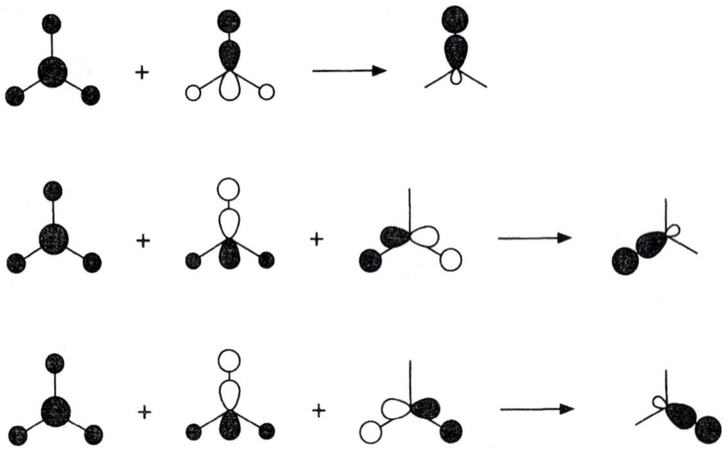

8.7 From **6-24**

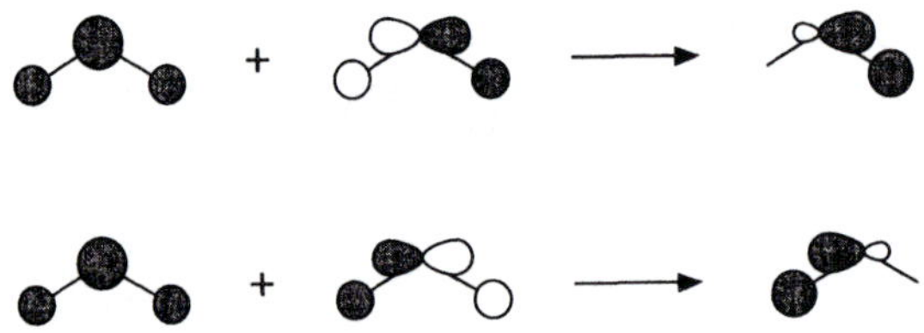

$2a_1$ needs to be included to get the s/p hybrids to properly point at the hydrogen atoms.

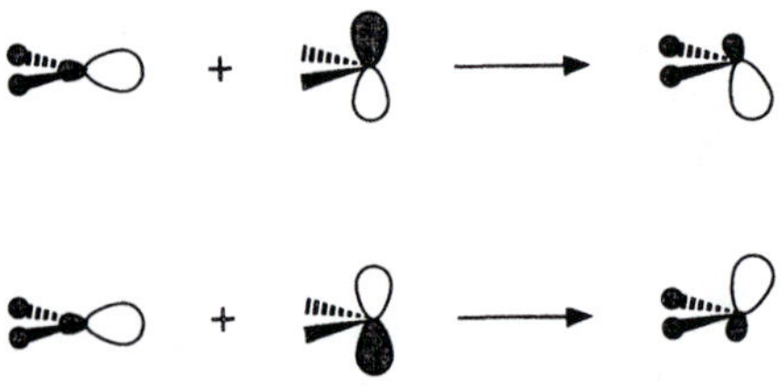

Chapter 9

9.1

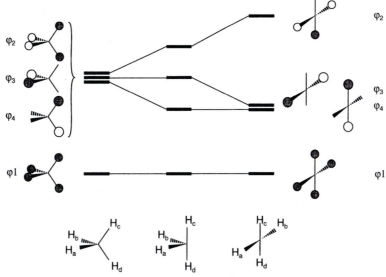

φ_1 is practically unchanged in energy (in the first step, c and d move further apart but get closer to a and b; in the second, a and b get further apart but get closer to c and d). φ_4 is stabilized during the first step (c and d get further apart) and unchanged in energy during the second (zero coefficients on a and b). φ_3 is unchanged during the first step (zero coefficients on c and d) but stabilized during the second (a and b get further apart). φ_2 is destabilized during both steps (one bonding interaction decreases and two antibonding interactions increase).

9.2 (i)

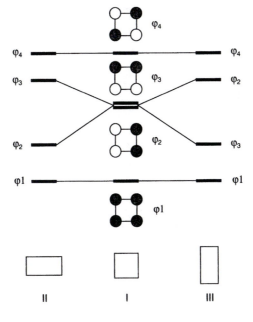

I → III: φ_1 remains practically unchanged (two bonding interactions increase but two decrease); φ_2 is destabilized (two bonding interactions decrease, two antibonding interactions increase); φ_3 is stabilized (two bonding interactions increase and two antibonding interactions decrease); φ_4 remains practically unchanged (two antibonding interactions increase and two decrease). Same type of analysis for I → II. (ii) Rectangle (iii) the rectangular geometry is more stable than the square for this electronic configuration.

9.3

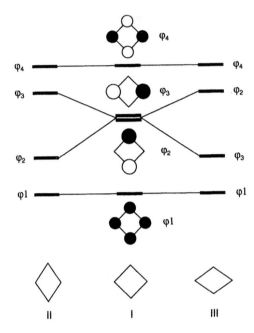

The rhombus structure is more stable than the square.

Chapter 10

10.1 (i) AlH_2: $1a_1^2$ $1b_2^2$ $2a_1^1$; SiH_2: $1a_1^2$ $1b_2^2$ $2a_1^2$; PH_2: $1a_1^2$ $1b_2^2$ $2a_1^2$ $1b_1^1$. (ii) The geometry is controlled by the $2a_1$ MO. (iii) A single electron in $2a_1$ for AlH_2, two in $2a_1$ for SiH_2 and PH_2. (iv) It is linear (geometry controlled by $1b_2$). (v) One electron remains in $2a_1$: these molecules are always bent but with an open angle.

10.2 (i) Four valence electrons: linear. (ii) They have 5 and 6 valence electrons. PH_2^{2+}: $124.8°$ (one electron in $2a_1$); SH_2^{2+}: $96.6°$ (two electrons in $2a_1$).

10.3 (ii) BH_2^+ (4 electrons) BH_2 (5 electrons) BH_2^- (6 electrons). (iii) BH_2^+ linear, BH_2 bent and BH_2^- bent more. (iv) Ground state: $1a_1^2$ $1b_2^2$ $2a_1^1$. First excited state: $1a_1^2$ $1b_2^2$ $1b_1^1$; bent → linear.

10.4 SO_2. First write this as $S^{2+}(O^-)_2$. Each O^- has a filled set of lone pair orbitals and one electron to form a σ-bond to sulfur. S^{2+} has four valence electrons. Overall there are then 6 σ-electrons (the same as in SiH_2) and the molecule will be nonlinear. BCl_3 also has 6 σ-electrons and the molecule is planar.

Chapter 11

11.1   (i) C_1C_2 and C_3C_4. (ii) Using the symmetry properties with respect to the plane running through the C_3—C_4 bond we have:

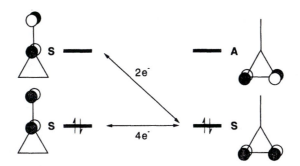

(iii) There is an electron transfer from C_3C_4 to C_1C_2 (two-electron interaction between π_{34} and π_{12}^*). A dipole moment is thus generated along the C_1—C_2 axis and directed towards the top of the molecule.

11.2   (i)

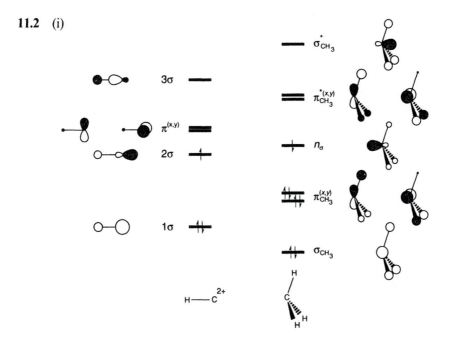

Interaction of 1σ, 2σ and 3σ with σ_{CH_3}, n_σ and $\sigma^*_{CH_3}$, and of each component of π, π_{CH_3} and $\pi^*_{CH_3}$ with its corresponding partner. (ii) The $2\sigma/n_\sigma$ interaction is a two-electron one; these orbitals have a large overlap since they point towards each other. They are close in energy. (iii) There are two stabilizing hyper-conjugative two-electron interactions (π^x, $\pi^x_{CH_3}$ and π^y, $\pi^y_{CH_3}$).

11.3 (i) The hyperconjugative (stabilizing) interaction is increased by the pivoting of the methyl group since the overlap between the π_{CH_3} and $2p$ orbitals becomes larger. In (**b**) this increase comes about by shortening the distance between H_2 and H_3 and the $2p$ orbital on the cationic center. In (**a**) it is the shortening of the distance between H_1 (where there is a large coefficient in π_{CH_3}) and the $2p$ orbital on the cationic center which determines the sense of the pivoting.

a   b

(ii) There is electron transfer from the π_{CH_3} orbital, bonding between carbon and hydrogen, towards $2p$ which weakens the C—H bonds (a loss of bonding electrons). C—H_1 is weaker than C—H_2 or C—H_3 in (**a**) because the coefficient on H_1 is twice that on H_2 or H_3. In (**b**) C—H_1 is unchanged (zero coefficient on H_1) and C—H_2 and C—H_3 are weakened. (iii) There is a bonding interaction between H_2 and H_3 in (**a**). The loss of electrons reduces this interaction so that the H_2—C—H_3 angle opens up. Contrarily in (**b**) one removes electrons that were antibonding between H_2 and H_3; the angle closes down.

11.4 (i) $\diagdown C{=}C\diagup$ and CH_3. (ii) π_{CC} and π^*_{CC} on the first fragment, $\pi^y_{CH_3}$ and $\pi^{y*}_{CH_3}$ on the second. (iii) $\pi_{CC}/\pi^y_{CH_3}$ (4 electrons); $\pi^*_{CC}/\pi^y_{CH_3}$ (2 electrons); $\pi_{CC}/\pi^{y*}_{CH_3}$ (2 electrons); $\pi^*_{CC}/\pi^{y*}_{CH_3}$ (0 electron). (iv) The analysis is analogous to that developed for butadiene.

11.5 (i) Cyclic C_6H_5 and CH_2^+. (ii) Only the π-orbitals of C_6H_5 (identical to those of benzene in Chapter 8.2.5), which are symmetric with respect to the plane, perpendicular to the plane of the molecule and containing $C_1C_4C_7$, can interact with the $2p$ AO on C_7. There are therefore two two-electron interactions which result in electron transfer from the ring to the cationic center: $\pi_1 \rightarrow 2p_7$ and $\pi_3 \rightarrow 2p_7$. The first tends to weaken all of the CC bonds in the same way since π_1 is bonding equally between each pair of linkages in the ring. The second reduces the number of bonding electrons between C_1C_6, C_1C_2, C_4C_5 and C_3C_4 and the number of antibonding electrons between C_5C_6 and C_2C_3. As a result, the last two linkages are shortened and the first four lengthened.

Chapter 12

12.1   (i) See Section 6.3.2. (ii) $1a_1^2\ 1b_2^2\ 2a_1^2$. (iii) HOMO: $2a_1$ LUMO: $1b_1$. Structure 1: the symmetry is defined with respect to the molecular plane.

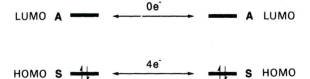

Structure 2: all the MOs are of the same symmetry, but the overlap between the HOMO and LUMO is important while that between the HOMOs is weak.

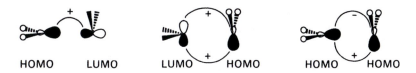

Essentially we have:

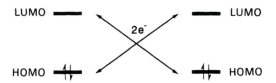

Structure **2** is therefore the most favorable.
(iv)

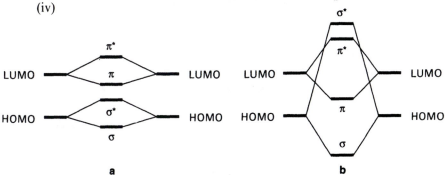

(v) (a) The ground state electronic configuration is $\sigma^2\sigma^{*2}$ in the first case and $\sigma^2\pi^2$ in the second. (b) In the first case, there is one occupied bonding MO and one occupied antibonding MO, leading overall to no bond. In the second, there are two bonding MOs which are occupied (double bond).
(vi) In the initial stages of the reaction (large distance) the methylene groups are arranged with their planes perpendicular. As the distance between them shortens the groups pivot to give a planar molecule.

12.2 Structure **4** is more favorable than structure **3** (two two-electron interactions versus one four-electron interaction.)

12.3 (i) (a) The interaction between $1s_H$ and π involves 2 electrons in **1** and 4 electrons in **6**. (b) The bridged structure is very unfavorable for **6**. (c) One has $4q + 2$ electrons in **5** ($q = 0$) and $4q$ electrons in **6** ($q = 1$). (ii) (a)

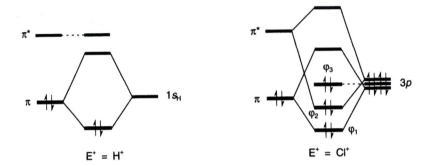

(b) An extra stabilizing interaction (leading to φ_2) occurs for the case $E^+ = Cl^+$ which favors the bridged form.

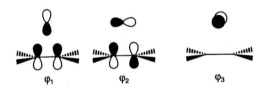

12.4 (i) See Section 6.3.3. (ii) (a) The interaction between the vacant orbital on E^+ and the two occupied MOs on water which describe the lone pairs.

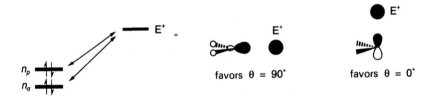

(b) An angle of attack of $45°$ constitutes a good compromise for the two interactions. (c) If the energy separation between the two lone pairs increases, then the interaction with the higher energy of the two (n_p) will predominate and θ will decrease.

12.5 (i) (ii)

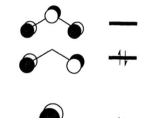

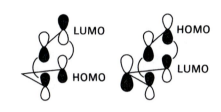

(iii)

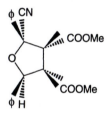

(iv)

Index

Entries in boldface show the principal reference